TWENTY FIRST CENTURY
science

GCSE Biology

Project Directors

Jenifer Burden   Andrew Hunt

John Holman   Robin Millar

Project Officers

Peter Campbell   John Lazonby

Angela Hall   Peter Nicolson

Authors and Editors

Jenifer Burden   Anna Grayson

Angela Hall   Pam Large

Contributors

Ann Fullick   Bill Indge

Jean Martin   Nick Owens

Linn Winspear

Nuffield
Curriculum Centre

THE UNIVERSITY of York

Contents

Introduction

Welcome to *Twenty First Century Science*

GCSE Biology

This is a course in three parts. The first three modules, B1–B3, look at questions that matter to everyone and that biology can help to answer. These are questions about ourselves, our bodies, and the world we live in. In these modules you will learn some of the most important biological explanations. An understanding of how information is inherited from one generation to the next helps to show how we grow and develop, and how we are affected by our environment. Biological explanations help people to make choices about how they can keep healthy. In these modules you will also learn about how science works to develop new explanations, and the different ways in which people make decisions that affect them and that may affect other people in our society.

Modules B4–B6 look more deeply at biological explanations. Here you see how biologists have discovered what happens as our bodies grow from a fertilized egg cell, developing systems to control conditions inside the body and patterns of behaviour to help us survive. You learn about the brain and the nervous system, and find out how scientists are just beginning to understand complex processes like memory.

The final module (B7) has six topics. The first explores how all living things on Earth are dependent on others for their survival. The second topic looks at some of the closest relationships between living things – symbiotes and parasites. The third topic is about new ways to tackle problems using our growing knowledge of inheritance – new DNA technologies. The last three topics take a deeper look at human anatomy – how blood provides a transport system for your body, how energy is released from food in your cells, and how your body uses this energy for movement. Stories of biology in action in this module show how biological ideas are researched and put to use in the twenty-first century.

How to use this book

If you want to find a particular topic, use the **Contents** and **Index** pages. You can also use the **Glossary**. This explains all the key words used in the book.

Each module has two introduction pages, which tell you the main ideas you will study. They look like this:

Why study homeostasis?
Why it is useful to know about this topic.

The science
The scientific information you will learn about in this module.

Biology in action
What you will learn from this module about how science works.

Find out about
The main ideas explored in this module.

Each module is split into sections. Pages in a section look like this:

Heading
Each section looks at a different part of the module.

Find out about
The key points explored in the section.

Higher Tier
The 'H' flag next to something on the page means that it refers to Higher Tier material in the specification.

Questions
Each section has questions for you to try. You can answer most of the questions using the book.

Circled questions
For a few questions, your teacher may give you some help. These questions have a circle around the question number.

B3 Life on Earth

Find out about
- how evolution happens – natural selection
- how humans have changed some species

Selective breeding has produced tulips with different coloured flowers.

Head lice are quite common. They feed on blood.

B Evidence for change NOW

Evolution did not just happen in the past. Scientists can measure changes in species which are happening now. Humans are causing many of these changes.

Selective breeding

Early farmers noticed that there were differences between individuals of the same species. They chose the crop plants or animals that had the features they wanted. For example, the biggest yield, or the most resistance to diseases. These were the ones they used for breeding. This way of causing change in a species is called **selective breeding**. It has been used for breeding wheat, sheep, dogs, roses, and many other species.

Some changes people don't want

People have been using poisons to kill head lice for many years. In the 1980s, doctors were sure that **populations** of head lice in the UK would soon be wiped out.

But a few headlice survived the poisons. Now parts of the country are fighting populations of 'superlice'.

So headlice are another example of change. But this wasn't selective breeding – no one *wanted* to cause superlice.

Natural selection

Head lice are changing because of human beings. But humans haven't been around on Earth for very long. Most changes to species happened before human beings arrived. Something else in the environment caused the change. This is called **natural selection**. Natural selection is how evolution happens.

70

B3 Life on Earth

Steps in natural selection

Treating head lice

Your Local Health Authority issues a directive, known as a rotational policy, every two to three years to inform everyone concerned which type of insecticide is currently recommended for use in your area.

The rotational policy is intended to prevent head lice becoming resistant to treatment — in other words, to help ensure that the treatments available continue to be effective in killing lice.

Key words
selective breeding
populations
natural selection

Questions

1 How does evolution happen?

2 Copy and complete the table to compare selective breeding and natural selection.

Steps in selective breeding	Steps in natural selection
Living things in a species are not all the same.	Living things in a species are not all the same.
Humans choose the individuals with the feature that they want.	
These are the plants or animals that are allowed to breed.	
They pass their genes on to their offspring.	
More of the next generation will have the chosen feature.	
If people keep choosing the same feature, even more of the following generation will have it.	

(3) Explain what is meant by a population.

(4) Read the extract from a leaflet about head lice. Explain how this rotational policy stops the evolution of resistant populations of head lice.

(5) Natural selection is sometimes described as 'survival of the fittest'. How good a description of natural selection do you think this is?

71

Each module ends with a summary, and some also have questions. Here is an example:

Summary
A checklist of key points explained in this module.

Questions
Some further questions to help bring together what you have learnt in this module.

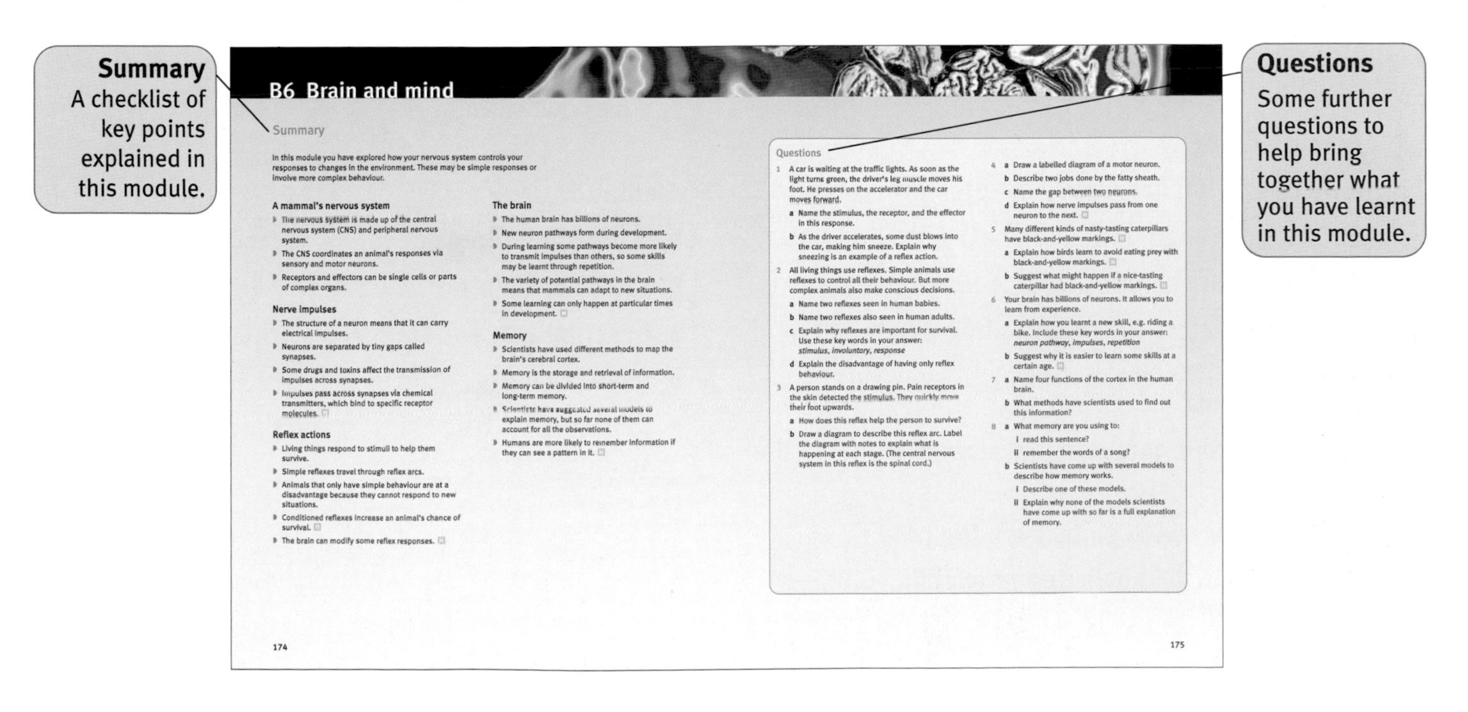

B6 Brain and mind

Summary

In this module you have explored how your nervous system controls your responses to changes in the environment. These may be simple responses or involve more complex behaviour.

A mammal's nervous system
- The nervous system is made up of the central nervous system (CNS) and peripheral nervous system.
- The CNS coordinates an animal's responses via sensory and motor neurons.
- Receptors and effectors can be single cells or parts of complex organs.

Nerve impulses
- The structure of a neuron means that it can carry electrical impulses.
- Neurons are separated by tiny gaps called synapses.
- Some drugs and toxins affect the transmission of impulses across synapses.
- Impulses pass across synapses via chemical transmitters, which bind to specific receptor molecules.

Reflex actions
- Living things respond to stimuli to help them survive.
- Simple reflexes travel through reflex arcs.
- Animals that only have simple behaviour are at a disadvantage because they cannot respond to new situations.
- Conditioned reflexes increase an animal's chance of survival.
- The brain can modify some reflex responses.

The brain
- The human brain has billions of neurons.
- New neuron pathways form during development.
- During learning some pathways become more likely to transmit impulses than others, so some skills may be learnt through repetition.
- The variety of potential pathways in the brain means that mammals can adapt to new situations.
- Some learning can only happen at particular times in development.

Memory
- Scientists have used different methods to map the brain's cerebral cortex.
- Memory is the storage and retrieval of information.
- Memory can be divided into short-term and long-term memory.
- Scientists have suggested several models to explain memory, but so far none of them can account for all the observations.
- Humans are more likely to remember information if they can see a pattern in it.

174

Questions

1 A car is waiting at the traffic lights. As soon as the light turns green, the driver's leg muscle moves his foot. He presses on the accelerator and the car moves forward.
   a Name the stimulus, the receptor, and the effector in this response.
   b As the driver accelerates, some dust blows into the car, making him sneeze. Explain why sneezing is an example of a reflex action.

2 All living things use reflexes. Simple animals use reflexes to control all their behaviour. But more complex animals also make conscious decisions.
   a Name two reflexes seen in human babies.
   b Name two reflexes also seen in human adults.
   c Explain why reflexes are important for survival. Use these key words in your answer: *stimulus, involuntary, response*
   d Explain the disadvantage of having only reflex behaviour.

3 A person stands on a drawing pin. Pain receptors in the skin detected the stimulus. They quickly move their foot upwards.
   a How does this reflex help the person to survive?
   b Draw a diagram to describe this reflex arc. Label the diagram with notes to explain what is happening at each stage. (The central nervous system in this reflex is the spinal cord.)

4 a Draw a labelled diagram of a motor neuron.
   b Describe two jobs done by the fatty sheath.
   c Name the gap between two neurons.
   d Explain how nerve impulses pass from one neuron to the next.

5 Many different kinds of nasty-tasting caterpillars have black-and-yellow markings.
   a Explain how birds learn to avoid eating prey with black-and-yellow markings.
   b Suggest what might happen if a nice-tasting caterpillar had black-and-yellow markings.

6 Your brain has billions of neurons. It allows you to learn from experience.
   a Explain how you learnt a new skill, e.g. riding a bike. Include these key words in your answer: *neuron pathway, impulses, repetition*
   b Suggest why it is easier to learn some skills at a certain age.

7 a Name four functions of the cortex in the human brain.
   b What methods have scientists used to find out this information?

8 a What memory are you using to:
      i read this sentence?
      ii remember the words of a song?
   b Scientists have come up with several models to describe how memory works.
      i Describe one of these models.
      ii Explain why none of the models scientists have come up with so far is a full explanation of memory.

175

Internal assessment

In *GCSE Biology* your internal assessment counts for 33.3% of your total grade. Marks are given for:

- *either* a practical investigation
- *or* a case study and a data analysis

Your school or college will decide on the type of internal assessment. You may be given the marking schemes to help you understand how to get the most credit for your work.

Internal assessment (33.3% of total marks)

EITHER: Investigation (33.3%)

Investigations are carried out by scientists to try and find the answers to scientific questions. The skills you learn from this work will help prepare you to study any science course after GCSE.

To succeed with any investigation you will need to:

- choose a question to explore
- select equipment and use it appropriately and safely
- design ways of making accurate and reliable observations

Your investigation report will be based on the data you collect from your own experiments. You may also use information from other people's work. This is called secondary data.

Marks will be awarded under five different headings.

Strategy

- Choose the task for your investigation.
- Decide how much data you need to collect.
- Choose a procedure to give you reliable data.

Collecting data

- Take careful, accurate measurements safely.
- Collect enough data and check its reliability.
- Collect data across a wide enough range.
- Control factors that might affect the results.

Interpreting data

- Present your data to make clear patterns in the results.
- State your conclusions from the results.
- Use chemical knowledge to explain your conclusion.

Evaluation

- Say how you could improve your method.
- Explain how reliable your evidence is.
- Suggest ways to increase the confidence in your conclusions.

Presentation

- Write a full report of your investigation.
- Lay out your report clearly and logically.
- Describe you apparatus and procedure.
- Show all units correctly.
- Take care with spelling, grammar, and scientific terms.

OR: Case study and data analysis (33.3%)

A **case study** is a report which weighs up evidence about a scientific question. You find out what different people have said about the issue. Then you evaluate the information and make your own conclusions.

You choose a topic from one of these categories:

- A question where the scientific knowledge is not certain.
- A question about decision-making using scientific information.
- A question about a personal issue involving science.

Selecting information

- Collect information from a range of sources.
- Decide how reliable each source is.
- Choose relevant information.
- Say where your information came from.

Understanding the question

- Use science knowledge to explain your topic.
- Report on the scientific evidence used by people with views on the issue.

Making your own conclusion

- Compare different evidence and points of view.
- Weigh the risks and benefits of different courses of action.
- Say what you think should be done based on the evidence.

Presenting your study

- Set out your report clearly and logically.
- Use an appropriate style of presentation.
- Illustrate your report.
- Take care with spelling, grammar, and scientific terms.

A **data analysis** task is based on a practical experiment which you carry out. You may do this alone or work in groups and pool all your data. Then you interpret and evaluate the data.

Interpreting data

- Present your data in tables, charts, or graphs.
- State your conclusions from the data.
- Use chemical knowledge to explain your conclusions.

Evaluation

- Say how you could improve your method.
- Explain how reliable your evidence is.
- Suggest ways to increase the confidence in your conclusions.

Why study genes?

What makes me the way that I am?
Your ancestors probably asked the same question. How are features passed on from parents to children? You may look like your relatives, but you are unique. Only in the last few generations has science been able to answer questions like these.

The science

Your environment has a huge effect on you, for example, on your appearance, your body, and your health. But these features are also affected by your genes. In this Module you'll find out how. You'll discover the story of inheritance.

Ideas about science

In the future, science could help you to change your baby's genes before it is born. Cloned embryos could donate cells to cure diseases. But, as new technologies are developed we must decide how they should be used. These can be questions of ethics – decisions about what is right and wrong.

You and your genes

Find out about:

- how do genes and your environment make you unique
- how and why do people find out about their genes
- how can we use our knowledge of genes
- should this be allowed

Find out about:

- what makes us all different
- what genes are and what genes do

A The same and different

New plants and animals look a lot like their parents. They have **inherited** information from them. This information controls how the new organisms develop.

A lot of information goes into making a human being. So inheritance does a big job pretty well. All people have most features in common. Children look a lot like their parents. If you look at the people around you, the differences between us are very, very small. But we're interested in them because they make us unique.

These sisters have some features in common.

Most features are affected by both the information you inherit and your environment.

Environment makes a difference

Almost all of your features are affected by the information you inherited from your parents. For example, your eye colour depends on this information.

But most of your features are also affected by your **environment**. For example, your skin colour depends on inherited information. But if you spend more time in the sun, your skin will get darker.

Questions

1. Choose two of the students in the photograph. Write down five ways they look different.
2. What two things can affect how you develop?
3. Explain what is meant by inherited information.

Key words

inherited
environment

Inheritance – the story of life

One important part of this story is where all the information is kept. Living organisms are made up of cells. If you look at a cell under a microscope you can see the **nucleus**. Inside the nucleus are long threads called **chromosomes**. Each chromosome contains thousands of **genes**. It is genes that control how you develop.

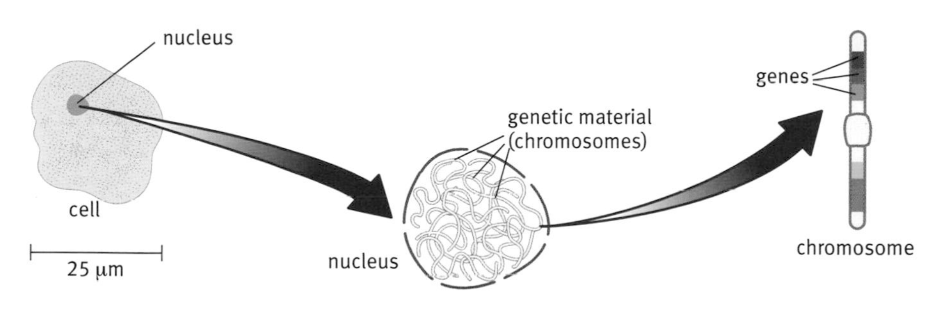

All the information needed to create a whole human being fits into the nucleus of a cell. The nucleus is just 0.006 mm across!

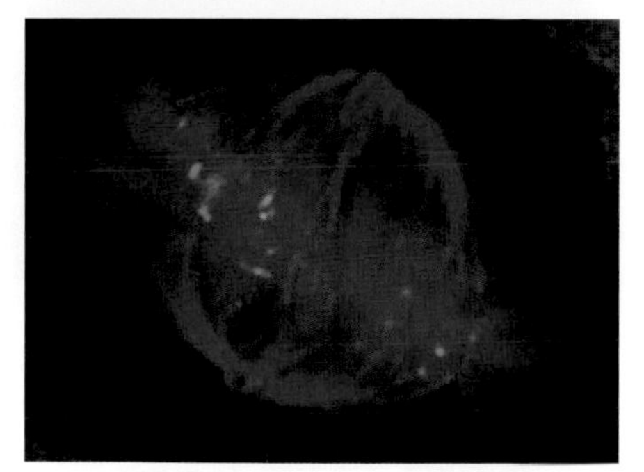

These cells have been stained to show up different parts. The long pink threads are the chromosomes. (Mag: × 6500 approx)

What are chromosomes made of?

Chromosomes are made of a chemical called **DNA**. DNA is short for deoxyribonucleic acid. Most kinds of living thing use DNA to make their chromosomes.

How do genes control your development?

A fertilized egg cell has the instructions for making every **protein** in a human being. That's what genes are – instructions for making proteins. Each gene is the instruction for making a different protein.

What's so important about proteins?

Proteins are important chemicals for cells. There are many different proteins in the body, and each one does a different job. They may be:

- **structural** proteins – to build the body
- **enzymes** – to speed up chemical reactions in cells

Genes control the proteins a cell makes. This is how they control what the cell does.

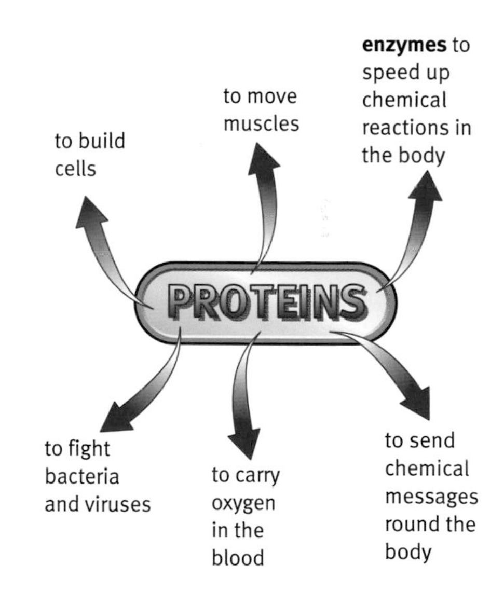

There are about 50 000 types of proteins in the human body.

Questions

4 Write these cell parts in order starting with the smallest:

chromosome, gene, cell, nucleus

5 Explain how genes control what a cell does.

6 **a** List two kinds of job that proteins do in the human body.

b Name two proteins in the human body and say what they do.

Key words

nucleus
chromosomes
structural
DNA
protein
genes
enzymes

Find out about:

- how you inherit genes
- Huntington's disorder (an inherited illness)

Craig and his grandfather, Robert

B Family values

It can be funny to see that people in a family look like each other. Perhaps you don't like a feature you've inherited - your dad's big ears or your mum's freckles. For some people, family likenesses are very serious.

Craig's story

My grandfather's only 56. He's always been well but now he's a bit off colour. He's been forgetting things – driving my Nan mad. No one's said anything to me, but they're all worried about him.

Robert's story

I'm so frustrated with myself. I can't sit still in a chair. I'm getting more and more forgetful. Now I've started falling over for no reason at all.
The doctor has said it might be **Huntington's disorder**. It's an inherited condition. She said I can have a blood test to find out, but I'm very worried.

Huntington's disorder

Huntington's disorder is an inherited condition. You can't catch it. The disorder is passed on from parents to their children. The symptoms of Huntington's disorder don't happen until middle age. First the person has problems controlling their muscles. This shows up as twitching. Gradually a sufferer becomes forgetful. They find it harder to understand things. After a few years people with Huntington's disorder can't control their movements. Sadly, the condition is fatal.

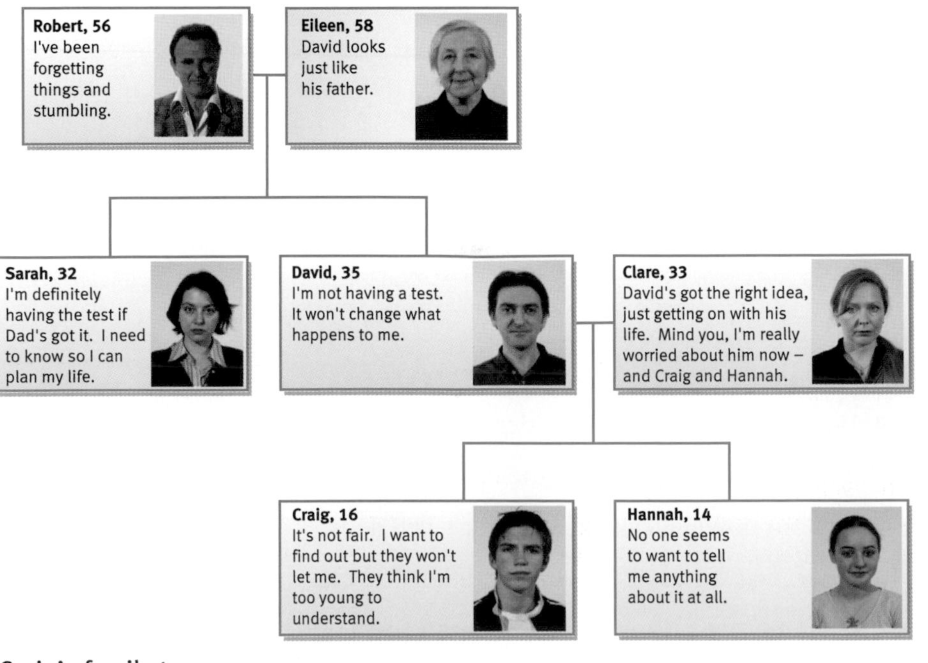

Craig's family tree

Key words

Huntington's disorder

Questions

1. List the symptoms of Huntington's disorder.
2. Explain why Huntington's disorder is called an inherited condition.

How do you inherit your genes?

Sometimes people in the same family look a lot alike. In other families brothers and sisters look very different. They may also look different from their parents. The key to this mystery lies in our genes.

Parents pass on genes in their **sex cells**. In animals these are sperm and egg cells (ova). Sex cells have copies of half the parent's chromosomes. When a sperm cell fertilizes an egg cell, the fertilized egg cell (ovum) gets a full set of chromosomes.

How many chromosomes does each cell have?

Chromosomes come in pairs. Every human body cell has **23 pairs** of chromosomes. The chromosomes in each pair are the same size and shape. They carry the same genes in the same place. This means that your genes also come in pairs.

Sex cells have single chromosomes

Sex cells are made with copies of half the parent's chromosomes. This makes sure that the fertilized egg cell has the right number of chromosomes – 23 pairs.

One chromosome from each pair came from the egg cell. The other came from the sperm cell.

Each chromosome carries thousands of genes. So the fertilized egg cell has a mixture of the parents' genes. Half of the new baby's genes are from the mother. Half are from the father.

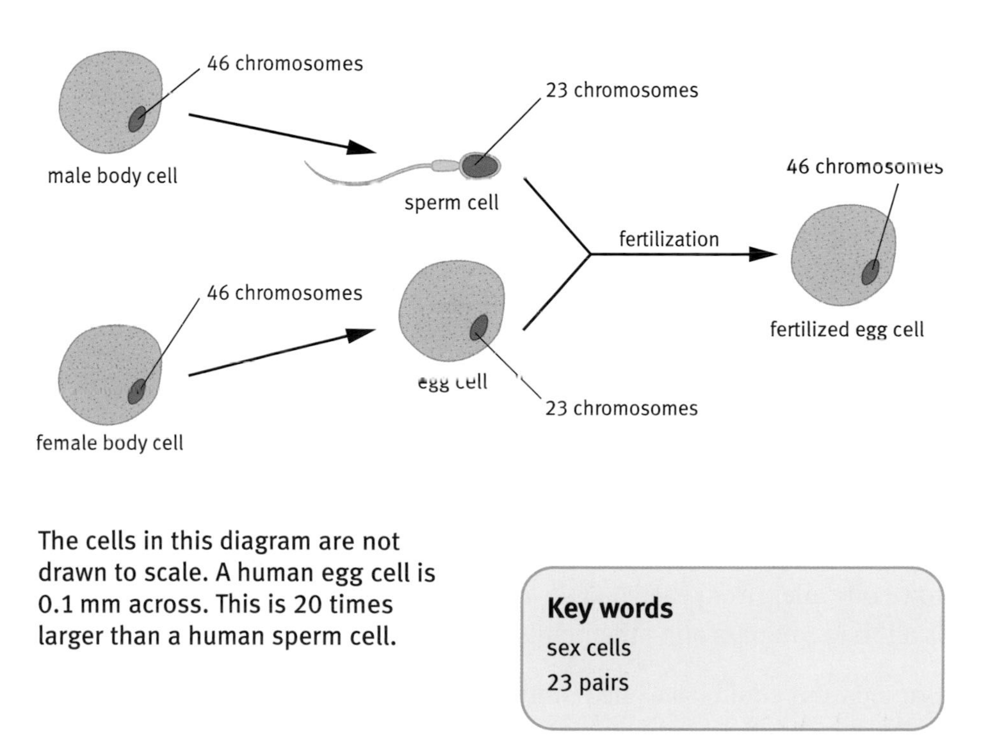

The cells in this diagram are not drawn to scale. A human egg cell is 0.1 mm across. This is 20 times larger than a human sperm cell.

Key words
sex cells
23 pairs

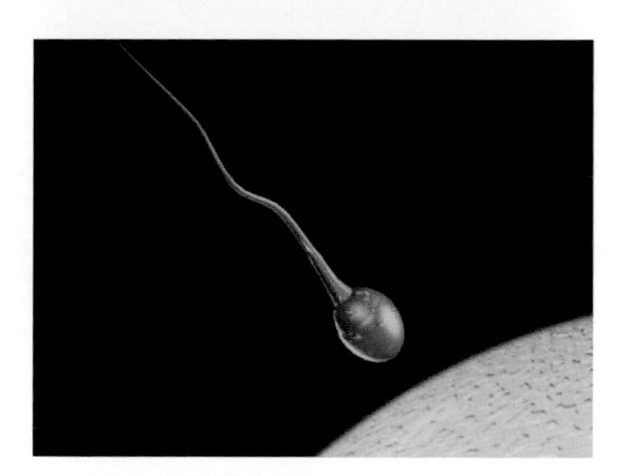

The fertilized egg cell will have genes from both parents. (Mag: × 2000 approx)

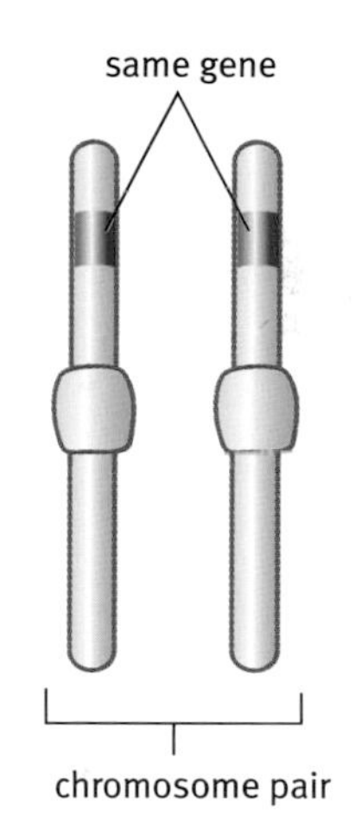

These chromosomes are a pair.

Questions

3 a Draw a diagram to show a sperm cell, an egg cell, and the fertilized egg cell they make.

b In each cell write down the number of chromosomes it has in the nucleus.

c Explain why the fertilized egg cell has pairs of chromosomes.

4 Explain why children may look a bit like each of their parents.

5 Two sisters with the same parents won't look exactly alike. Explain why you think this is.

Find out about:

- how pairs of genes control some features
- cystic fibrosis (an inherited illness)
- testing a baby's genes before they are born

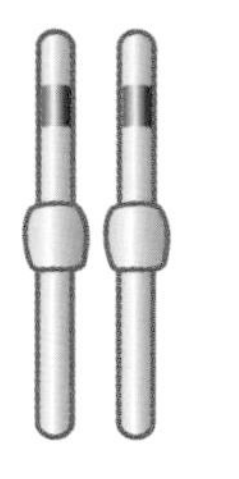

This diagram shows one pair of chromosomes. The gene controlling dimples is coloured in.

C The human lottery

Will this baby be tall and have red hair? Will she have a talent for music, sport – even science?! Most of these features will be affected by her environment. Most features are affected by more than one gene. A few are controlled by just one gene. We can understand these more easily.

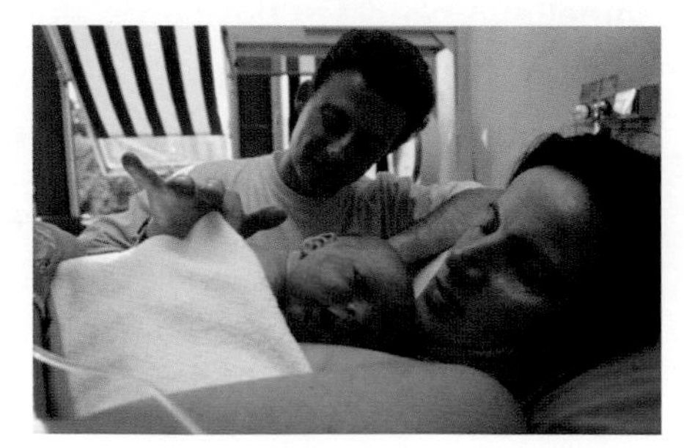

This baby has inherited a unique mix of genetic information.

Genes come in different versions

Both chromosomes in a pair carry genes that control the same features. Each pair of genes is in the same place on the chromosomes.

But genes in a pair may not be exactly the same. They may be slightly different versions. You can think about it like football strips – a team's home and away strips are both based on the same pattern, but they're slightly different. Different versions of the same genes are called **alleles**.

Dominant alleles – they're in charge

The gene that controls dimples has two alleles. The D allele gives you dimples. The d allele won't cause dimples.

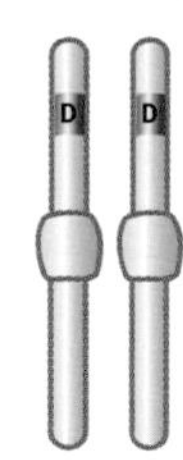

This person inherited a D allele from both parents. They have dimples.

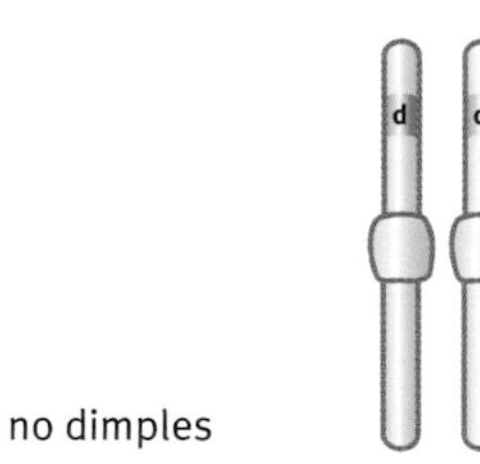

This person inherited a d allele from both parents. They don't have dimples.

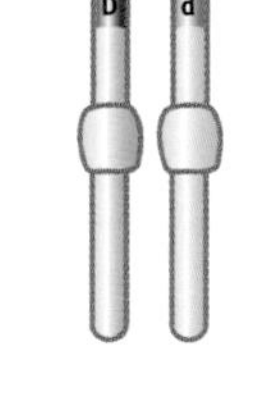

This person inherited one D and one d allele. They have dimples.

The D allele is **dominant**. You only need one copy of a dominant allele to have its feature. The d allele is **recessive**. You must have two copies of a recessive allele to have its feature – in this case no dimples.

Which alleles can a person inherit?

Sex cells get one chromosome from each pair the parent has. So they only have one allele from each pair. If a parent has two D or two d alleles, that is all they can pass on to their children.

But a parent could have one D and one d allele. Then half of their sex cells will get the D allele and half will get the d allele.

The human lottery

We cannot predict which egg and sperm cells will meet at fertilization. The diagram shows all the possibilities for one couple.

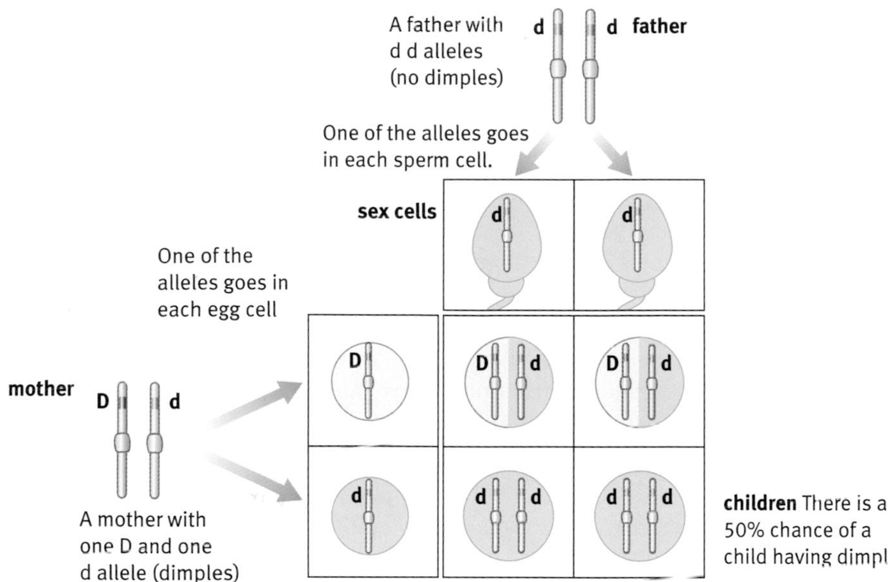

The allele that gives you straight thumbs is dominant (T). The allele for curved thumbs is recessive (t).

Why don't brothers and sisters look the same?

Human beings have about 30 000 genes. Each gene has different versions – different alleles.

Brothers and sisters are different because they each get a different mixture of alleles from their parents. Except for identical twins, each one of us has a unique set of genes.

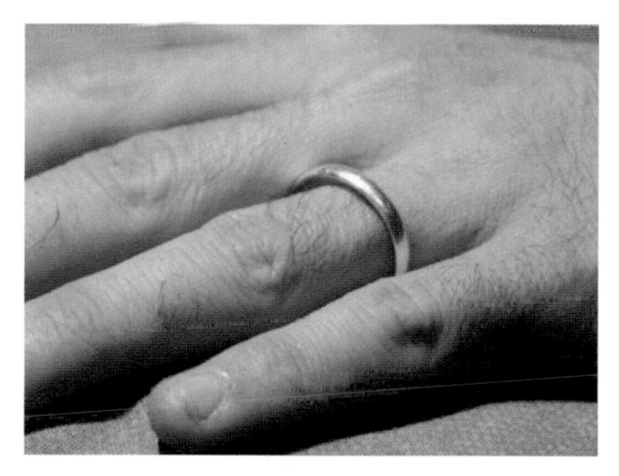

The allele that gives you hair on the middle of your fingers is dominant (R). The allele for no hair is recessive (r).

What about the family?

Huntington's disorder is caused by a dominant allele. You only need to inherit the allele from one parent to have the condition. Craig and Hannah's grandfather, Robert, has Huntington's disorder. So their dad, David, may have inherited this faulty allele. At the moment he has decided not to have the test to find out.

Key words

alleles
dominant
recessive

Questions

1 Write down what is meant by the word allele.

2 Explain why you inherit two alleles for each gene.

3 Explain the difference between a dominant and a recessive allele.

4 What are the possible pairs of alleles a person could have for:

a dimples

b straight thumbs

c no hair on the second part of their ring finger

5 Use diagrams to explain why a couple who have dimples could have a child with no dimples.

⑥ Use diagrams to work out the chance that David has inherited the Huntington's disorder allele.

Dear Clare,

Please help us. My husband and I have just been told that our first child has cystic fibrosis. No one in our family has ever had this disease before. Did I do something wrong during my pregnancy? I'm so worried.

Yours sincerely

Emma

Dear Emma,

What a difficult time for you all. First of all, nothing you did during your pregnancy could have affected this, so don't feel guilty. Cystic fibrosis is an inherited disorder ...

Dear Doctor

*We've had a huge postbag in response to last month's letter from Emma. So this month we're looking in depth at **cystic fibrosis**, a disease which one in twenty-five of us carries in the UK ...*

What is cystic fibrosis?

You can't catch cystic fibrosis. It is a genetic disorder. This means it is passed on from parents to their children. The disease causes big problems for breathing and digestion. Cells that make mucus in the body don't work properly. The mucus is much thicker than it should be, so it blocks up the lungs. It also blocks tubes that take enzymes from the pancreas to the gut. People with cystic fibrosis get breathless. They also get lots of chest infections. The shortage of enzymes in their gut means that their food isn't digested properly. So the person can be short of nutrients.

How do you get cystic fibrosis?

Most people with cystic fibrosis (CF) can't have children. The thick mucus affects their reproductive systems. So babies with CF are usually born to healthy parents. At first glance this seems very strange – how can a parent pass a disease on to their children when they don't have it themselves?

The answer lies with one of the thousands of genes responsible for producing a human being. There are two versions (or alleles) of this gene. The first is dominant and instructs cells to make normal mucus. The second is a faulty recessive allele, which leads to the symptoms of CF.

So a person who has one dominant allele and one recessive allele will not have CF. But they are a **carrier** of the faulty allele. When parents who are carriers make sex cells, half will contain the normal allele – and half will contain the faulty allele. When two sex cells carrying the faulty allele meet at fertilization, the baby will have CF. One in every 25 people in the UK carries the CF allele.

This diagram shows how healthy parents who are both carriers of the cystic fibrosis allele can have a child affected by the disease.

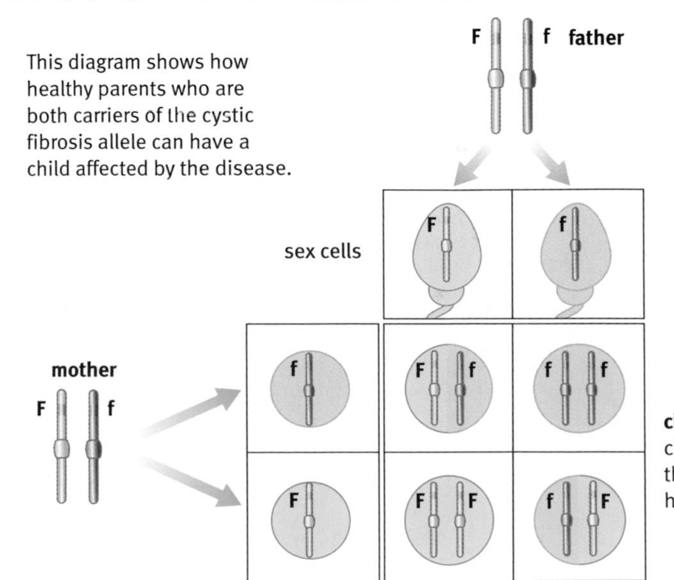

children There is a 25% chance that a child from the carrier parents will have cystic fibrosis.

Key words
cystic fibrosis
termination
carrier

Can cystic fibrosis be cured?

Not yet. But treatments are getting better, and life expectancy is increasing all the time. Physiotherapy helps to clear mucus from the lungs. Sufferers take tablets with the missing gut enzymes in. Antibiotics are used to treat chest infections. And an enzyme spray can be used to thin the mucus in the lungs, so it is easier to get rid of. New techniques may offer hope for a cure in the future.

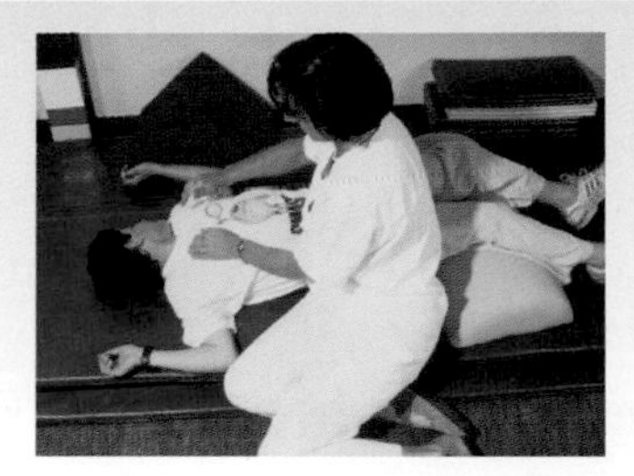

Tom has cystic fibrosis. He has physiotherapy every day to clear thick mucus from his lungs.

Can it be prevented?

Yes, if a couple know they are at risk of having a child with cystic fibrosis, but this involves a very hard decision for the parents. During pregnancy cells from the developing fetus can be collected, and the genes examined. If the fetus has two alleles for cystic fibrosis, the child will have the disease. The parents may choose to end the pregnancy. This is done with a medical operation called a **termination** (an abortion).

How do doctors get cells from the fetus?

The fetal cells can be collected two ways:

- an amniocentesis test
- a chorionic villus test

The diagrams show how each of these tests is carried out.

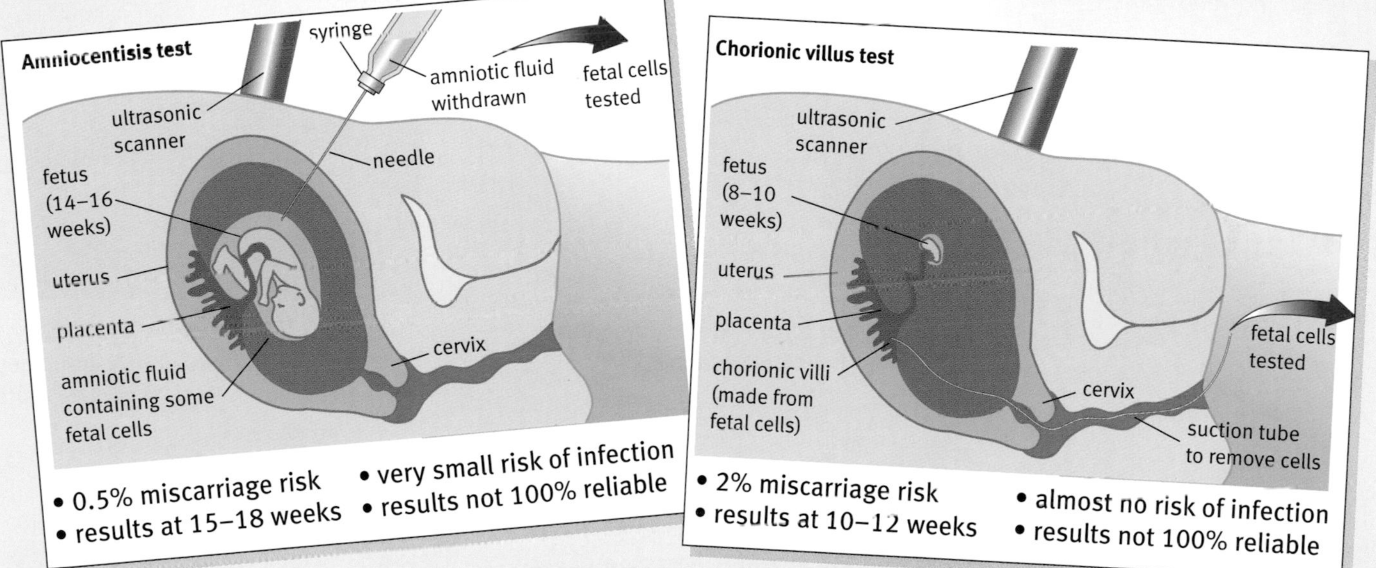

Questions

7 The magazine doctor is sure that nothing Emma did during her pregnancy caused her baby to have cystic fibrosis. How can she be so sure?

8 People with cystic fibrosis make thick, sticky mucus. Describe the health problems that this may cause.

9 Explain what it means when someone is a 'carrier' of cystic fibrosis.

10 Two carriers of cystic fibrosis plan to have children. Draw a diagram to work out the chance that they will have:

a a child with cystic fibrosis

b a child who is a carrier of cystic fibrosis

c a child who has no cystic fibrosis alleles

Find out about:

- what's gender
- how hormones change a person's sex

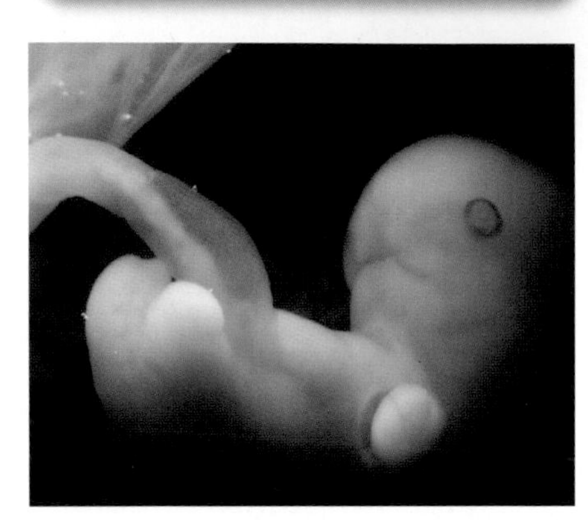

This embryo is six weeks old.

D Male or female?

Ever wondered what it would be like to be the opposite sex? Well, if you are male there was a time when you were – just for a short while. Male and female babies are very alike until they are about six weeks old.

What decides an embryo's sex?

A fertilized human egg cell has 23 pairs of chromosomes. Pair 23 are the sex chromosomes. Males have an X chromosome and a Y chromosome – **XY**. Females have two X chromosomes – **XX**.

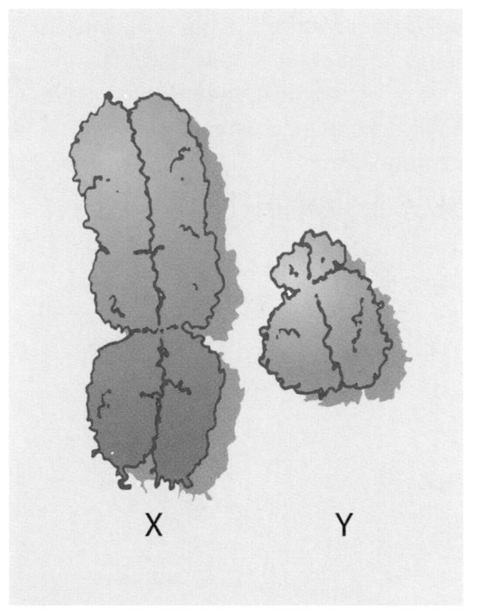

Women have two X chromosomes. Men have an X and a Y.

These chromosomes are from the nucleus of a woman's body cell. They have been lined up to show the pairs.

What's the chance of being male or female?

A parent's chromosomes are in pairs. When sex cells are made they only get one chromosome from each pair. So half a man's sperm cells get an X chromosome and half get a Y chromosome. All a woman's egg cells get an X chromosome.

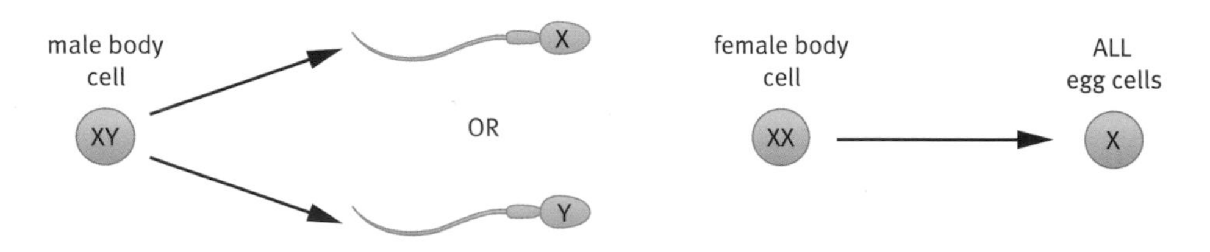

When a sperm cell fertilizes an egg cell the chances are 50% that it will be an X or a Y sperm. This means that there is a 50% chance that the baby will be a boy or a girl.

How does the Y chromosome make a baby male?

A male embryo's testes develop when it is about six weeks old. This is caused by a gene on the Y chromosome – the SRY gene. SRY stands for 'sex-determining region of the Y chromosome'.

Testes produce the male sex **hormone** called androgen. Androgen makes the embryo develop into a male. If there is no male sex hormone present, the sex organs develop into the ovaries, clitoris, and vagina of a female.

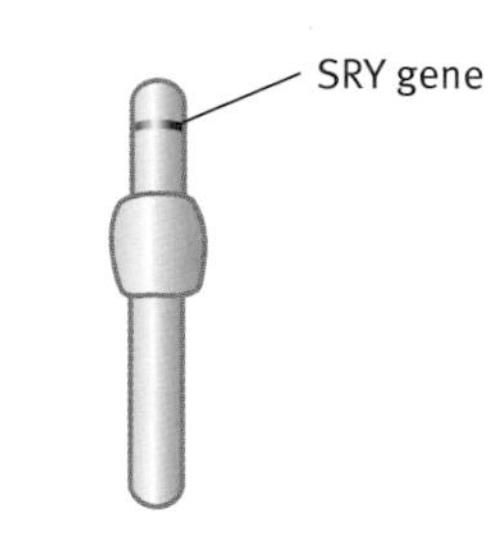

The Y chromosome

What are hormones?

Hormones are another group of proteins. They control many processes in the cells. Tiny amounts of hormones are made by different parts of the body. You can read more about hormones in Module B3 *Life on Earth.*

Key words
XY
XX
hormone

Jan's story

At eighteen Jan was studying at college. She was very happy, and was going out with a college football player. She thought her periods hadn't started because she did a lot of sport.

Then in a science class Jan looked at the chromosomes in her cheek cells. She discovered that she had male sex chromosomes – XY.

Sometimes a person has X and Y chromosomes but looks female. This is because their body makes androgen but the cells take no notice of it. About 1 in 20 000 people have this condition. They have small internal testes and a short vagina. They can't have children.

Jan had no idea she had this condition. She found it very difficult to come to terms with. But she has now told her boyfriend and they have stayed together.

Jan on holiday, aged eighteen.

Questions

1 Why do you think sex chromosomes are called X and Y?

2 What sex chromosome(s) would be in the nucleus of:

  a a man's body cell

  b an egg cell

  c a woman's body cell

  d a sperm cell

3 Draw a diagram to show the chance of a baby being male or female.

4 Imagine you are Jan or her boyfriend. How would you have felt about her condition?

5 What is a hormone?

⑥ How do hormones get around the body?

⑦ Name one human hormone and say what it does.

Find out about:

- how people make ethical decisions
- how genetic information could be used

E Ethics - making decisions

Elaine's nephew has cystic fibrosis. When they found out, Elaine and her husband Peter became worried about any children they might have. They both had a genetic test. The tests showed that they were both carriers for cystic fibrosis. Elaine and Peter decided to have an amniocentesis test when Elaine was pregnant.

'We had an amniocentesis test for each of my pregnancies,' says Elaine. 'Sadly we felt we had to terminate the first one, because the fetus had CF. We are lucky enough now to have two healthy children – and we know we haven't got to watch them suffer.'

Elaine and Peter made a very hard decision when they decided to terminate their first pregnancy. When a person has to make a decision about what is the right or wrong way to behave, they are thinking about **ethics**. Deciding whether to have a termination is an example of an ethical question.

Ethics – right and wrong

For some ethical questions, the right answer is very clear. For example, should you feed and care for your pet? Of course. But in some situations, like Elaine and Peter, people may not agree on one right answer. People think about ethical questions in different ways.

For example, Elaine and Peter felt that they had weighed up the consequences of either choice. They thought about how each choice – continuing with the pregnancy or having a termination – would affect all the people involved. They had to make a judgement about the difficulties their unborn child would face with cystic fibrosis.

In order to consider all the consequences they also had to think about the effects that an ill child would have on their lives, and on the lives of any other children they might have. Some people feel that they could not cope with the extra responsibility of caring for a child with a serious genetic disorder.

Key words
ethics

Different choices

Not everyone weighing up the consequences of each choice would have come to the same decision as this couple did.

Some people feel that any illness would have a devastating effect on a person's quality of life. But people lead very happy, full lives in spite of very serious disabilities.

Jo has a serious genetic disorder. Her parents believe that termination is wrong. They decided not to have more children, rather than use information from an amniocentisis test.

Elaine and Peter made their ethical decision only by thinking about the consequences that each choice would have. This is just one way of dealing with ethical questions.

When you believe that an action is wrong

For some people having a termination would be completely wrong in itself. They believe that an unborn child has the right to life, and should be protected from harm in the same way as people are protected after they are born. Other people believe that terminating a pregnancy is unnatural, and that we should not interfere. People may hold either of these viewpoints because of their own personal beliefs, or because of their religious beliefs.

A couple in Elaine and Peter's position who felt that termination was wrong might decide not to have children at all. This would mean that they could not pass on the faulty allele. Or they could decide to have children, and to care for any child that did inherit the disease.

Questions

1 Explain what is meant by 'an ethical question'.

2 Describe three different points of view that a couple in Elaine and Peter's position might take.

3 What is your viewpoint on genetic testing of a fetus for a serious illness? Explain why you think this.

Not everyone in Elaine and Peter's position would have made the same decision.

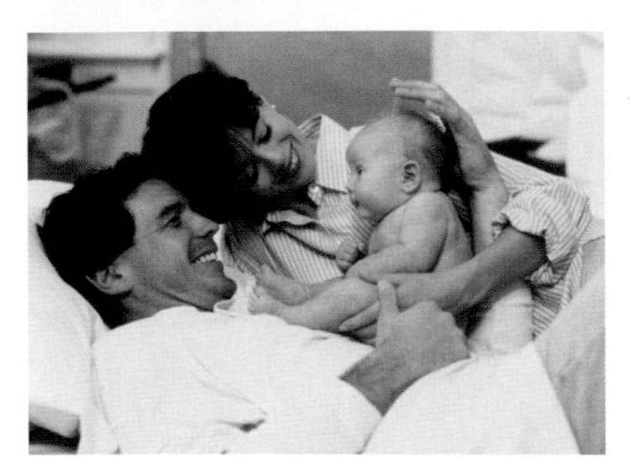

This couple are both carriers of cystic fibrosis. They had an amniocentesis test during their pregnancy. The results were unclear. When their daughter was born she was completely healthy.

How reliable are genetic tests?

Genetic testing is used to look for alleles that cause genetic disorders. People like Elaine and Peter use this information to make decisions about whether to have children. Genetic tests can be used to make a decision about whether a pregnancy should be continued or not.

So, it is important to realize that the tests are not completely reliable. Current tests for CF detect about 90% of cases. A genetic test on an embryo is even more accurate. In a few cases only will it not detect CF. This is called a **false negative**. **False positive** tests are not as common, but they can happen. The photo describes one such result.

Why do people have genetic tests?

Usually people only have a genetic test because they know that a genetic disorder runs in their family. Most parents who have a child with cystic fibrosis did not know that they were carriers. So, they would not have had a genetic test during pregnancy.

There have been small studies to find out what would happen if everyone was tested for the cystic fibrosis allele. Testing the whole population for an allele is called **genetic screening**.

Who decides about genetic screening?

The decision about whether to use genetic screening is taken by governments and local NHS trusts. People in the NHS have to think about different things when they decide if genetic screening should be used:

- what are the costs of testing everyone for the allele?
- what are the benefits of testing everyone for the allele?
- is it better to spend the money on other things, e.g. hip replacement operations, treating people with cancer, and treating people who already have cystic fibrosis?

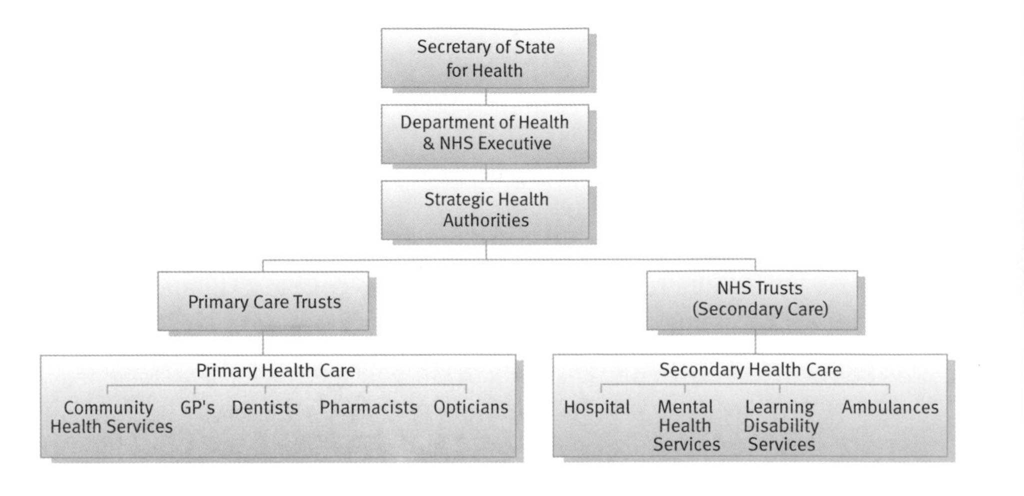

NHS trusts are responsible for the health care of their local people. They are given funds from the Department of Health.

Is it right to use genetic screening?

It is easy to see why people may want genetic screening:

- when two people decided to have children, they would know if their children were at risk of inheriting the disorder

At first glance, genetic screening may seem like the best course of action for everyone. But the best decision for the majority is not always the right decision. There are ethical questions to consider about genetic screening for cystic fibrosis, including:

- who should know the test results?
- what effect could the test result have on people's future decisions?
- should people be made to have screening, or should they be able to opt out?, is it right to interfere?

About 1 in 25 people in the UK carry the allele for cystic fibrosis. Some people think that having this information is useful, but there are also good reasons why not everyone agrees. A decision may benefit many people. But it may not be the right decision if it causes a great amount of harm to a few people.

People have different ideas about whether genetic screening for cystic fibrosis would be a good thing.

Questions

4 What are 'false negative' and 'false positive' results?

5 Why is it important for people to know about false results?

6 Explain what is meant by the term 'genetic screening'.

7 Give two arguments for and two against genetic screening for cystic fibrosis.

8 Which argument do you agree with? Explain why.

(9) Give an example of a decision about a different issue which was made because it caused less harm to a few people, instead of being the most benefit to a few. It could be a non-science issue.

Key words

false negative
false positive
genetic screening

Can we, should we?

If you could have more information about your genes, would you want it? In future it may be possible to screen children at birth for many different alleles. People would know if they had genes that increased their risk of a particular disease. But remember that most diseases are affected by many genes - and your lifestyle.

> One change in a particular gene can lead to higher risk of heart disease, scientists claim. Following detailed family histories, a group based at Yale University think that genetic tests could tell people if they were at greater risk. Other scientists disagree. 'We simply don't know enough about how different genes work together to be able to say that one gene can be the most important cause.'

Biobank, a new research project, began in 2005 to investigate genes linked to common diseases.

> It will lead to new drugs to treat cancers and heart disease, say Biobank supporters. Over the next 20 years, 500,000 people will donate genetic samples and answer questions about their lifestyles. Scientists hope to find out which versions of genes make you more likely to get certain diseases. And also what lifestyle factors may trigger the diseases. But critics of Biobank are worried that people's private genetic information could be used for other purposes – which the volunteers would not want. 'No one should know the names of the volunteers. We're concerned about what would happen if insurance companies, employers or the police had access to the information.'

Scientists already use information about people's DNA to help them solve crimes. They produce DNA profiles from cells left at a crime scene. There is only a 1 in 50 million chance of two people having the same DNA profile – unless they are identical twins.

> A senior judge has called for a national DNA database recording everyone living in or entering the country. At the moment police can only keep samples from people who've been arrested. But already this has helped them link people to crimes that had been committed many years before. But human rights campaigners argue that the database is unnecessary. 'It is an invasion of privacy, and puts innocent people on the same level as criminals.' 2004

Questions

(10) Suggest some pros and cons of knowing about your genes from birth.

11 Explain why many scientists think that Biobank will benefit society.

12 Give one argument that people have given against Biobank.

13 Give different arguments for and against the government setting up a DNA database.

(14) Explain why you agree or disagree with setting up a DNA database.

Who should know about your genes?

Many people think that only you and your doctor should know information about your genes. They are worried that it could affect a person's job prospects and chances of getting life insurance.

How does life insurance work?

People with life insurance pay a regular sum of money to an insurance company. This is called a premium. In return, when they die, the insurance company pays out an agreed sum of money. People buy life insurance so that there will be money to support their families when they die.

CONDITION	PERCENTAGE OF DEATHS CAUSED BY SMOKING IN 1995	
	Men	Women
CANCERS		
Lung	90	73
Throat & mouth	74	47
Oesophagus	71	62
HEART AND CIRCULATION DISEASES	28	19

People use information like this to decide the premium each person should pay for insurance. The higher the risk, the bigger the premium.

Should insurance companies know about your genes?

Insurance companies assess what a person's risk is of dying earlier than average. If they believe that the risk is high, they may choose to charge higher premiums than average. Some people think that insurance companies might use the results of genetic tests in the wrong way. Individual people might do the same. Here are some of the arguments:

- Insurers may not insure people if a test shows that they are more likely to get a particular disease. Or they may charge a very high premium.
- Insurers may say that everyone must have genetic tests for many diseases before they can be insured.
- People may not tell insurance companies if they know they have a genetic disorder.
- People may refuse to have a genetic test because they fear that they will not be able to get insurance. They may miss out on medical treatment which could keep them healthy.

In 2001 insurance companies in the UK agreed not to collect and share genetic information about people. This was to give the government time to regulate how information about people's genes can be used. The agreement runs out in October 2006.

Questions

15 'People are already asked lots of information about their family history and their lifestyle when they get life insurance. Genetic testing is just another way of getting information.'

Do you agree or disagree with this point of view? Explain your answer.

16 Science can give us a lot of information about our genes. But that doesn't mean other people should be allowed to know about it. Give another example of where something can be done in science, but society does not allow it.

Find out about

- how new techniques can allow people to select embryos
- how people think this technology should be used

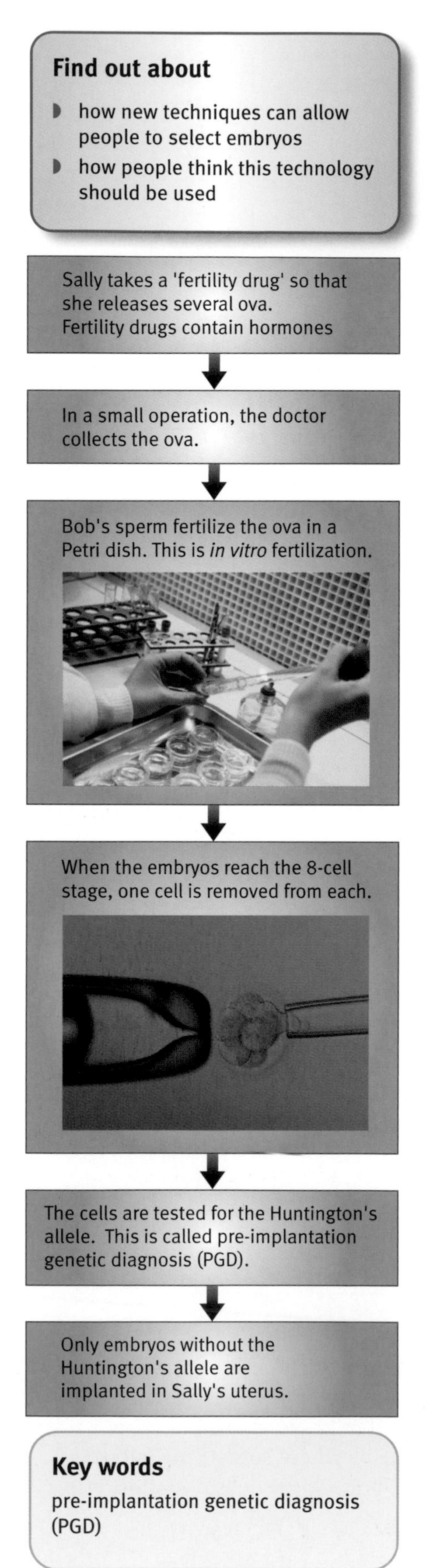

Key words

pre-implantation genetic diagnosis (PGD)

F Can you choose your child?

Many people do not agree with termination. If they are at risk of having a child with a genetic disease, they may have decided not to have children. Now doctors can offer them another treatment. It uses *in vitro* fertilization (IVF).

How does *in vitro* fertilization work?

In this treatment the mother's egg cells are fertilized outside her body. IVF has been used since 1977 to help couples who could not conceive a child naturally. Since then over 300 000 women world-wide have become pregnant by IVF treatment. Now doctors can also use this treatment to help couples whose children are at risk from a serious genetic disorder.

Pre-implantation genetic diagnosis

Bob and Sally want children, but Bob has the allele for Huntington's disorder. Sally has become pregnant twice. Tests showed that both the fetuses had the Huntington's allele and the pregnancies were terminated. They were keen to have a child, so their doctor suggested that they should use **pre-implantation genetic diagnosis (PGD)**. Sally's treatment is explained in the flow chart. The first use of PGD to choose embryos was in the UK in 1989. At the moment, PGD is only allowed for families with particular inherited conditions.

New technology – new decisions

New technologies like PGD often give us new decisions to make. In the UK, Parliament makes laws to control research and technologies to do with genes. Scientists are not free to do whatever research they may wish to do. From time to time Parliament has to update the law.

But Parliament can't make decisions case by case. So the Government has set up groups of people to decide which cases are within the law on reproduction. One of these groups is the Human Fertilisation and Embryology Authority (HFEA).

The HFEA interprets the laws we already have about genetic technologies. It also takes into account public opinion, as well as practical and ethical considerations. One of its jobs is to decide when PGD can be used.

Questions

1 Draw a flow chart to show the main steps in embryo selection using PGD.

Case One

Early 2002: Zain Hashmi has a serious inherited blood disorder. He needs a bone marrow transplant to give him normal blood-making cells. His body will reject a transplant unless the donor's tissue is a good match. No suitable donor can be found. His only hope is for a new brother or sister with a matching tissue type. Blood from their umbilical cord could be used to make the cells that he needs.

Zain's parents can have permission to use PGD to select embryos without the blood disorder. The HFEA already allows PGD to be used for this disease. But Zain's parents also want to check the embryos to see if they are a tissue match for their son. No one has asked for PGD to be used in this way before. The HFEA agree.

December 2002: An anti-abortion group takes the case to the High Court. They believe that creating embryos to benefit another human being is wrong. The High Court reverses the HFEA's decision. The judges say that it is against current law, and any change must be made by Parliament.

April 2003: The HFEA appeals against the High Court's decision. It wins. The family can use PGD to select an embryo with a matching tissue type to their son's. Zain's parents hope a new brother and sister will be able to save his life.

Case Two

October 2003: A couple applies to the HFEA to use PGD for tissue matching an embryo. Their son also has a blood disorder. But his disease is not inherited. PGD would not normally be used to test embryos for this disease. The HFEA rules that they cannot use PGD just to find a tissue match.

July 2004: A second couple whose son has the same blood disorder apply to the HFEA to use PGD. The HFEA reconsiders its decision, looking at evidence from all cases of PGD. It decides that PGD can be used for tissue typing, because babies born after PGD do not seem to suffer any more harmful effects than normal IVF babies. The risks to the new brother and sister of having their cells taken are also very small.

People had different opinions about the decision in Case One.

Questions

2 Write down one viewpoint that embryo selection:

  a should not be done because it is wrong

  b should be done because it is the best decision for all involved

3 Make a list of other viewpoints for embryo selection.

4 Which viewpoint do you agree with? Give your reasons.

(5) Cases One and Two are similar in many ways.

  a Why did the HFEA at first decide that PGD could not be used in Case Two?

  b What evidence changed its mind?

Find out about:
- treatment to replace faulty genes

G Gene therapy

Paul and Kamni look fine, but they have health problems. Kamni's own white blood cells will get rid of her cold. Cystic fibrosis is an inherited disorder caused by faulty alleles. But there is no cure for Paul's cystic fibrosis.

Paul has cystic fibrosis.

Kamni has a bad cold.

Finding a new treatment for cystic fibrosis

Key words
gene therapy
genetic modification

The cell nuclei of cystic fibrosis patients contain two faulty, recessive alleles. So one of the proteins the cells make is faulty. The faulty protein causes the cystic fibrosis symptoms. Some scientists have been trying to develop a new treatment for cystic fibrosis. Their plan is to put copies of the normal allele into the cells of cystic fibrosis patients. This kind of treatment is called **gene therapy**.

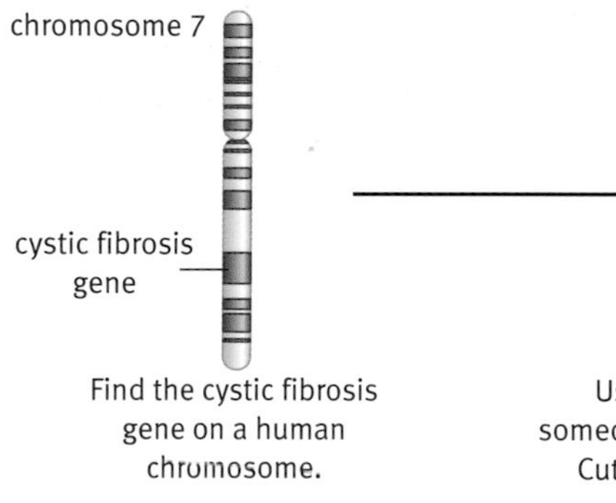

Find the cystic fibrosis gene on a human chromosome.

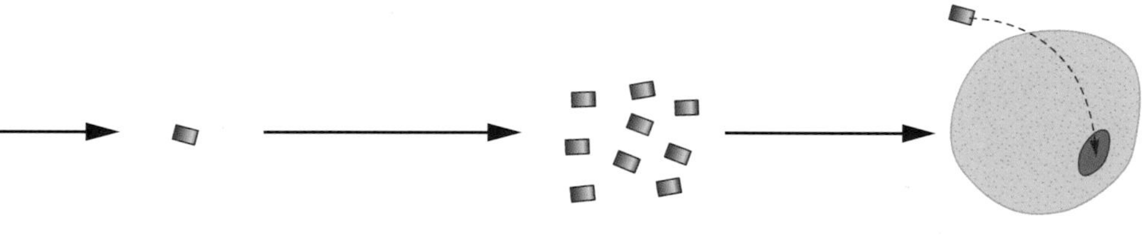

Use chromosomes from someone without cystic fibrosis. Cut out the normal alleles.

Make lots of copies.

Put these copies into cells of the cystic fibrosis patient (genetic modification).

The final step, **genetic modification**, was very difficult. Scientists in the mid-1990s did trials on human patients. They trapped the alleles in fat droplets and used nose sprays to get them into the air passages. Scientists thought that if they had been successful, some of the symptoms should disappear. So it was very exciting when the health of some of the patients did improve a little.

Unfortunately the improvements didn't last. Cells lining the lungs die and are replaced all the time. New cells only contain the patients' original alleles. Scientists are continuing their research, but in some countries all gene therapy trials on humans are closed.

Questions

1 Draw a flow chart to explain the steps in gene therapy.

2 In the 1990s some people thought that gene therapy would soon be able to treat cystic fibrosis. Explain the main problem scientists have had trying to do this.

A gene therapy success story?

Rhys Evans is a happy, healthy four year old. But when he was just four months old Rhys got a chest infection which didn't get better. Doctors soon realized that Rhys was very seriously ill.

Rhys's mother, Marie, remembers this frightening time: 'It was a bit of a mystery really, he lived on a knife edge. The doctors told us "We don't know if your child is going to live today".'

Eventually Rhys was referred to Great Ormond Street Hospital. Doctors at the hospital found that Rhys's immune system was missing an important protein. He wasn't able to fight off diseases by himself.

Rhys was given gene therapy treatment to replace the faulty gene. The diagram shows how this was done.

Rhys's disease was caused by a faulty gene. The disease is called Severe Combined Immunodeficiency Disease (SCID).

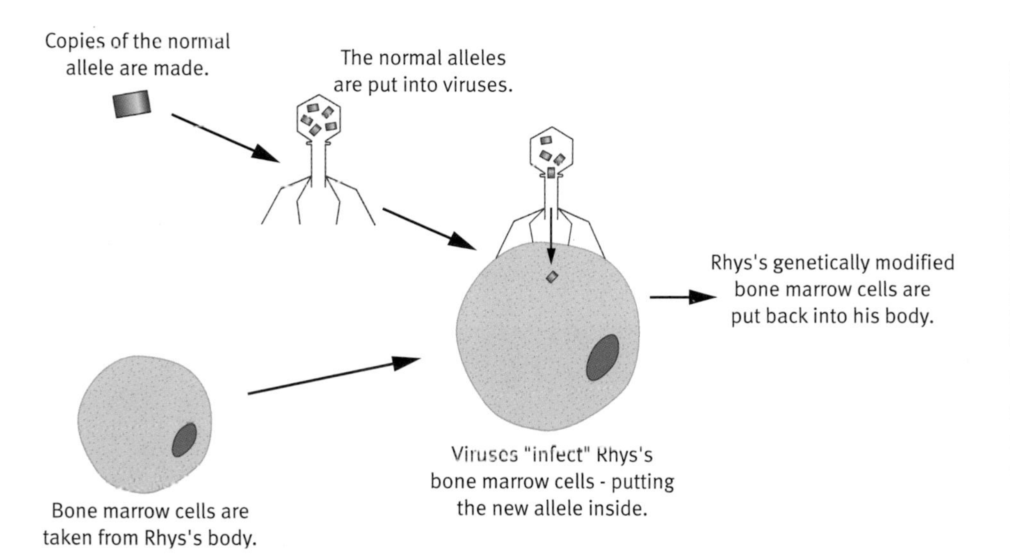

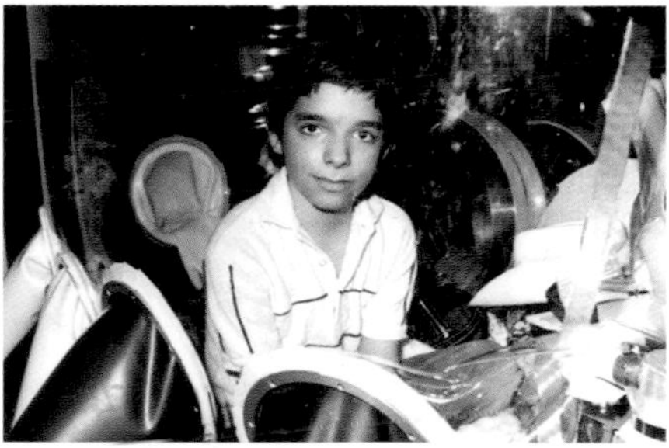

People with SCID have to live in a sterile 'bubble'. This protects them from microorganisms which could kill them.

Future possibilities?

Cells in the body are called 'body cells' – except for egg and sperm cells. These are called sex cells. At the moment gene therapy treatments only put new genes into body cells. The person's sex cells are not changed. So even if the person gets better, they could still pass the faulty gene on to their children.

In the future it may be possible to use gene therapy to prevent known genetic diseases. New genes could be put into the sex cells or fertilized egg cells. All the new person's cells would have correct versions of their faulty genes. At the moment gene therapy of sex cells is illegal. Many people are worried that replacing any genes in sex cells would be a dangerous step. The same method could be used to control other features, for example eye colour. People think this could be a step on the road to 'designer babies'.

Questions

3 Why did Rhys's faulty gene make him so ill?

4 How did doctors get the normal gene into Rhys's cells?

(5) The media often use the term 'designer babies'.

- **a** What do they mean by this term?
- **b** Why are some people worried that gene therapy could be misused in this way?

Find out about:
- asexual reproduction
- cloning and stem cells

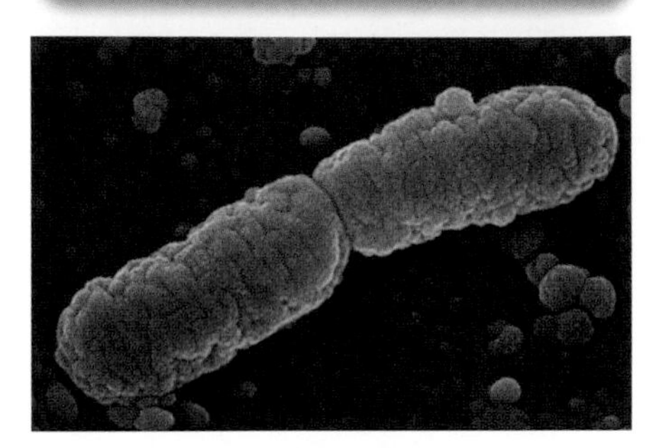

The bacterium cell grows and then splits into two new cells. (mag: × 7500 approx)

H Cloning – science fiction or science fact?

Cloning: a natural process

Many living things only need one parent to reproduce. This is called **asexual reproduction**. Single-celled organisms like the bacterium in the picture use asexual reproduction.

The new bacteria only inherit genes from one parent. So their genes are identical to each other's and their parent's. We call genetically identical organisms **clones**. The only variation between them will be caused by differences in their environment.

Asexual reproduction

Larger plants and animals have different types of cells for different jobs. As an embryo grows, cells become specialized. Some examples are blood cells, muscle cells, and nerve cells.

Plants keep some unspecialised cells all their lives. These cells can become anything that the plant may need. For example, they can make new stems and leaves if the plant is cut down. These cells can also grow whole new plants. So they can be used for asexual reproduction.

The unspecialized cells in this strawberry plant have produced all the different types of cells needed by the new plants.

Hydra

Some simple animals, like the *Hydra* in the picture alongside, also use asexual reproduction. Larger animals do not have unspecialized cells after they have grown. So cloning is very uncommon in animals.

Questions

1 How many clones are shown in the *Hydra* picture?

2 Why is natural cloning more common in plants than animals?

3 Why are a pair of identical twins genetically identical to each other, but not to their parents?

Key words
asexual reproduction
clones

Sexual reproduction

Most animals use sexual reproduction. The new offspring have two parents so they are not clones. But clones are sometimes produced – we call them identical twins.

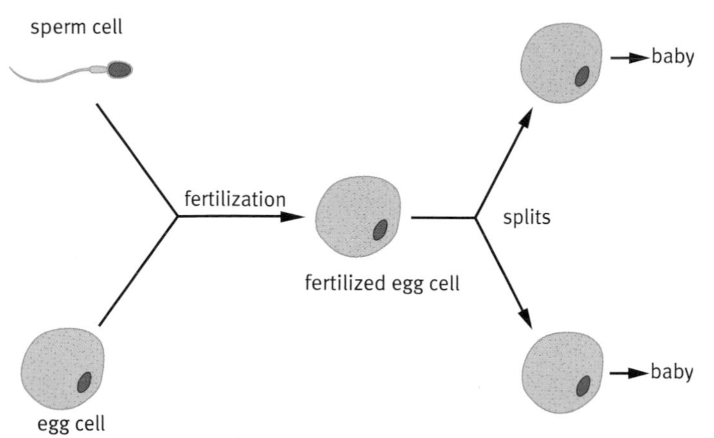

Identical twins have the same genes. But their genes came from both parents. So they are clones of each other, but not of either parent.

Cloning Dolly

Scientists can also clone animals. But this is much more difficult. Dolly the sheep was the first cloned sheep to be born.

- The nucleus was taken from an unfertilized sheep egg cell.
- The nucleus was taken out of a body cell from a different sheep.
- This body cell nucleus was put into the empty egg cell.
- The cell grows to produce a new animal. Its genes will be the same as those of the animal that donated the nucleus. So it will be a clone of that animal.

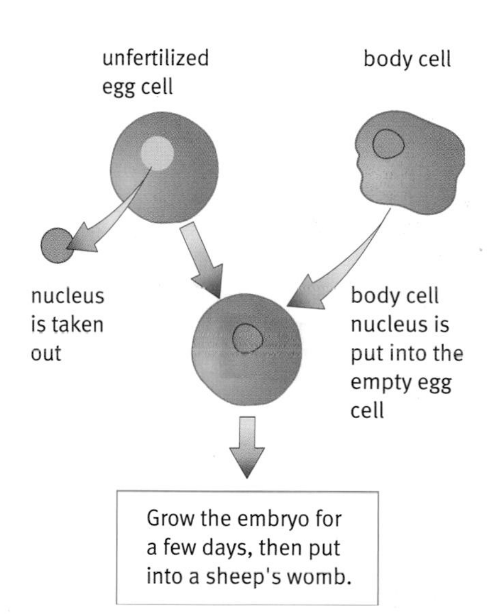

Is it safe to clone mammals?

Dolly died in 2003, aged 6. The average lifespan for a sheep is 12–14 years. Perhaps Dolly's illness had nothing to do with her being cloned. She might have died early anyway. One case is not enough evidence to decide either way.

But it took 277 attempts before Professor Wilmut's team managed to clone Dolly. Many other cloned animals have suffered unusual illnesses. So scientists think that more research needs to be done before cloned mammals will grow into healthy adults.

Professor Ian Wilmut and his team at the Roslin Institute, Edinburgh, cloned Dolly.

Questions

4 Describe how Dolly the sheep was cloned.

5 Where did Dolly inherit her genes from?

Who would you clone? Most scientists don't want to clone adult humans.

Cloning humans

Cloning humans – what does that make you think of? A double of you, or someone else? Scientists are trying to improve methods for cloning animals. So in the future it may be possible to clone humans. But most scientists don't want to clone adult human beings.

However, some scientists do want to clone human embryos. They think that some cells from cloned embryos could be used to treat diseases. The useful cells are called **stem cells**.

What are stem cells?

Stem cells are unspecialized cells. They can grow into any type of cell in the human body.

Stem cells can be taken from embryos that are a few days old. Researchers use human embryos that are left over from fertility treatment.

Scientists want to grow stem cells to make new cells to treat patients with some diseases. For example, new brain cells could be made for patients with Parkinson's disease.

But these new cells would need to have the same genes as the person getting them as a treatment. When someone else's cells are used in a transplant they are rejected.

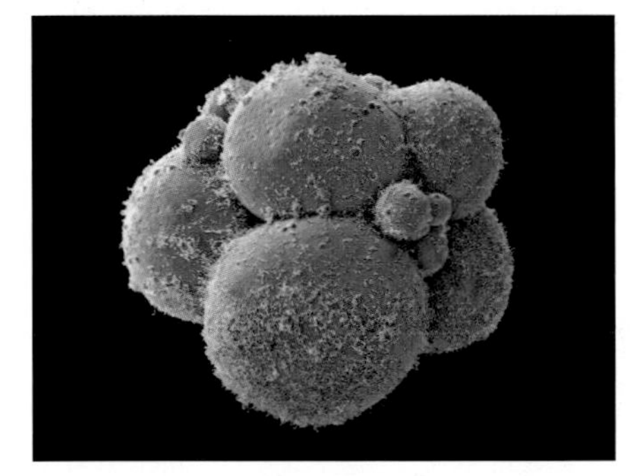

Cells from eight-cell embryos like this one can develop into any type of body cell. They start to become specialized when the embryo is five days old. (Mag: × 500 approx)

What's cloning got to do with this?

Cloning could be used to produce an embryo with the same genes as the patient. Stem cells from this embryo would have the same genes as the patient. So cells produced from the embryo would not be rejected by the patient's body. This is called **therapeutic cloning**.

Doctors have only started to explore this technology. Success is still years away, but millions of people could benefit if it is made to work.

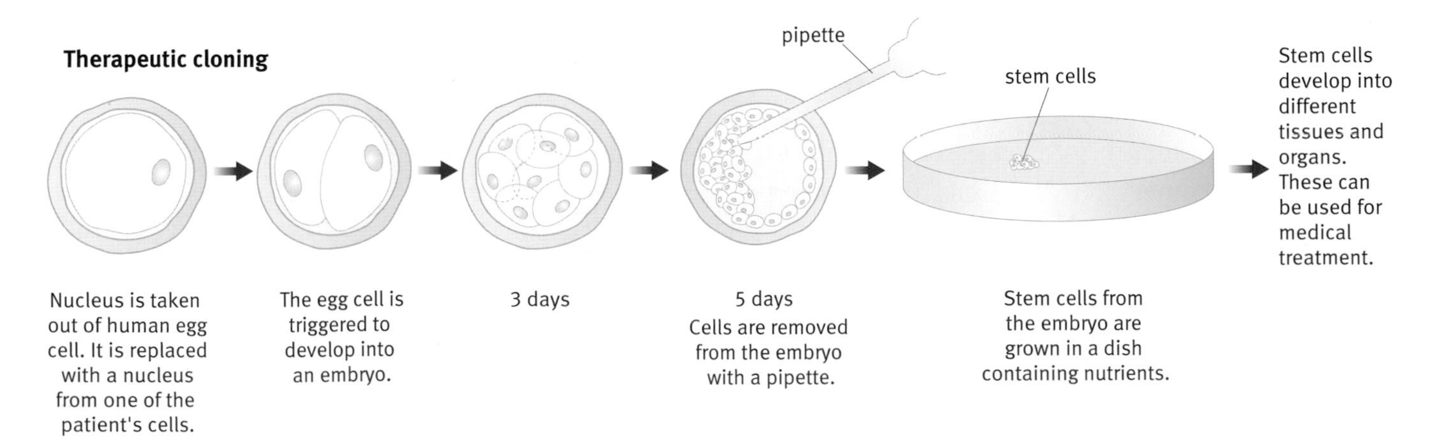

Should human cloning be allowed?

James has Parkinson's disease. His brain cells do not communicate with each other properly. He cannot control his movements.

Key words

stem cells
therapeutic cloning

Questions

6 How are stem cells different from other cells?

7 Explain why scientists think stem cells would be useful in treating Parkinsons's disease.

8 Explain how this is different from cloning an adult.

9 For each of these cells, say whether ot not your body would reject it:

a bone marrow from your twin

b your own skin cells

c a cloned embryo stem cell

10 For embryo cloning to make stem cells:

a describe one viewpoint in favour

b describe two different viewpoints against

11 People often make speculations when they are arguing for or against something. This is something they think will happen, but may not have evidence for. Write down a viewpoint that is a speculation.

B1 You and your genes

Science explanations

How living things develop is one of the most complex explanations. In this Module you've begun to explore the science behind what makes you the way you are.

You should know:

- most of your features are affected by your environment and your genes
- genes are found in the nuclei of cells and are instructions for making proteins
- your chromosomes, and genes, are in pairs
- genes have different versions, called alleles
- the difference between dominant and recessive alleles
- men and women have different sex chromosomes
- how a gene on the Y chromosome causes an embryo to develop as a man
- why you may look like your parents
- why you may look like your brothers and sisters, but not be identical
- how to interpret family trees
- how to complete genetic cross diagrams
- the symptoms of cystic fibrosis and Huntington's disorder
- why people can be carriers of cystic fibrosis, but not of Huntington's disorder
- doctors can test embryos, fetuses, and adults for certain alleles by genetic tests
- what happens during embryo selection (pre-implantation genetic diagnosis)
- how gene therapy could be used to treat some genetic disorders
- that some organisms use asexual reproduction and have offspring that are clones
- how animal clones are produced naturally and artificially
- that cells in multicellular organisms become specialized very early on in the organism's development
- what stem cells are, and how they could be used to treat certain diseases

Ideas about science

It is difficult to make decisions about some uses of science. Many of the issues in this module have ethical questions. Ethics is about deciding whether something is a right or wrong way to behave. Just because science can help us to do something doesn't mean it's right or that it should be allowed.

People may disagree about some ethical questions. Often they agree about the facts of an issue, but disagree about what should be done. For example, it is possible to test for some alleles that cause disease. People disagree about whether these tests should be done. They disagree about whether people should be allowed to terminate a pregnancy if a fetus has a genetic disorder.

There are different viewpoints presented for each of the issues discussed in this Module:

- some people think that certain actions are wrong whatever the circumstances
- some people think that you should weigh up the benefit and harm for everyone involved and then make a decision

People may make different decisions because of their beliefs, and/or their personal circumstances. When you consider an ethical issue you should be able to:

- say clearly what they issue is
- describe some different viewpoints people may have
- say what you think and why

You've looked at different issues in this Module:

- should we test fetuses for particular genetic disorders?
- should other people, like insurance companies and employers, be allowed to have information about a person's genes?
- should we use genetic tests to choose embryos without certain genetic disorders?
- should doctors be allowed to use gene therapy to treat people with some genetic disorders?
- should doctors be allowed to clone stem cells from embryos to treat certain illnesses (therapeutic cloning)?

Why study keeping healthy?

Good health is something everyone wants. Stories about keeping healthy are all around you, for example, news reports about what to eat, new viruses and 'superbugs'. New evidence is reported everyday. So the message about how to stay healthy often changes. It's not always easy to know which advice is best.

The science

Some diseases are caused by harmful microorganisms. Your body has ways to stop them getting in. If you are infected it has amazing ways of fighting back. Vaccines and drugs can help you survive many diseases, and doctors are always trying to develop new ones. But, not all diseases are caused by microorganisms. Your lifestyle may also put you at risk. Media reports often warn about the dangers of smoking, eating badly, and not exercising.

Ideas about science

So, how do you decide which health reports are reliable? Knowing about correlation and cause and peer review will help. There are also ethical questions (arguments about right and wrong) to consider when deciding how we should use vaccines and drugs.

Keeping healthy

Find out about:

- how your body fights infections
- arguments people may have about vaccines
- where 'superbugs' come from
- how new vaccines and drugs are developed and tested
- how scientists can be sure what causes heart disease

Find out about:

- how some microorganisms make you ill
- how your body keeps MOs out
- infections and lifetyle diseases

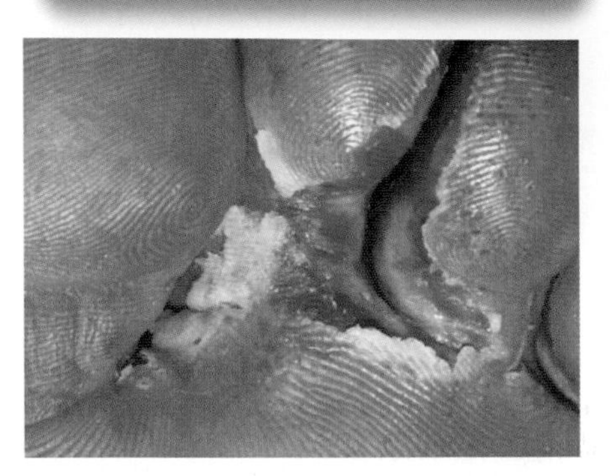

The fungus that causes athlete's foot grows on the skin.

A What's up, Doc?

Most days you don't think about your health. It's only when you're ill that you realize how important good health is. Everyone has some health problems during their lives. Usually it is minor – like a cold. But sometimes it is more serious. Some illnesses may be life-threatening, like heart disease or cancer.

There are lots of reasons for feeling ill. In the doctor's waiting room:

- the man with the painful knee has arthritis
- the young woman feeling sick and tired doesn't know that she's pregnant
- the man having his monthly check-up has had heart disease

None of these conditions can be passed on to other people. But the other patients all have **infectious** diseases. Infections can be passed from one person to another.

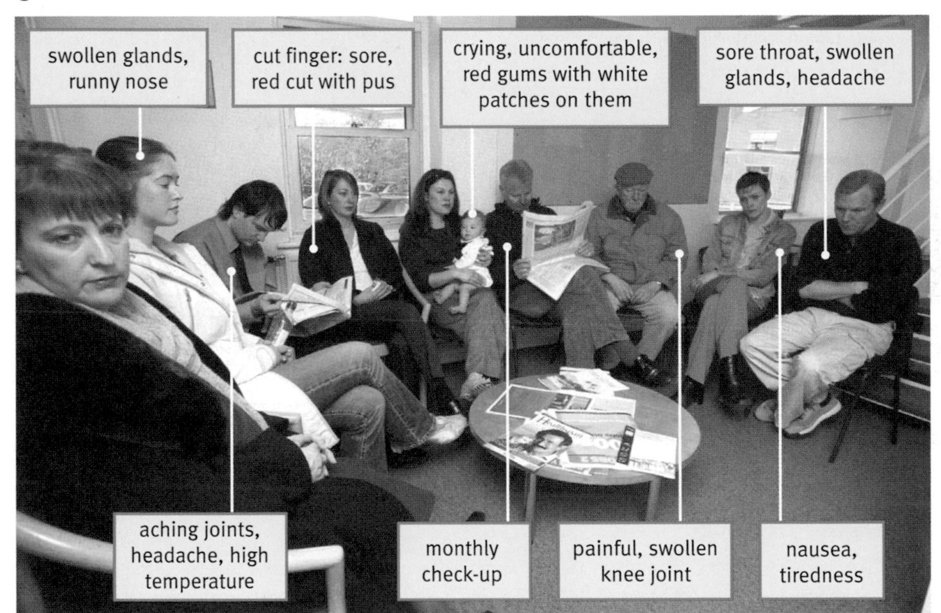

Passing it on

Infections are caused by some **microorganisms** that invade the body. Microorganisms (MOs) are **viruses**, **bacteria**, and **fungi**.

When disease MOs get inside your body, they reproduce very quickly. This causes **symptoms** – the ill feelings you get when you are unwell. Symptoms can be caused by:

- damage done to your cells when the MOs reproduce
- poisons made by the MOs

There are medicines that can cure many diseases caused by bacteria and fungi. But we still don't have many good treatments for diseases caused by viruses. Instead we take medicines that help us feel better until our bodies get rid of the viruses. You will learn more about this later in this chapter.

What are microorganisms like?

Microorganisms are very small. To see bacteria you need a microscope. Viruses are even smaller. They are measured in nanometres, and one nanometre is only one millionth of a millimetre.

Every breath of air you take has billions of MOs in it. And every surface you touch is covered with them. But most of the time you stay fit and healthy. This is because:

- most MOs do not cause human diseases
- your body has barriers that keep most MOs out

	Virus	Bacterium	Fungus
Size	20–300 nm	1–5 μm	50+ μm
Appearance			
Examples of diseases caused	flu, polio, common cold, AIDS, measles	tonsillitis, tuberculosis, plague, cystitis	athlete's foot, thrush, ringworm

Lifestyle diseases

One hundred years ago infectious diseases killed most people in the UK. Today better hygiene and health care means these illnesses are more controlled. Now **lifestyle diseases** are much more common than they were. These include heart disease and some cancers.

Lifestyle diseases aren't caused by infections. For example, most things that increase the risk of a heart attack are to do with a person's lifestyle – a high-fat diet, smoking, and lack of exercise.

One hundred years ago people often died of an infection before reaching old-age. Today the average lifespan is longer. So the way people live has more time to affect their bodies.

But we must not forget the power of MOs. Some infections are becoming more common, for example, food poisoning. We cannot prevent the common cold. And new infectious diseases are developing. These are all strong reminders of how vulnerable we are to attack by some of the smallest organisms on Earth.

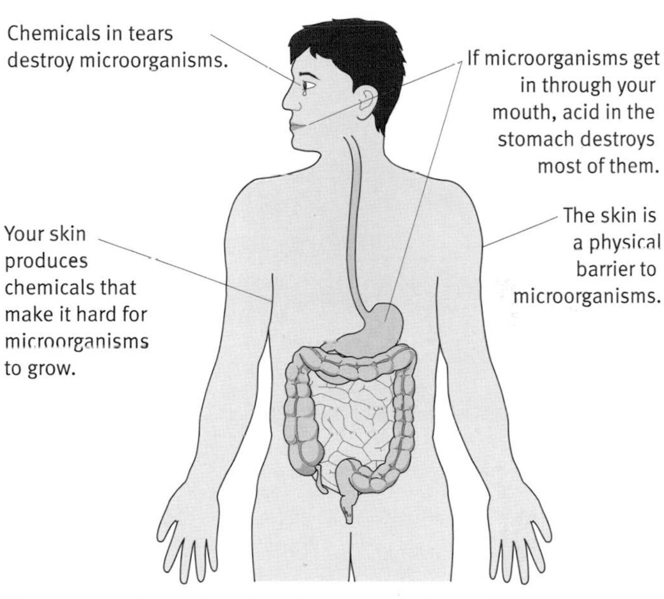

The human body has barriers to stop harmful MOs getting inside.

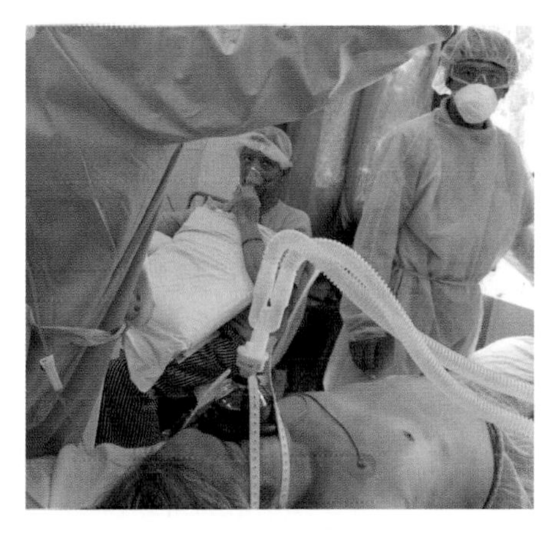

In 2003 a new infection called SARS appeared.

Questions

1. Name three types of MOs that can cause disease.
2. Write down two different diseases caused by each type of MO you have named.
3. Explain two ways that MOs make you feel ill.
4. Describe three defences your body has to stop MOs from getting in.
5. Write a few sentences to explain to someone why people don't usually 'catch' heart disease.

Key words

infectious
microorganisms (MOs)
fungi
viruses
bacteria
symptoms
lifestyle diseases

Find out about:

- how white blood cells fight infections
- what antibiotics do

B Microbe attack!

Jolene cut her finger when she was gardening. She didn't wash it quickly, so bacteria on her skin and in the soil invaded her body. Once inside they started to reproduce. And when bacteria reproduce, they do it in style.

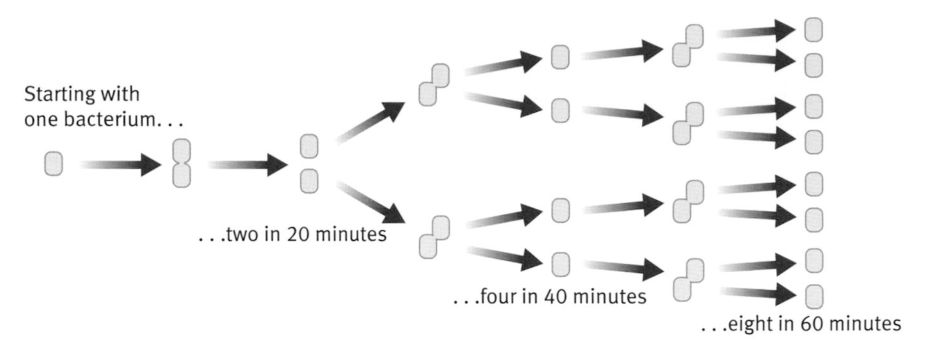

Bacteria can reproduce rapidly inside the body.

Reproduction in bacteria is simple. Each bacterium splits into two new ones. These grow for a short time before splitting again. If conditions are right – warmth, nutrients, moisture – they can split every 20 minutes.

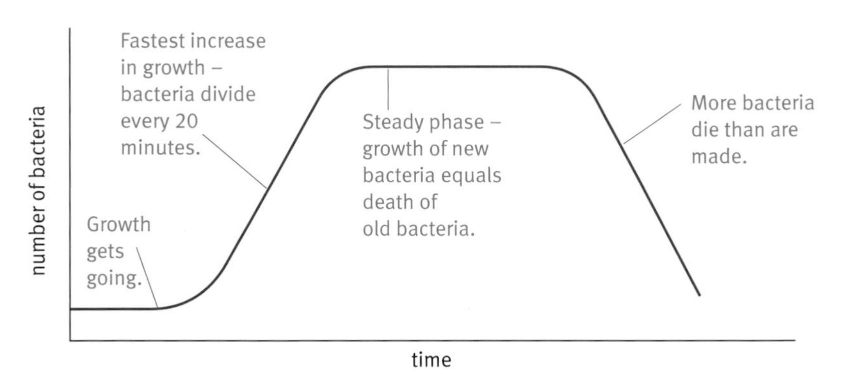

In ideal conditions in a sealed container bacteria can't keep up their fastest growth. Food starts to run out, or waste products kill them off.

Jolene's body responds by sending more blood to the area.

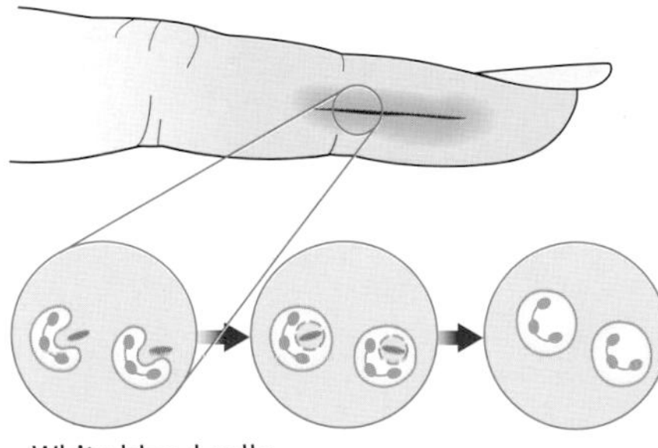

White blood cells surround the bacteria and digest them.

The battle for Jolene's finger

Conditions inside Jolene's body are ideal for the bacteria. But they don't have everything their own way.

The redness and swelling in Jolene's finger is called inflammation. Extra blood is being sent to the wounded area, carrying with it the body's main defenders – the **white blood cells**. One type of white blood cell surrounds the bacteria and **digests** them.

The worn-out white blood cells, dead bacteria, and broken cells collect as pus. So redness and pus show that your body is fighting infection. As the bacteria are killed, the inflammation and pus get less until the tissue heals completely.

Your body army – fighting infection

The parts of your body that fight infections are called your **immune system**. White blood cells are an important part of your immune system.

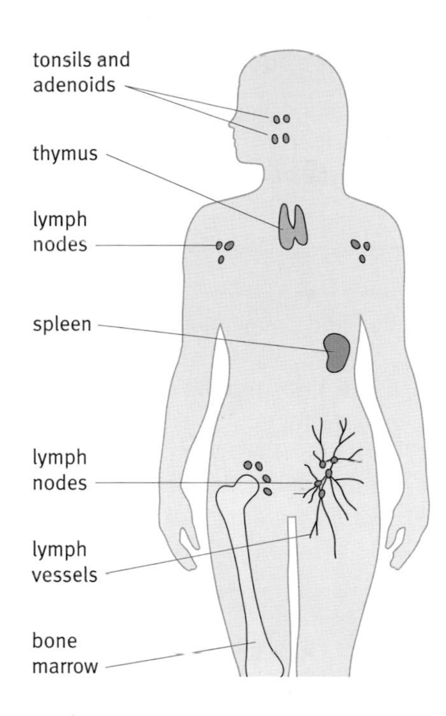

The main parts of your immune system.

What's the verdict?

In most cases the body will overcome invading bacteria. Keeping the cut clean and using antiseptic is usually enough treatment. But, Jolene's cut is quite deep, so her doctor gives her a course of **antibiotics**. These are chemicals that kill bacteria and fungi. Different antibiotics affect different bacteria or fungi.

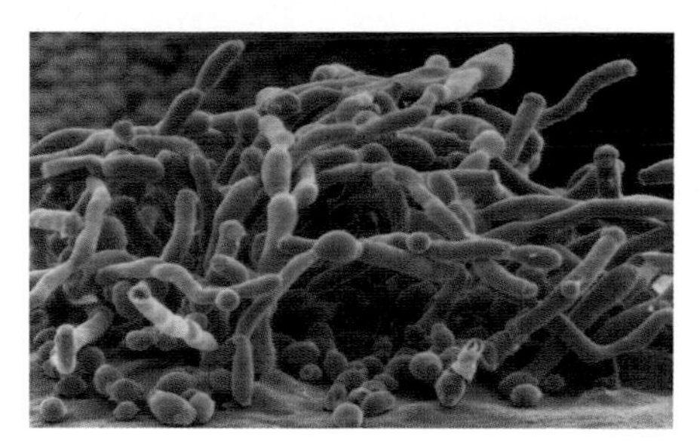

Candida albicans is a common fungus that causes thrush. It lives on warm, moist body surfaces. It usually infects the vagina or mouth.

Do antibiotics have side effects?

Soon after her finger has healed, Jolene is back at the doctor's. She has a common disease called thrush which infects the mouth and reproductive passage. Jolene's friend has told her that taking antibiotics gives you thrush. But she's not got the story quite right.

There is a **correlation** between some antibiotics and thrush. A person is more likely to get thrush if they have had a course of antibiotics than if they have not. But they won't definitely develop thrush. The diagram explains the problem.

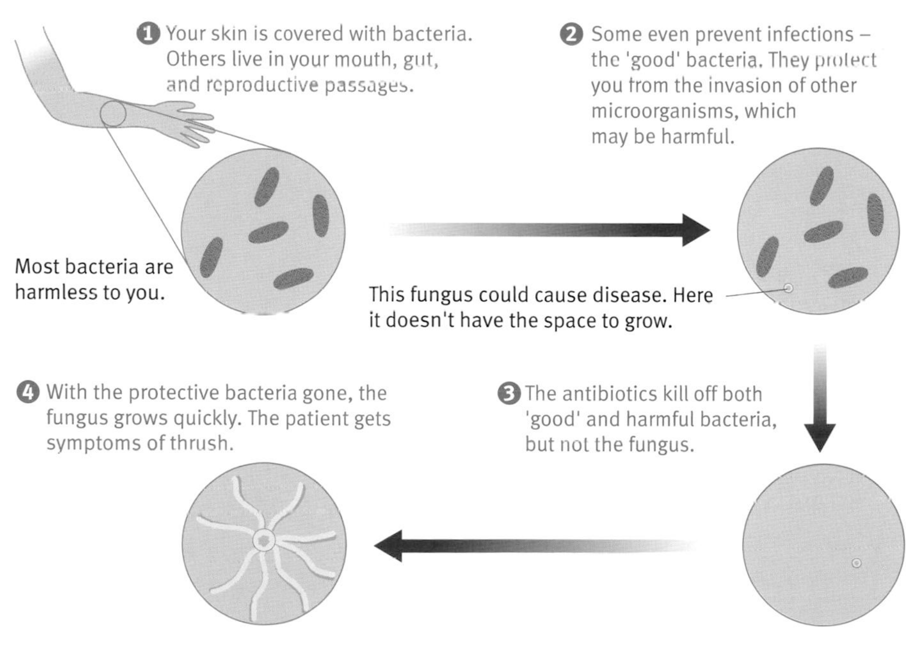

Fortunately for Jolene there are other antibiotics that kill fungi. It can also be helpful to eat 'bio yoghurt' when you are taking antibiotics. These contain 'good' bacteria to help replace the lost ones as quickly as possible.

Key words

white blood cells
digests
immune system
antibiotics
correlation

Questions

1 What are ideal conditions for bacteria to reproduce?

2 You have a small cut. How can you reduce your risk of infection?

3 Write down one sentence to describe the job of the immune system.

4 What types of MOs do we treat with antibiotics?

(5) Explain why taking antibiotics may lead to thrush.

Find out about

- how white blood cells fight infection
- how you become immune to a disease

C Everybody needs antibodies – not antibiotics!

A bad cold is something we've all had. All you feel like doing is curling up in bed. And there's not usually much sympathy – 'What's all the fuss about? It's just a cold!'

Natalie has been ill for a few days. Her doctor explains that he won't be giving her any antibiotics. Her cold is caused by a virus, which antibiotics cannot treat. Natalie's own body is fighting the infection by itself.

Fighting the virus

Natalie's neck glands are swollen because millions of new white blood cells are being made there. These white blood cells are fighting the virus in her body.

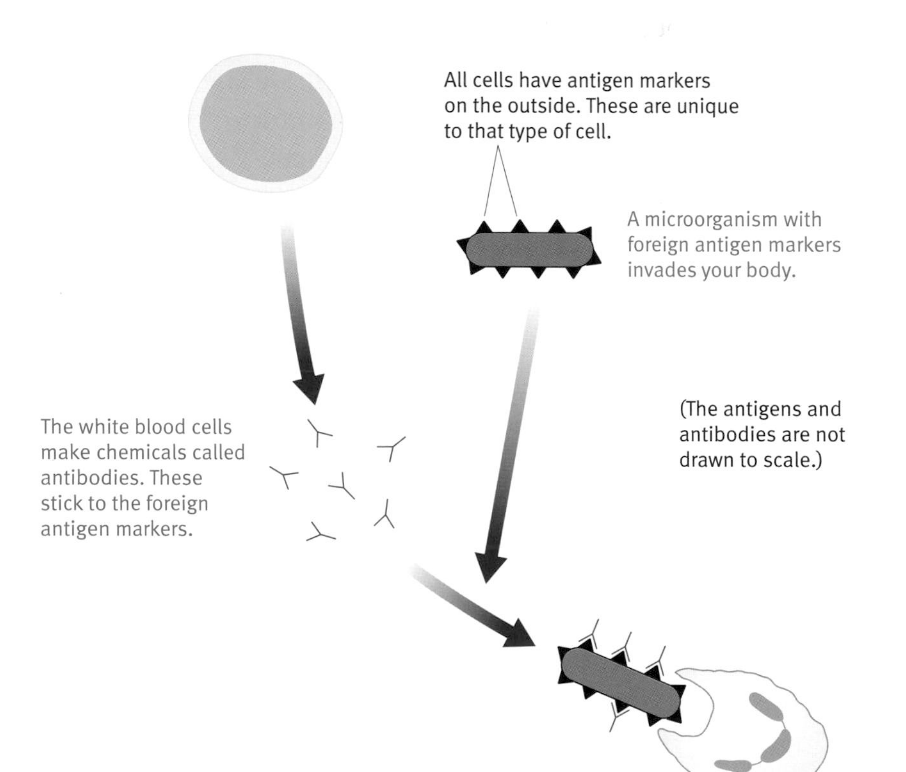

One type of white blood cell makes **antibodies** to label MOs. A different type digests the MOs.

If antibodies are so good, why do I get ill?

The **antigens** on every MO are different. So your body has to make a different antibody for each new kind of MO. This takes a few days, so you get ill before your body has destroyed the invaders.

This doesn't really matter for diseases like a cold. But for more serious diseases this is a problem. The disease could kill a person before their body has time to destroy the MOs.

Why do you get some diseases just once?

Once your body has made an antibody it is not forgotten. Some of the white blood cells that make the antibody stay in your blood. If the same MO invades again, these white blood cells reproduce very quickly and start making the right antibody. This means that the body reacts much faster the second time you meet a particular MO. Your body destroys the invaders before they make you feel ill. So you are **immune** to that disease.

Not another one!

Natalie's cold soon got better, but she had only been back at school for about three weeks before she caught another one. If you have an illness like chickenpox, you are very unlikely to catch it again, because you are immune. So why do we catch an average of three to five colds every year?

The problem is that there are hundreds of different cold viruses. So every cold you catch is caused by a different virus. To make things worse, the viruses have a very high **mutation** rate (more about this on page 45). This means that their DNA changes regularly. So do the markers on their surface. Your body needs to make a different antibody to fight the virus. So, we suffer the symptoms of a cold all over again.

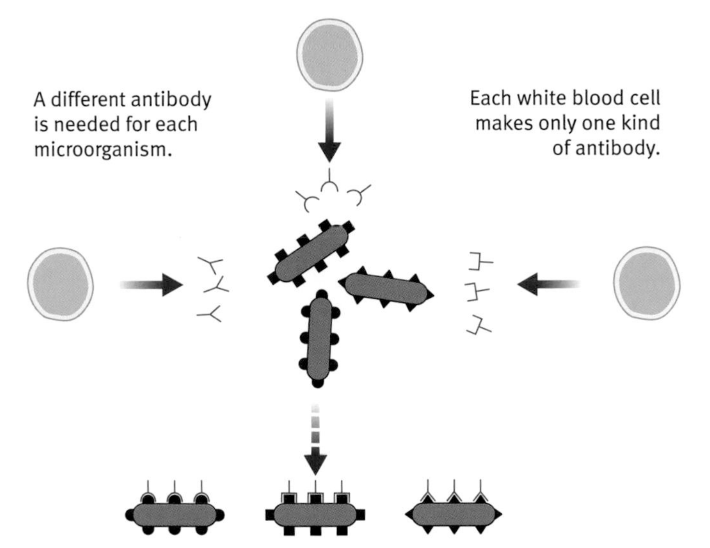

Only the correctly shaped antibody can attach to each kind of MO.

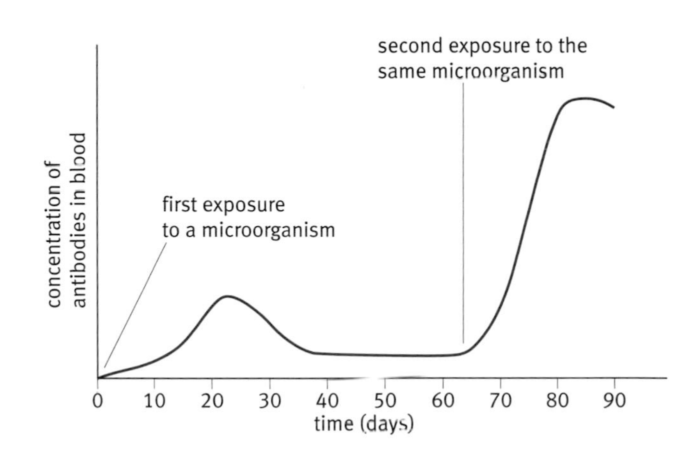

A person is infected twice by a disease MO. Their white blood cells make antibodies much faster the second time.

Questions

1. Why are antibiotics not given to patients infected with a virus?
2. Explain two ways that white blood cells protect the body from invading MOs. You could do this with a diagram.
3. Explain why different antibodies have to be made for every MO.
4. (4) Draw a flowchart to explain how you can become immune to chickenpox.
5. (5) Write a few sentences to explain to Natalie why she will never be immune to catching colds.

Key words

antibodies
antigens
immune
mutation

Find out about:

- how vaccines work
- deciding if vaccines are safe to use

Small amounts of disease MOs are put into your body. Dead or inactive forms are used so you don't get the disease itself. Sometimes just parts of the MOs are used.

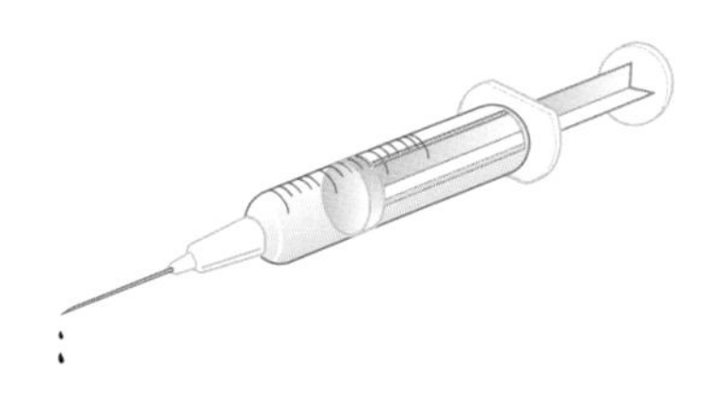

White blood cells recognise the foreign MOs. They make the right antibodies to stick to the MOs.

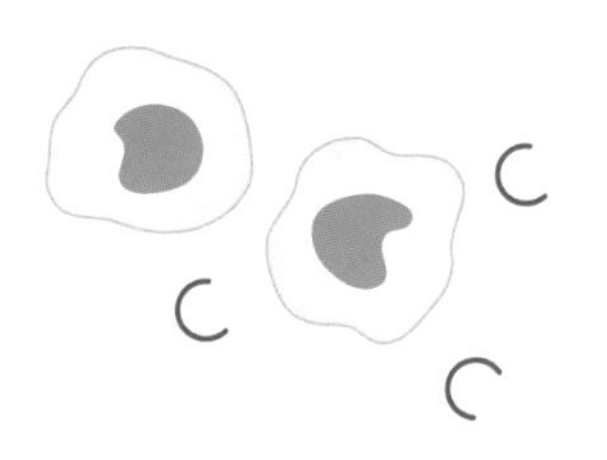

The antibodies make the MO's clump together. White blood cells digest the clump.

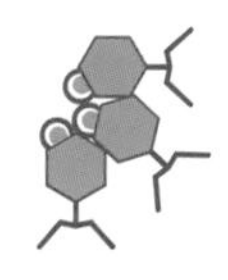

If you meet the real disease MO, the antibodies you need are made very quickly.

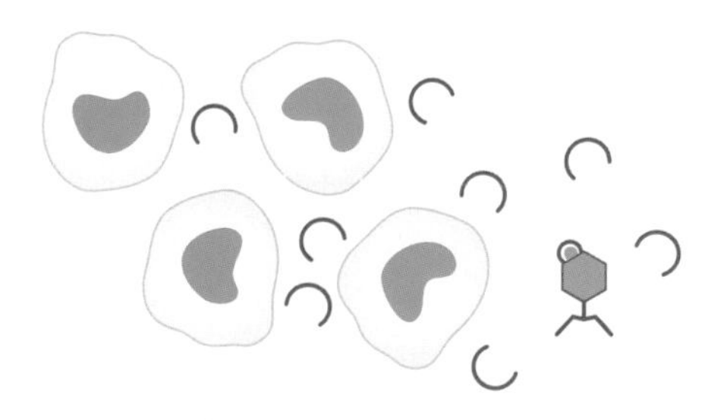

The MOs are destroyed before they can make you ill.
(Not to scale)

Key words
vaccinations

D Vaccines

In the UK we are lucky to be able to get medicines for many diseases. But it would be even better not to catch a disease in the first place. **Vaccinations** aim to do just that.

Vaccinations make use of the body's own defence system. They kick-start your white blood cells into making antibodies. So you become immune to a disease without having to catch it first.

Age	Immunisation
2, 3, and 4 months	polio, DTP-Hib (diphtheria, tetanus, pertussis, and hib – causes pneumonia and meningitis), meningitis C
13 months	MMR (measles, mumps, and rubella)
3–5 years	polio, DTaP (diphtheria, tetanus, and acellular pertussis), and MMR
10–14 years	BCG (against tuberculosis)
13–18 years	tetanus and polio

Many childhood diseases are very rare in the UK because of vaccination programmes.

Are vaccines safe?

Any medical treatment you have should do two things:

- improve your health
- be safe to use

Vaccines can improve your health by protecting you from disease. They are tested to make sure that they are safe to use. But it is important to remember that no action is ever completely safe. People react differently to medical treatments, including vaccines.

Doctors decide that a treatment is safe to use when:

- the risk of serious harmful effects is very small
- the benefits outweigh any risk

You can read more about this in Section F.

Questions

1 What is a vaccine made of?

2 Describe how a vaccine can stop you from catching an infectious disease.

3 Explain why a vaccine can never be 'completely safe'.

Should Tom have a flu vaccine?

Tom is weighing up the pros and cons of having a flu vaccine. Flu is a serious disease that kills thousands of people each year. Most of them are elderly or suffering from other illnesses. People most at risk of dying from flu are advised to have a new vaccine each year.

Costs	Benefits
small risk of a reaction to the vaccine – Tom might feel a bit ill for a few days	much smaller risk of suffering from flu
cost of providing Tom's flu vaccine is about £3.70	saving to the NHS if Tom does not get flu could be £1000s

Why are new flu vaccines made each year?

The **influenza** (flu) virus reproduces very quickly. It also has a very high mutation rate. Mutation means that a small change happens to the DNA. So new kinds of flu virus develop regularly. A different vaccine is needed against each new flu virus.

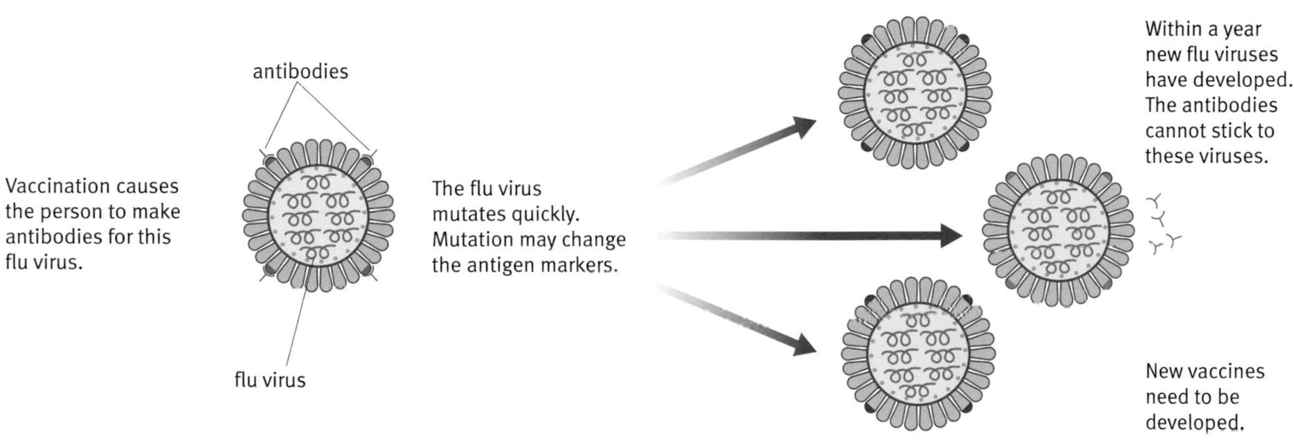

Fighting AIDS

AIDS kills millions of people worldwide each year. It is caused by the **HIV** virus. A major problem in fighting AIDS is that the virus damages the immune system itself. This makes the immune system poor at fighting off other infections. A person with AIDS can become very ill from an infection that a healthy person would quickly fight off.

Another big problem is that we don't have a vaccine against HIV. Unfortunately HIV also mutates very quickly. So a vaccine would probably be useless before it had been fully tested.

Key words

influenza
HIV
AIDS

Questions

4 Explain why a new flu vaccine must be produced every year.

5 Why is it difficult to produce a vaccine against HIV?

⑥ An elderly relative or friend has been offered a 'flu jab' by their doctor. They are worried it may not be safe. What would you advise them to do? Explain your reasons.

Whose choice is it?

To stop a large outbreak of a disease, almost everyone in the population needs to be vaccinated. If they are not, large numbers of the disease-causing MOs will be left in infected people. If the vaccination rate drops just a little, lots of people will get ill.

The vaccination rate is 98%. Unvaccinated people are unlikely to catch the disease.

The vaccination rate has dropped to 90%. Unvaccinated people are much more likely to catch the disease.

The MMR vaccine protects against three diseases – measles, mumps, and rubella. In 2001 the media wanted to know if Prime Minister Tony Blair's baby son, Leo, had been given the MMR vaccine.

Why does the government encourage vaccinations?

Doctors encourage parents to have their children vaccinated at an early age. In the UK there are mass vaccination programmes for some diseases, such as measles. This means that few people suffer from these diseases. Parents have to balance the possible harm from the disease against the risk of possible side-effects from the vaccine.

- Almost everyone who has a vaccine notices no harmful effects.
- Harmful effects from MMR can be mild (3 in every 100 000 children), or produce a serious allergic reaction (1 in every million children).
- Some children who catch measles are left severely disabled (1 in every 4000 cases).
- Measles can be fatal (1 in 100 000 cases).

For society as a whole, vaccination is the best choice. But for each parent, it is a difficult choice, with their child at the centre of it. It is important that people have clear and unbiased information to help them make their decision.

Recent news stories about the MMR vaccine have worried many parents. In the 1970s there were similar worries about the whooping cough vaccine.

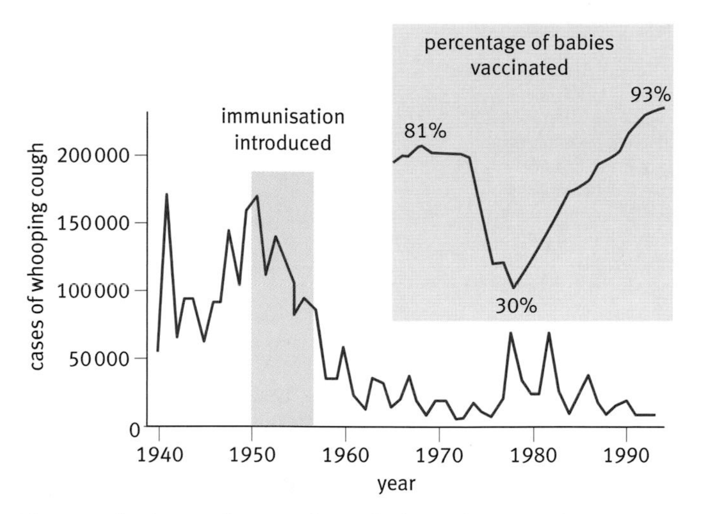

The graph shows the number of whooping cough cases in the UK each year between 1940 and 1992.

The graph shows how the number of cases of measles (children under 15) changed between 1980 and 2004.

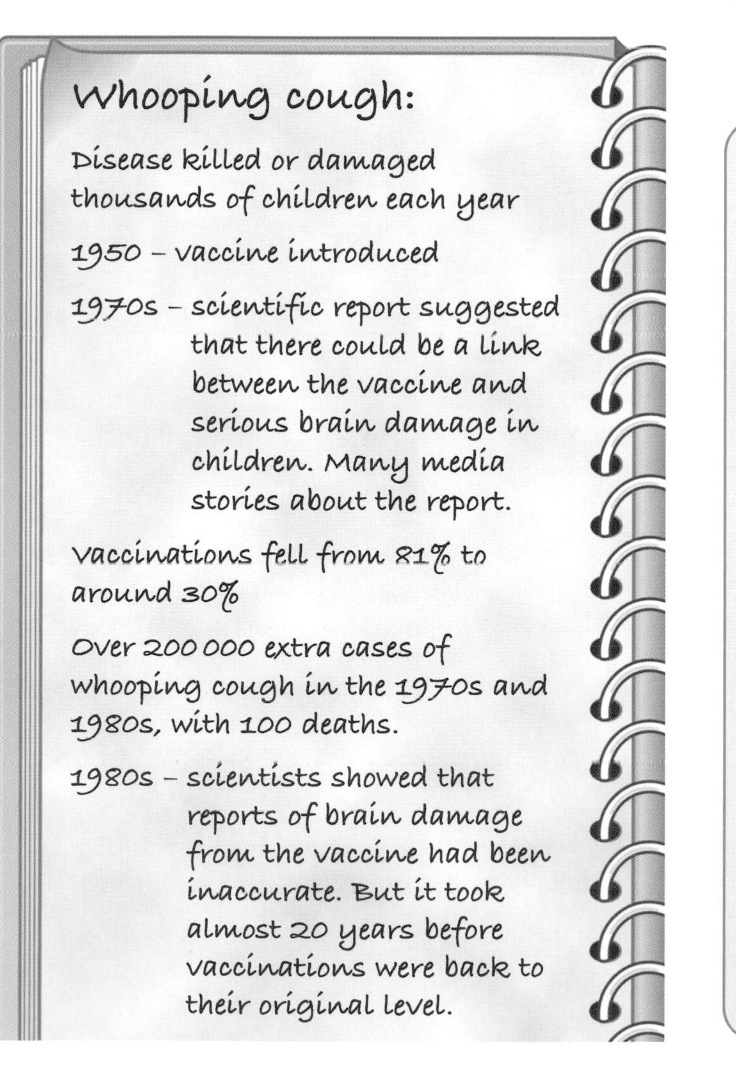

Questions

7 To stop a large outbreak of a disease almost all of the population must be vaccinated against it. Explain why.

8 a Estimate the number of whooping cough cases one year before vaccination began.

b Describe what happened to the number of cases between 1950 and 1970.

c What happened to the percentage of babies vaccinated between 1973 and 1979?

d Explain why this change happened.

9 Look at the number of whooping cough cases between 1965 and 1990. Is there any correlation with the percentage of babies vaccinated?

10 Suggest why the number of deaths from measles peaked in 1988.

11 Scientific data is more likely to be accepted when other scientists have been able to repeat it. Use the example of whooping cough to explain why.

MMR – what's the story?

27 February 1998

MMR LINKED TO AUTISM?

A scientist continues to claim his work shows a connection between the measles, mumps, and rubella jab (MMR) and autism. His comments have spread panic in parents throughout the UK. New figures show that fewer children are having the MMR vaccine.

Health officials are seriously worried that a measles epidemic could be the result. Two children in the Republic of Ireland have already died, and many were left disabled in a recent measles epidemic involving 1500 cases after vaccination rates dropped.

The World Health Organisation suggests that ideally 95% of children should receive the vaccination. In the UK in some areas, that figure has sunk as low as 61%, leaving the door open to the diseases and all the problems they bring.

February 2002

EVIDENCE AGAINST MMR LOOKS THIN

One of the scientists who originally backed the claim that MMR might be linked to autism today explained his change of mind. 'There is now evidence that MMR is not a risk factor for autism. There is a massive amount of medical information from around the world to support this conclusion.'

Unfortunately the symptoms of autism tend to appear at the same age as the first MMR vaccine is given. But medical studies show that although autism levels have risen, this is not linked to the introduction of MMR. Autism is no more common among children who have been vaccinated than in those who have not.

More than 500 million doses of MMR have been given in more than 90 countries with no evidence of a link.

24th January 2004

'No link' between MMR and autism

Scientists have reported the strongest evidence yet that MMR does not cause autism.

Researchers looked at number of autism cases in a city in Japan, before and after the MMR vaccine was withdrawn in 1993.

Autism rates kept on rising, even after the vaccine was withdrawn. 'These results rubbish the claim that MMR has an effect on the rate of autism' said a leading scientist. He also suggested that cases of autism are going up because doctors are better at diagnosing it.

28 January 2004

MMR JABS RISE FOR FIRST TIME IN YEAR

The number of young children having the controversial MMR jab has risen for the first time in more than a year, figures showed yesterday. The percentage of two-year-olds who had MMR rose by 0.9% to 79.8% between July and September. The Health Protection Agency welcomed the news but said it could be partly due to a change in the way that the information was collected. This follows a drop in use of the vaccine, amid concerns that it might cause autism, a link that has not been proved.

What is autism?

People with autism often have difficulty communicating with others. They may have difficulty with language skills and some thinking skills. A person with autism describes their condition:

> Reality to an autistic person is a confusing, interacting mass of events, people, places, sounds and sights. There seems to be no clear boundaries, order or meaning to anything. A large part of my life is spent just trying to work out the pattern behind everything.

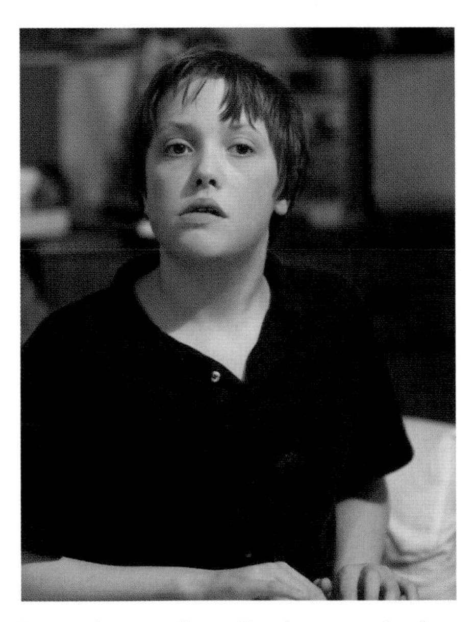

James has autism. He does not look disabled. This can make it harder for other people to understand his condition.

Questions

12 Describe three pieces of evidence about the safety of the MMR vaccine.

13 Describe two points of view parents may have about having their child vaccinated against MMR.

14 For each of these points of view list the main benefits and drawbacks for:

a the child

b other children

Smallpox

Smallpox was a devastating disease. In the 1950s there were 50 million cases worldwide. This fell to 10–15 million cases by 1967 because of vaccination by some countries. But 60 percent of the world's population were still at risk.

In 1967 the World Health Organisation (WHO) began a campaign to wipe out smallpox by vaccinating people across the world. In 1977 the last natural case of smallpox was recorded, in Somalia, Eastern Africa.

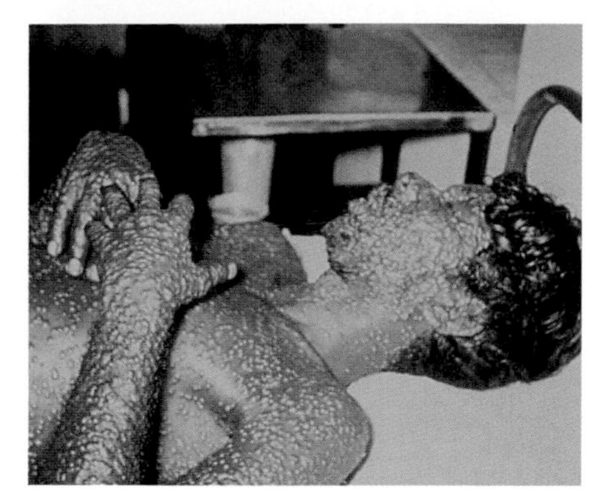

Smallpox killed every fourth victim. It left most survivors with large scars, and many were also blinded.

Why could smallpox be wiped out?

The smallpox virus has a much lower mutation rate than, for example, the flu virus. This meant that the vaccine was effective all through the WHO campaign. The WHO also had the cooperation of governments across the world.

Should people be forced to have vaccinations?

There is enough measles vaccine for every child in the UK. If everyone had to be vaccinated by law, there would be a much lower risk of any child catching the disease. A few children would still get the disease, because vaccinations don't have a 100% success rate.

So it would be possible for measles vaccination to be compulsory – but it isn't. Society does not think it is right to force anyone to have this particular treatment. There is a difference between what *can* be done with science, and what people think *should* be done.

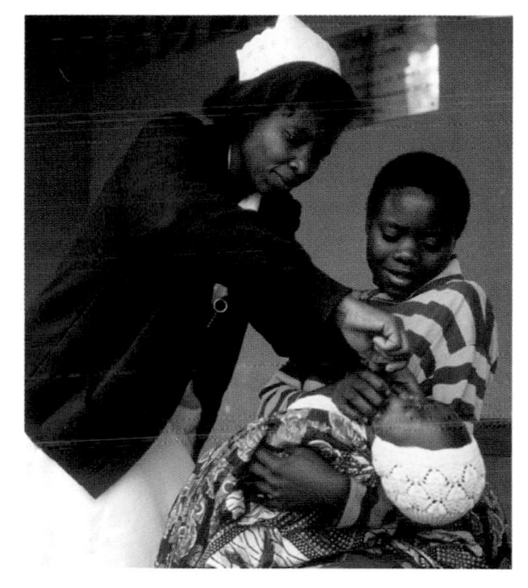

People have become concerned about the safety of vaccines for their children. But for some, the decision is easy.

Different decisions

Where you live may make a difference to your choice about vaccination.

People in poorer countries are more likely to catch a disease due to poor hygiene, or overcrowded housing. They will also suffer more if they catch a disease because they may:

- be weaker because of poor diet or other diseases
- have less access to medicines and other health care

So people from poorer countries may make different decisions about vaccinations, compared with people in better-off communities.

Questions

15 Describe why measles vaccination is not compulsory in the UK.

16 Explain why smallpox could be wiped out by vaccination but flu cannot be.

17 Give two reasons why people in different parts of the world may feel differently about having vaccinations.

18 Scientists cannot make vaccines against every disease. How would you decide which diseases to target?

Find out about:
- where 'superbugs' come from
- how *you* can help fight them

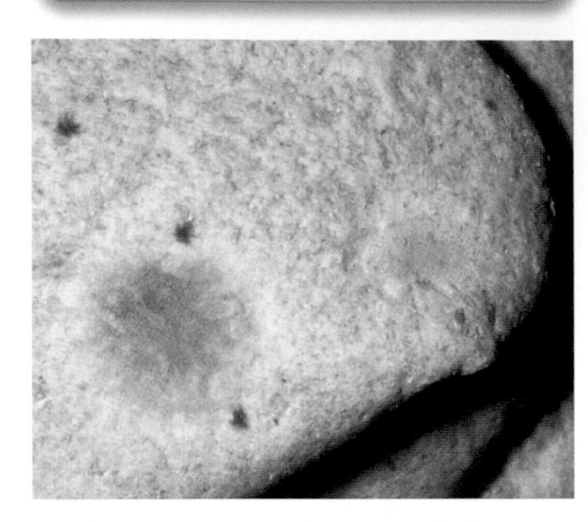

Antibiotics are made naturally by bacteria and fungi to destroy other MOs. The fungus growing on this bread makes penicillin.

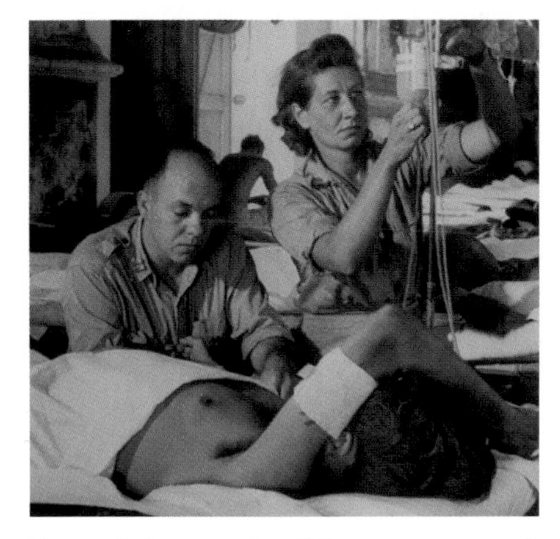

Tens of thousands of lives were saved during World War II by penicillin.

E The end for antibiotics?

The first antibiotics

The Ancient Egyptians may have been the first people to use antibiotics. They used to put mouldy bread onto infected wounds. Scientists now know that the mould is a fungus that makes **penicillin**. In the 1940s scientists started to grow the fungus to make larger amounts of penicillin.

The bugs fight back

To begin with penicillin was called a 'wonder drug'. Before the 1940s bacterial infections had killed millions of people every year. Now they could be cured by antibiotics. Antibiotics were also used to treat animals. They were even added to animal feed, to stop farm animals from getting infections.

But within ten years one type of bacteria was no longer killed by penicillin. It had become resistant. New antibiotics were discovered, but each time resistant bacteria soon developed. The 'superbugs' we are dealing with now are resistant to all known antibiotics, except one. How long that will last, we don't know.

Where have 'superbugs' come from?

You won't be surprised to learn that it's the genes of a 'superbug' that make it resistant to an antibiotic. A tiny change in one gene – a mutation – can turn a bacterial cell into a 'superbug'. Just one 'superbug' on its own won't do much damage. But if it reproduces rapidly, it could produce a large population of bacteria, all resistant to an antibiotic.

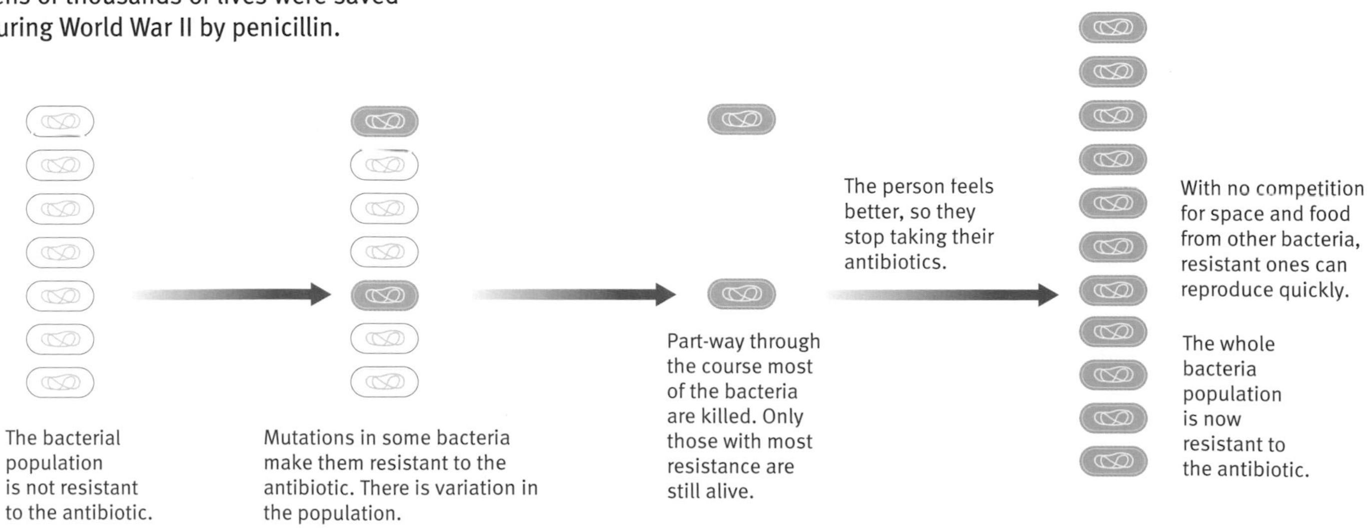

A few mutations can result in antibiotic resistant bacteria.

Why are superbugs developing so quickly?

Two things increase the risk of **antibiotic resistant** superbugs:

- people taking antibiotics they don't really need
- people not finishing their course of antibiotics

If you are given a course of antibiotics and take them all, it is likely that all the harmful bacteria will be killed. But if you stop taking the antibiotics because you start to feel better, the MOs that survive will be those which are most resistant to the antibiotic. They will live to breed another day – and so a population of antibiotic resistant bacteria soon grows.

How can we stop the superbugs?

Scientists cannot stop antibiotic resistant bacteria from developing. The mutations that produce these bacteria are part of a natural process. For now, we can only hope that scientists can develop new antibiotics fast enough to keep us one step ahead of the bacteria.

But as well as new drugs, there are other ways of tackling the problem:

- having better hygiene in hospitals to reduce the risk of infection
- only prescribing antibiotics when a person really needs them
- making sure people understand why it is important to finish all their antibiotics (unless side effects develop)

New drugs in strange places?

Scientists are always looking out for sources of new drugs. For example, crocodile blood might be the source of the next family of antibiotics. A chemical found in crocodile blood is a powerful antibacterial agent. It was discovered by a scientist who wondered why crocodiles didn't die of infections when they bit each other's legs off.

'SUPERBUGS' MRSA *ON THE RAMPAGE*

These killer bacteria are resistant to almost all known antibiotics.
The bad news is that they have broken out of hospitals.
People are dying of MRSA 'superbug' infections picked up at work, out shopping, and even at home. And the cause? The very antibiotics we've been using to kill them!

The bacteria MRSA is resistant to almost all antibiotics.

Crocodile blood could be the source of important new antibacterial drugs.

Questions

1 What are antibiotic resistant bacteria?

2 Write bullet-point notes to explain how antibiotic-resistant bacteria can develop.

3 Describe two things that you can do to reduce the risk of antibiotic-resistant bacteria developing.

Key words

penicillin
antibiotic resistant

Find out about:
- how new drugs are developed
- how they are tested

From painkillers to vaccines, antibiotics to antihistamines, medicines are part of everyday life.

F Where do new medicines come from?

Most of us take medicines prescribed by our doctor without asking many questions. We assume that they will do us good. But what if you could speak to the scientist who developed the medicine?

Scientists around the world are trying to develop new drugs. New antibiotics, new treatments for asthma, cancer ...

Developing a new drug takes years of research, and lots of money. The rewards for a successful discovery can be huge improvements in human health. For drug companies there may also be large profits.

A scientist explains how a new drug is developed:

Sian is a cancer research scientist.

First we study the disease to understand how it makes people ill. This helps us work out what we need to treat it – for example a chemical to kill a microorganism, or a chemical to replace one the body isn't making properly.

We search through many natural sources to find a chemical that may be the correct shape to do this. We look at computer models of the molecules to test our ideas.

When we find a chemical that could work, there are many tests that must be done. It's also important that we could make lots of it without too many problems. Only a very small number of possible drugs get through all these stages.

Stage 1: human cells

Early tests are done on human cells grown in a laboratory. Scientists try out different concentrations of a possible new drug. They test it on different types of body cells with the disease. These tests check how well the chemical works against the disease – how effective it is. They also give the scientists data about how safe the drug is for the cells.

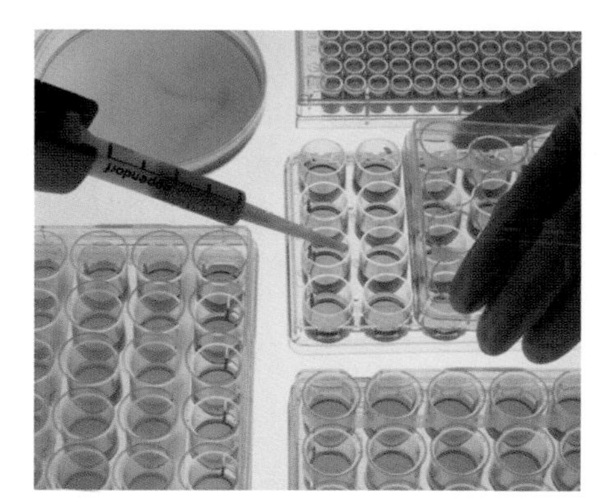

Drugs are tested on cells in the laboratory. These are called *in vitro* tests.

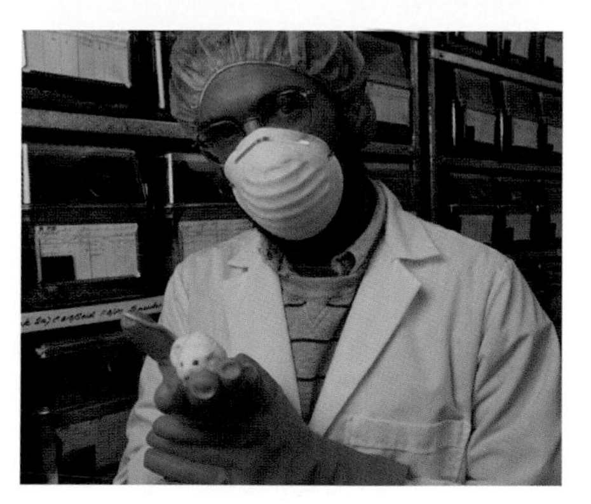

Trials using animals or human volunteers are called *in vivo* tests.

Stage 2: animal tests

If the drug passes tests on human cells, it is tried on animals. Animal trials are carried out to make sure that the drug works as well in whole animals as it does on cells grown in the laboratory.

Not everybody agrees that it is right to test drugs on animals. The British Medical Association (BMA) believes that animal experimentation is necessary at present to develop a better understanding of diseases and how to treat them, but says that alternative methods should be used whenever possible.

If animal trials go well, we apply for a patent. It costs a lot of money to develop a new drug. If we have a patent, no other company can sell the medicine for 20 years. But because clinical trials take many years, we often only have about 10 years when we're the only people making the drug.

Stage 3: clinical trials

If the drug passes animal trials, then it can be tested on people. these tests are called **human trials** or **clinical trials**. They give scientists more data about the effectiveness and safety of the drug.

Questions

1 Copy and complete the table:

Stage	Testing	To find out
one	Drug is tested on human cells grown in the laboratory.	• how safe the drug is for human cells • how well it works against the disease
two		
three		

2 Developing a new drug is usually very expensive. Suggest why.

Key words
human trials
clinical trials

Clinical trials – crunch time

Five years ago Anna was diagnosed with breast cancer. Fortunately her treatment worked and she recovered. Now Anna has been asked to take part in the trial of a new drug. Doctors hope it will reduce the risk of the cancer coming back.

Anna talks to her doctor:
The problem is I won't know if I'm getting any treatment or not. Could I be risking my health? I know the trial could help people in the future – but what about me? Can you tell me if I will be given the real drug or not?

Before the trial Anna would sign a patient consent form. She signs it to say that all of her questions have been answered. She can also leave the trial at any time. Anyone taking part in a drug trial must give their 'informed consent'.

What treatment would Anna get?

People who agree to take part in this trial will be put randomly into one of two groups. Having **random** groups is very important in making sure the results of the study are reliable.

One group of people in the trial will be given the new drug, another group will not. This is the **control** group. The results from both groups will be compared.

Anna's doctor wouldn't know if she was getting the new drug or not. Neither would Anna. Someone else would prepare the treatments. This is because Anna would be part of a **double-blind** trial.

If Anna or her doctor knew what treatment she was getting, it could affect the way they report her symptoms. A random double-blind trial is considered the best type of clinical trial.

What treatment will the control group be given?

The drug being tested is a new treatment. In almost all clinical trials the control group are given the treatment that is currently being used. So comparing the results from both groups shows whether the new treatment is an improvement.

Sometimes there isn't any current treatment for an illness. In these cases the control group can be given a **placebo**. This looks exactly like the real treatment but has no drug in it. Using a placebo in a clinical trial is very rare. The control group in Anna's trial will be given a placebo.

Human trials – ethical questions

Taking the placebo would not increase Anna's risk of cancer returning. Taking the new drug may bring other risks. But her doctor will be looking out for any harmful effects. And the new drug may increase her chance of staying well.

It may seem unfair that the control group could miss out on any benefits of the new drug. But remember that not all drugs pass clinical trials. Proper testing is needed to find out if a new drug has real benefits. Tests also give doctors data about the risk of unwanted harmful effects.

- If the trial shows that the risks are too great it will be stopped.
- If the trial shows that the drug has benefits it will immediately be offered to the control group.

Questions

3 Explain why drug trials must be random.

4 Explain the difference between blind and double blind trials.

5 Describe a situation in which it would be wrong to use placebos in a trial.

⑥ What do you think Anna should do? Explain why you think this.

Blind trials

In some trials the doctor is told which patients are being given the drug. This may be because they need to look very carefully for certain unwanted harmful effects. The patient still should not know. This method is called a **blind trial**.

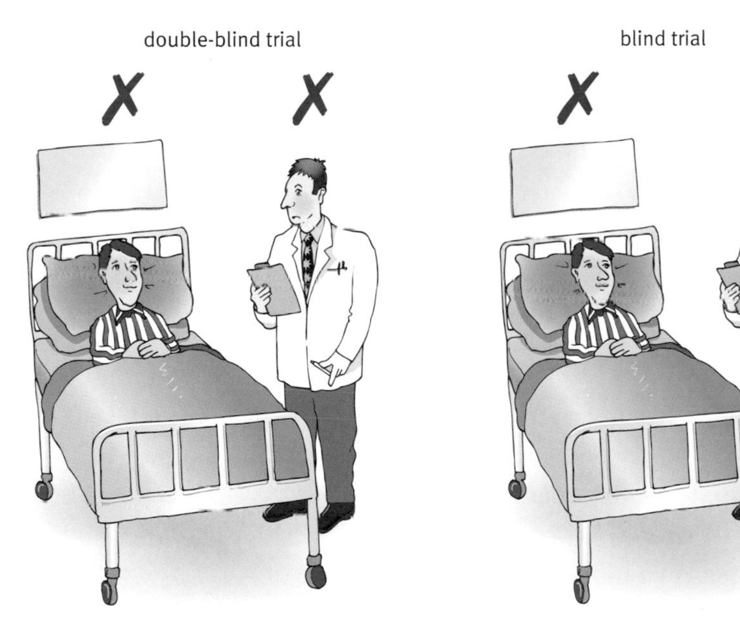

In a drug trial the doctor and/or patient may (✓) or may not (✗) know if the treatment is the new drug.

Trials without a control group

In rare cases a new drug is given to all the patients in a trial. This happens when there is no treatment, and patients are so ill that doctors are sure they will not recover themselves. The risk of possible harmful effects from the drug is outweighed by the possibility that it could extend their lifespan or be a cure. No-one is given a placebo. It would be wrong not to offer the hope of the new drug to all the patients. Penicillin is one example where this happened.

Key words

random
control
blind trial
placebo
double-blind

Find out about:
- what causes a heart attack
- how to look after your heart

I'll never forget. I went cold and clammy, covered in sweat. And the pain – it wasn't just in my chest. It was down my arm, up my neck and into my jaw. I don't remember much else until I woke up in intensive care. I never want to go through that again.

G Circulation

Three weeks ago 45-year-old Oliver suffered a serious **heart attack**. He was very lucky to survive. Now he wants to try and make sure it doesn't happen again.

Your body's supply route

Your heart is a bag of muscle. It pumps blood around your body. When you are sitting down your heart beats about 70 times each minute.

Tubes carry the blood around your body:

- **arteries** take blood from the heart to your body
- **veins** bring blood back to the heart

The diagram shows the flow of blood around your body.

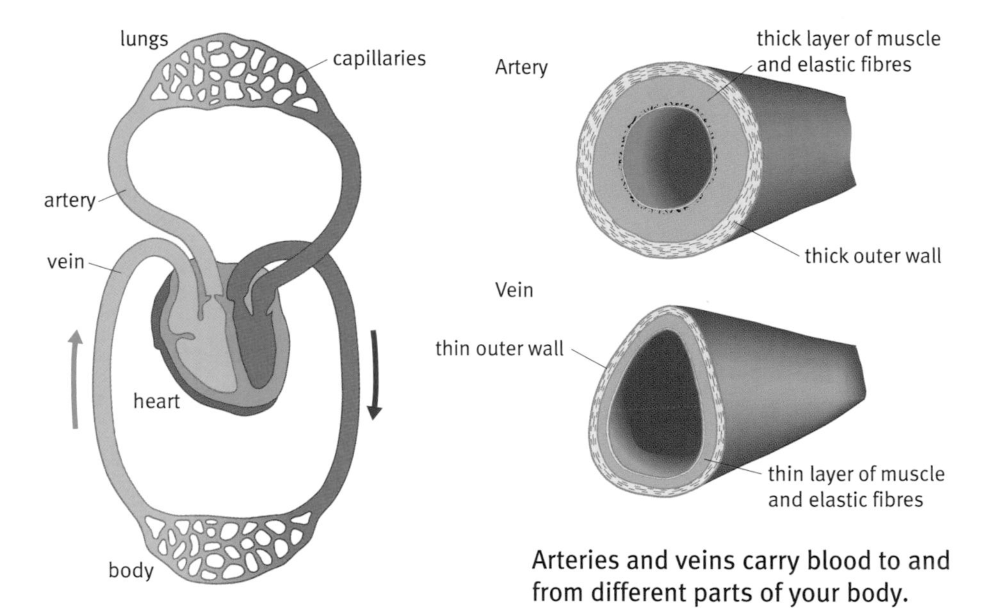

Arteries and veins carry blood to and from different parts of your body.

What is a heart attack?

Blood brings oxygen and food to cells. Cells use these raw materials for a supply of energy. Without energy the heart would stop. So heart muscle cells must have their own blood supply.

Sometimes fat can build up in the coronary arteries. A blood clot can form on the fatty lump. If this blocks an artery, some heart muscle is starved of oxygen. The cells start to die. This is a heart attack.

Fat build-up in a coronary artery.

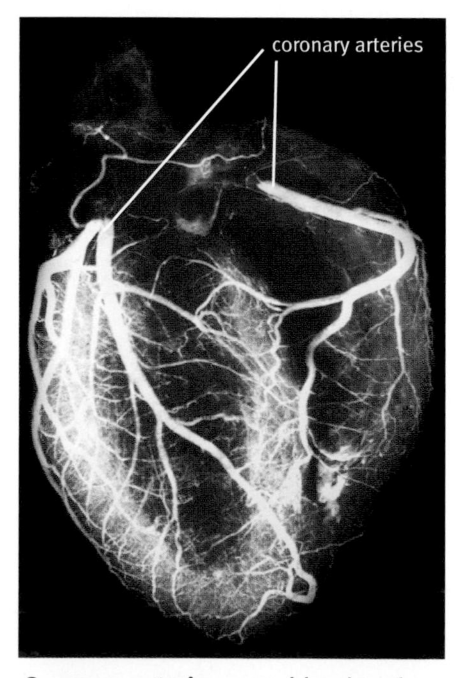

Coronary arteries carry blood to the heart muscle.

Key words
heart attack
arteries
veins
coronary arteries
risk factors

How serious is the problem of heart disease?

Heart disease is any illness of the heart, for example a blocked coronary artery and a heart attack.

Oliver survived because only a small part of his heart was damaged. He was given treatment to clear the blocked artery. If the blood supply to more of his heart had been blocked it could have been fatal.

In the UK 270 000 people have a heart attack every year. This is one every two minutes. Coronary heart disease is more common in the UK than in non-industrialized countries. This is because people in the UK do less exercise – most people travel in cars and have machines to do many jobs. And a typical UK diet is high in fat.

What causes heart disease?

Heart attacks are not normally caused by an infection. Your genes, your lifestyle, or most likely a mixture of both, all affect whether you suffer a heart attack. There isn't one cause of heart attacks – there are many different **risk factors**. Your own risk increases the more of these risk factors you are exposed to.

Is Oliver at risk of another heart attack?

Oliver has a family history of coronary heart disease. He is also overweight and often eats high-fat, high-salt food. This diet has given Oliver high blood pressure and high cholesterol levels. All these factors increase his risk of a heart attack. Oliver does like sport – but he'd rather watch it on TV than do exercise himself. Oliver's doctor has given him advice about reducing his risk.

HEALTHY HEART

- Cut down on fatty foods to lower blood cholesterol.
- If you smoke, stop.
- Lose weight to help reduce blood pressure and the strain on your heart.
- Take regular exercise (such as 20 minutes of brisk walking each day) to increase the fitness of the heart.
- Reduce the amount of salt eaten to help lower blood pressure.
- If necessary, take drugs to reduce blood pressure and/or cholesterol level.

Questions

1 Draw a diagram to show the inside of an artery and a vein.

2 Label your diagram to explain how these two blood vessels are suited to their job.

3 Explain why heart cells need a good blood supply.

4 Explain how too much fat in a person's diet can lead to a heart attack.

5 List four lifestyle factors that increase a person's risk of heart disease.

6 Heart disease is more common in the UK than in non-industrialised countries. Suggest why.

7 Your next-door neighbour wants to do more exercise. But she gets bored easily, and doesn't want to spend money going to a gym. Suggest some ways that she could get more exercise into her daily life.

Find out about

- how scientists identify risk factors for lifestyle diseases
- the evidence needed to prove a causal link

H Causes of disease – how do we know?

It's usually easy for doctors to find the cause of infectious diseases. The MO is always in the patient's body. It is harder to find the causes of lifestyle diseases, like heart disease or cancer.

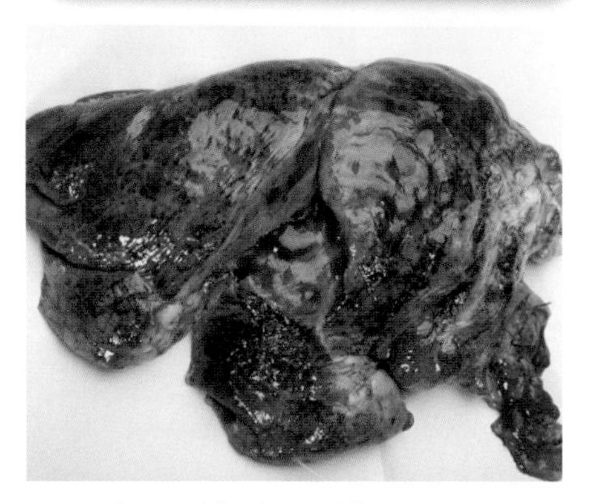

Lung tissue blackened by tar from cigarette smoke.

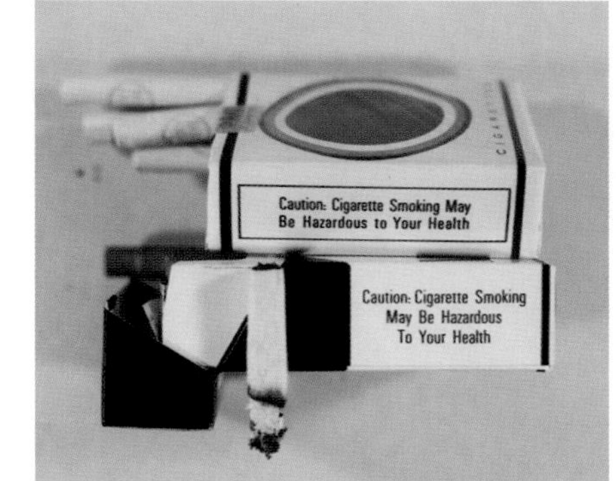

Health warning in 1971.

Health warning in 2003

Smoking and lung cancer

Government health warnings have been printed on cigarette packets since 1971. There was evidence showing a link – a *correlation* – between smoking and lung cancer. But in 2003 the message was made much stronger. How did doctors prove that smoking *caused* lung cancer?

An early clue

In 1948 a medical student in the USA, Ernst Wynder, observed the autopsy of a man who had died of lung cancer. He noticed that the man's lungs were blackened. There was no evidence that the man had been exposed to air pollution from his work. But his wife told Wynder that he had smoked 40 cigarettes a day for 30 years. Wynder knew that one case is not enough to show a link between any two things.

In 1950, two British scientists, Richard Doll and Austin Bradford Hill, started a series of scientific studies. First, they compared people admitted to hospital with lung cancer, to another group of people in hospital for other reasons. Smoking was very common at the time, so there were lots of smokers in both groups. But the percentage of smokers in the lung cancer group was much greater.

This data showed a link – a correlation – between smoking and lung cancer. Doll and Hill suggested smoking caused lung cancer. But, a correlation doesn't always mean that one thing causes another.

Cigarettes smoked per day	Number of cases of cancer per 100 000 men
0 – 5	15
6 – 10	40
11 – 15	65
16 – 20	145
21 – 25	160
26 – 30	300
31 – 35	360
36 – 40	415

The data shows how the number of cases of lung cancer in men is affected by the number of cigarettes smoked.

How reliable was the claim?

Doll and Hill published their results in a medical journal so that other scientists could look at them. This is called 'peer review'. Other scientists look at the data, and how it was gathered. They look for faults. If they can't find them, then the claim is more reliable.

The claim is also more reliable if other scientists can produce data that suggests the same conclusions.

A major study

In 1951 Doll and Hill started a much larger study. They followed the health of more than 40 000 British doctors for over 50 years. The results were published in 2004 by Doll and another scientist, Richard Peto. They showed that:

- smokers die on average 10 years younger than non-smokers
- stopping smoking at any age reduces this risk

The last piece of the puzzle – an explanation

Lung cancer rates in the USA rose sharply after 1920. The same pattern was seen in the UK.

Many doctors were now convinced that smoking caused lung cancer. But cigarette companies did not agree. They said other factors could have caused the increase in lung cancer, for example more air pollution from motor vehicles.

The missing piece of the puzzle was an explanation of *how* smoking caused cancer. In 1998 scientists discovered just this. They were able to explain *how* chemicals in cigarette smoke damage cells in the lung, causing cancer. This confirmed that smoking *causes* cancer.

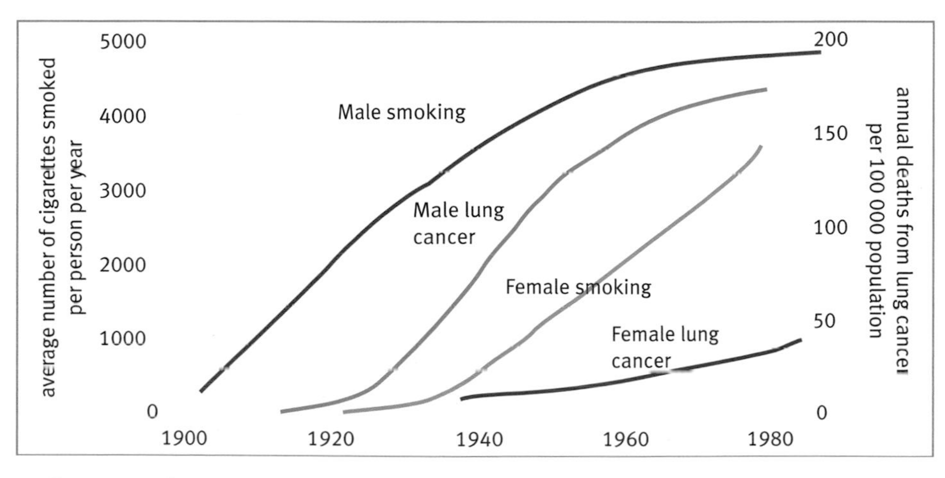

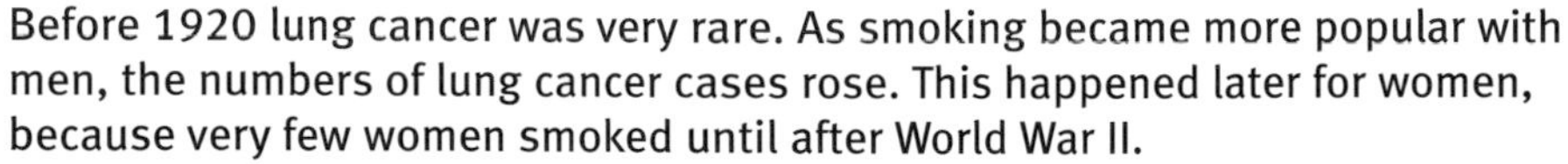
Before 1920 lung cancer was very rare. As smoking became more popular with men, the numbers of lung cancer cases rose. This happened later for women, because very few women smoked until after World War II.

Questions

1. Write down one example of an everyday correlation.
2. Draw a graph to show how the number of cases of lung cancer in men is affected by the number of cigarettes smoked.
3. Explain briefly what happens during 'peer review'.
4. Explain why scientists think it is important that a scientific claim can be repeated by other scientists.
5. It's unlikely that many people would have agreed with Wynder if he'd reported the case he saw in 1948. Suggest two reasons why.
6. If a man smokes 20 cigarettes a day from age 16 to 60, will he definitely develop lung cancer? Explain your answer.

Looking at the health of lots of people can show scientists the risk factors for different diseases.

What makes a good study?

There are many reports in the media about studies of health risks. These studies look for diseases caused by different risk factors. For example, the possible harmful effects from using a mobile phone.

You may want to use this information to make a decision about your own health. So it's important to know if the study has been done well. There are several things you can look for.

How many people were involved in the study?

A good study usually looks at a large sample of people. This means that the results are less likely to be affected by chance. In 1948, a study of heart disease began in the town of Framingham, Massachusetts, USA. The study recruited 5209 men and women between the ages of 30 and 62.

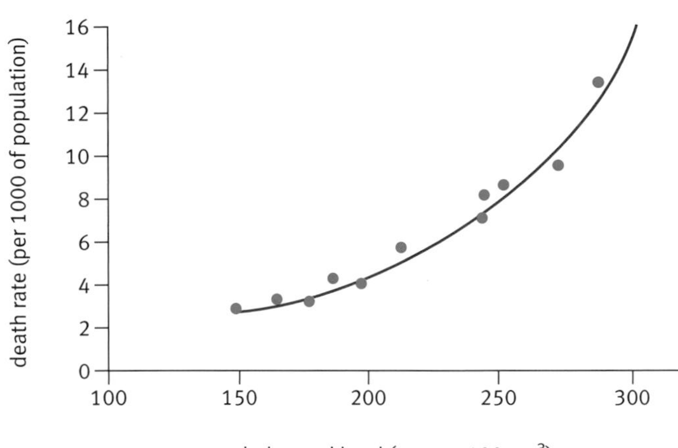

The graph shows some of the data from the Framingham study.

In 1971, their children were also recruited – another 5124 people. Now the third generation – the grandchildren – are joining the study.

Every two years the researchers record details such as:

- body mass
- blood pressure
- cholesterol level
- lifestyle factors, for example, if they smoke, how much exercise they do

In total the Framingham study will have looked at over 13 000 people. This study has been hugely important for heart-disease research. It has led to the identification of all known major risk factors for heart disease.

How well matched are the people in the study?

Health studies sometimes compare two groups of people. One group has the risk factor, the other doesn't. For example, a study that compares people who use mobile phones with people who do not.

In these studies it is important to **match** the people in the two groups as closely as possible.

In many studies, like Framlingham, people are not matched at the start of the study. The researchers are following the health of a particular group of people. When the results of these studies are analysed, researchers check for differences between the people who have a disease and those who don't. For example, you are researching the risk factor a certain disease. You find that all the people who get the disease were older at the start of the study than those who did not get the disease. If the two groups are not of the same age the researchers must make allowances for this when drawing their conclusions.

The British Regional Heart Study

The British Regional Heart Study began in 1975 in 24 towns across the UK. The researchers could not study everyone in the towns. They randomly selected 8000 men. Over 25 years the researchers took measurements of medical data and lifestyle factors.

The people in the study were all:

- men
- middle-aged at the start of the study

So, the results of the study gave a true picture of heart-disease risks for this group. But they could not be used to decide about risk factors for other people. Age and gender both affect your risk of heart disease.

In 2001 The British Regional Heart Study was expanded to look at heart disease in groups of women, and also groups of children.

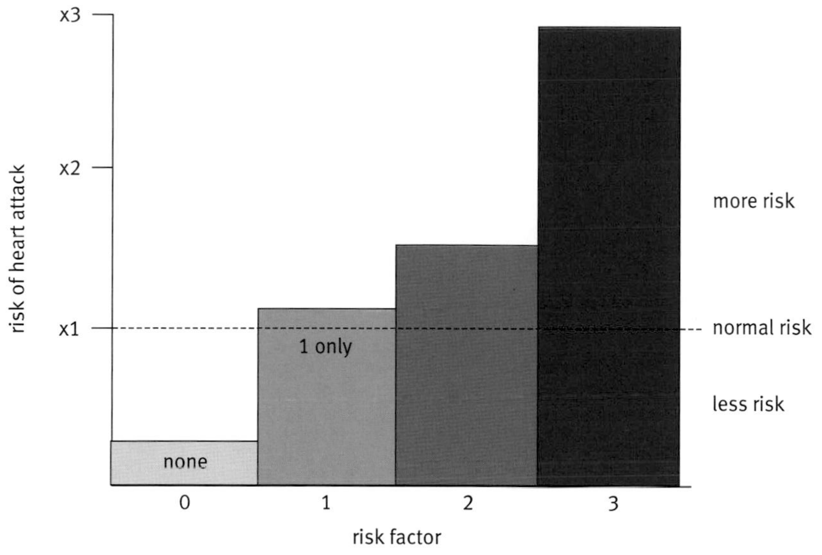

Data from the British Regional Heart Study shows that the more risk factors you are exposed to, the greater your risk of heart attack.

How big is the risk?

There's one other thing to look for when using data from health studies to make decisions. Imagine a headline like 'Risk of disease is two times greater'. It's important to check how big the original risk is. For example, what if the risk of an outcome is that it will happen to one person in a million. An increase of two times is still only two in one million – or one in every 500 000 people. This is still a very small risk.

Key word
match

Questions

7 a Name one factor that increases a person's risk of heart disease.

b Use information from the graphs on these pages to support your answer.

8 Suggest two things you should look for when deciding whether a study was well planned.

9 Your teenage daughter has started smoking. She says 'I don't believe smoking causes heart disease or lung cancer. Grandad has smoked all his life, and he's fine.' How would you explain to her that she may not be so lucky?

B2 Keeping healthy

Science explanations

In this Module you have found out how your body fights disease. You have also seen how scientists learn about the causes of diseases.

You should know:

- diseases are caused by some microorganisms, and by a person's lifestyle, for example, smoking, poor diet
- natural barriers help to stop harmful microorganisms entering the body
- these microorganisms may reproduce very quickly in good conditions, damaging cells or producing poisons which cause symptoms of disease
- white blood cells are part of the immune system to fight infections
- white blood cells can destroy microorganisms by digesting them or producing antibodies
- different antibodies are needed to fight every different microorganism
- once you have made one type of antibody you can make it again very quickly, so you are immune to that disease
- vaccines trigger the body to make antibodies before it is infected with a particular microorganism
- vaccines contain a harmless form of the microorganism
- no action can be completely safe, including vaccinations and other medical treatments
- why a very high percentage of people must be vaccinated to prevent an epidemic
- new vaccines must be made against flu every year because the virus changes quickly
- why it is difficult to make a vaccine against the HIV virus
- antibiotics are chemicals that kill bacteria and fungi
- an antibiotic may stop working because the bacteria or fungi has become resistant to it
- antibiotic resistant microorganisms are made because of mutations in their genes
- to slow down antibiotic resistant bacteria you should:
  - only use antibiotics when really needed
  - always finish the course
- new drugs are tested for safety and how well they work on:
  - human cells grown in the laboratory
  - animals
  - healthy human volunteers
  - people with the illness
- how blind and double-blind trials are different
- heart muscle needs its own blood supply to bring food and oxygen to the cells
- how the structure of arteries and veins is suited to the jobs they do
- fatty deposits in blood vessels supplying the heart can produce a heart attack
- heart disease is usually caused by lifestyle factors

Ideas about science

It is not always easy to make decisions about personal health. It can be difficult to decide whether information about health risks is reliable.

You should also be able to:

- correctly use the ideas of correlation and cause when discussing the issues in this module
- suggest factors that might increase the chance of an outcome
- explain that individual cases do not provide convincing evidence for or against a correlation
- evaluate a health study by commenting on sample size or sample matching
- explain why a correlation between a factor and an outcome doesn't definitely mean that one thing causes the other, and give an example to show this
- use data to argue whether a factor does or does not increase the chance of something happening
- know that having a good explanation for how a factor may cause something to happen makes it more likely that scientists will accept that it does
- describe what happens in 'peer review'
- know that scientific claims which have not been evaluated by other scientists are less reliable than ones which have
- know that if data cannot be repeated by other scientists it makes any scientific claim based on the data less reliable

People may have different viewpoints for personal and social decisions:

- some people think that certain actions are wrong whatever the circumstances
- some people think that you should weigh up the benefit and harm for everyone involved and then make your decision

People may make different decisions because of their beliefs, and their personal circumstances. When you consider an ethical issue such as vaccination policy you should be able to:

- say clearly what they issue is
- describe some different viewpoints people may have
- say what you think and why

Why study life on Earth?

Life on Earth - so many different kinds of living thing it's almost unbelievable. 'How did life begin?', and, 'Where do we come from?' are two of the biggest questions we ask science to answer.

Scientists think life began on Earth 3500 million years ago. Modern humans have only been around for about 40 000 years. And since then many other species have become extinct. We can learn to look after life on Earth better for future generations.

The science

Fossils are evidence for how life on Earth has evolved. Simple organisms have gradually developed and changed, forming new, larger species.

All life forms depend on their environment and on other species for survival. Larger organisms have evolved communication systems (nerves and hormones), that help them to survive.

Ideas about science

Today, most scientists agree that evolution happens. But 200 years ago they didn't. And not all scientists agree about how life on Earth started. Developing new explanations takes a lot of evidence and imagination. Even then, people may have reasons not to accept them.

Life on Earth

Find out about:

- how life on Earth may have begun and is evolving
- how scientists developed an explanation for evolution
- how humans evolved
- why some species become extinct, and whether this matters

Find out about

- why living things are all different
- what a species is
- evidence for evolution

A The variety of life

You can usually see the differences between different kinds of living things on Earth. But there are also a lot of similarities, even between living things that don't look the same. For example, almost all living things use DNA to pass on information from one generation to the next.

Human skin cells and cells in these butterfly wings use the same chemical reaction to make pigment.

Classification – working out where we belong

Scientists use the similarities and differences between living things to put them into groups. You've probably come across this idea before. It's called classification. The biggest group that humans belong to is *Animalia* (animals). The smallest is *Homo sapiens*, or human beings. *Homo sapiens* is our **species** name.

Classification names are in Latin, so that everyone can use the same name for something. It doesn't matter what languages two people speak, they can always use the same Latin name.

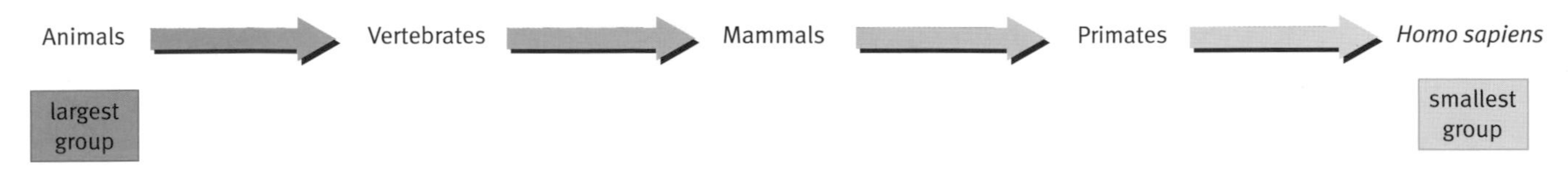

You are most closely related to other members of *Homo sapiens*. But you belong to these other groups as well.

What makes a species?

Scientists define a species as a group of organisms so similar that:

- they can breed together
- their offspring can also breed (they are **fertile**)

Horses and donkeys are good examples to explain species. They can breed together and produce offspring called mules. But mules are **infertile**. Horses and donkeys look pretty similar, but they are different species.

Horses and donkeys do look very similar. But their offspring are infertile. So horses and donkeys are different species.

horse

donkey

mule

Are all members of a species the same?

Look at the photo of people and their dogs. It's easy to see which belong to *Canis familiaris*, the dog species, and which are human. But you can also see that the dogs are not identical to each other. Neither are the people. Members of a species are different from each other. This **variation** is very important in evolution. You'll find out more about this later.

What causes variation?

There are different causes of variation.

- The photograph above shows both men and women. This difference is controlled by some of their genes. It is **genetic** variation.
- One of the women in the front has pierced ears. Other people don't. This difference has been caused by something other than genes. It is **environmental** variation.
- People have different skin colours. This is partly genetic variation. But it is also affected by environment – how much sun their skin is exposed to.

Almost all variation is caused by a mixture of genes and environment.

Key words
species
fertile
infertile
variation
genetic
environmental

Questions

1 What species do you belong to?

2 Explain why horses and donkeys are different species.

3 Explain what the word 'variation' means. Use examples in your answer.

(4) Write down one difference in people that is caused by

a genes only

b genes and environment

c environment only

Explaining similarities – the evidence for evolution

Most scientists agree that life on Earth started from a few simple living things. This explains why living things have so many similarities.

These simple living things changed over time to produce all the kinds of living things on Earth today. The changes also produced many species that are now extinct. This process of change is called **evolution**, and it is still happening today.

Key words
evolution
fossils

What evidence is there for evolution?

Fossils are made from the dead bodies of living things. They are very important as evidence for evolution. Almost all fossils found are of extinct species. This is more than 99% of all species that have ever lived on Earth.

How reliable is fossil evidence?

Conditions have to be just right for fossils to develop. Only a very few living things end up as fossils. So there are gaps in the fossil record. Sometimes a new species seems to appear without an in-between link to an earlier species.

Although there are gaps in the record, scientists have collected millions of fossils. This huge amount of evidence has helped to build up a picture of evolution.

Why are there gaps in the fossil record?

Evolution doesn't happen at the same speed all the time. It happens in spurts. But a 'spurt' of evolution may still take tens of thousands of years. It is quite possible that the right conditions for fossil-making didn't happen

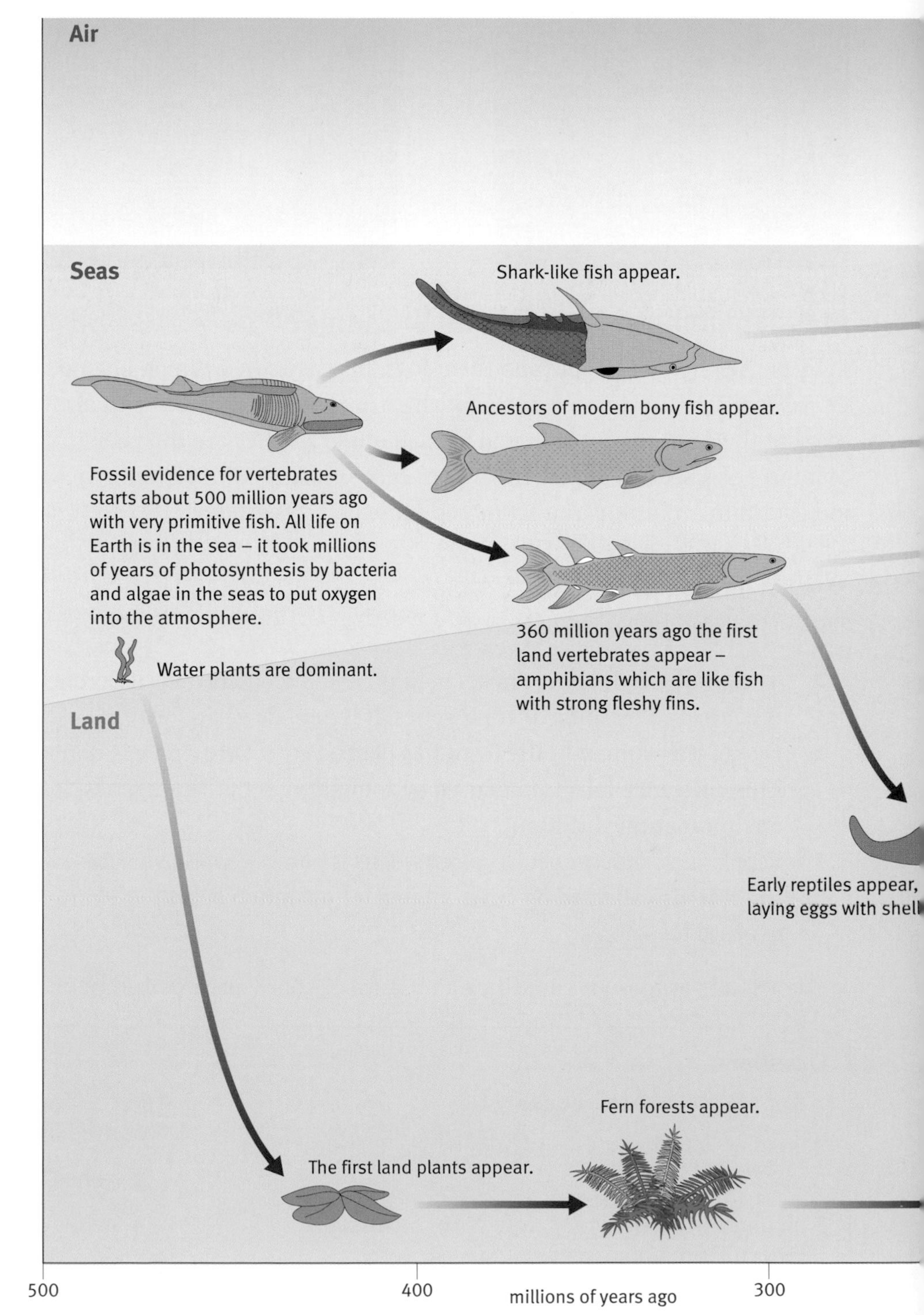

during that time. So there would be no fossil evidence of the small changes that happened as the new species evolved.

What other evidence do we have for evolution?

Scientists can also compare the genes from different living things. The more genes two living things share, the more closely related they are. This helps scientists to work out where different species fit on the evolutionary tree.

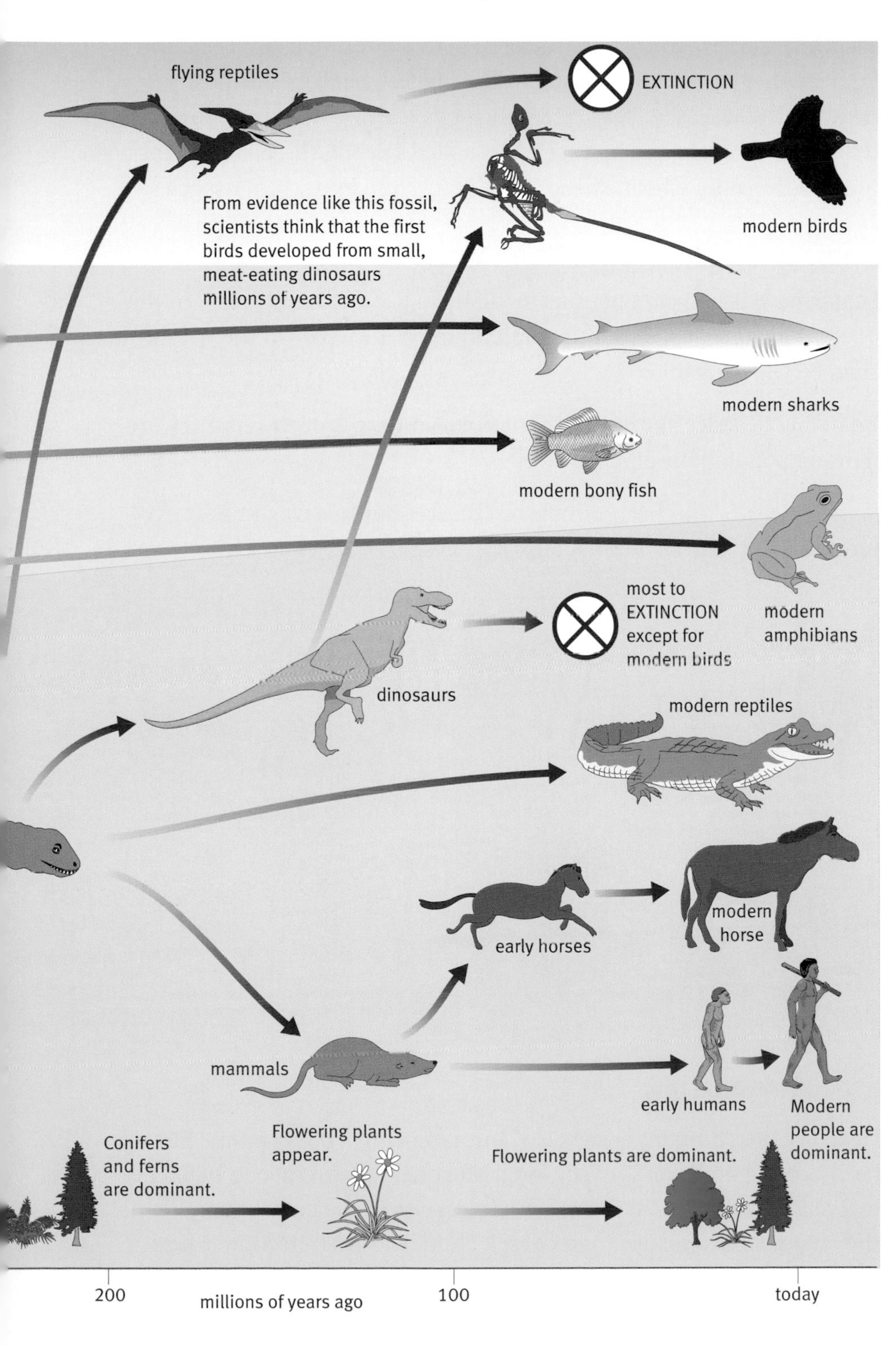

Over 98% of human genes are the same as those of a chimpanzee, but only 85% are the same as those of a mouse.

Questions

5 What percentage of all life on Earth is alive now?

6 Name two types of evidence that scientists use as evidence for evolution.

Find out about
- how evolution happens – natural selection
- how humans have changed some species

B Evidence for change NOW

Evolution did not just happen in the past. Scientists can measure changes in species which are happening now. Humans are causing many of these changes.

Selective breeding

Early farmers noticed that there were differences between individuals of the same species. They chose the crop plants or animals that had the features they wanted. For example, the biggest yield, or the most resistance to diseases. These were the ones they used for breeding. This way of causing change in a species is called **selective breeding**. It has been used for breeding wheat, sheep, dogs, roses, and many other species.

Selective breeding has produced tulips with different coloured flowers.

Some changes people don't want

People have been using poisons to kill head lice for many years. In the 1980s, doctors were sure that **populations** of head lice in the UK would soon be wiped out.

But a few headlice survived the poisons. Now parts of the country are fighting populations of 'superlice'.

So headlice are another example of change. But this wasn't selective breeding – no one *wanted* to cause superlice.

Head lice are quite common. They feed on blood.

Natural selection

Head lice are changing because of human beings. But humans haven't been around on Earth for very long. Most changes to species happened before human beings arrived. Something else in the environment caused the change. This is called **natural selection**. Natural selection is how evolution happens.

Steps in natural selection

① *Living things in a species are not identical. They have variation.*

Ancestors of modern giraffes had variation in the length of their necks.

② *They compete for things like food, shelter, and a mate. But what if something in the environment changes?*

Food supply became scarce. The giraffes competed for food.

③ *Some will have features that help them to survive. They are more likely to breed. They pass their genes on to their offspring.*

Taller giraffes were more likely to survive and breed. They passed on their features to the next generation.

④ *More of the next generation have the useful feature. If the environment stays the same, even more of the following generation will have the useful feature.*

Over many generations, more giraffes with longer necks were born.

Treating head lice

Your Local Health Authority issues a directive, known as a rotational policy, every two to three years to inform everyone concerned which type of insecticide is currently recommended for use in your area.

The rotational policy is intended to prevent head lice becoming resistant to treatment – in other words, to help ensure that the treatments available continue to be effective in killing lice.

Key words

selective breeding
populations
natural selection

Questions

1 How does evolution happen?

2 Copy and complete the table to compare selective breeding and natural selection.

Steps in selective breeding	Steps in natural selection
Living things in a species are not all the same.	Living things in a species are not all the same.
Humans choose the individuals with the feature that they want.	
These are the plants or animals that are allowed to breed.	
They pass their genes on to their offspring.	
More of the next generation will have the chosen feature.	
If people keep choosing the same feature, even more of the following generation will have it.	

3 Explain what is meant by a population.

4 Read the extract from a leaflet about head lice. Explain how this rotational policy stops the evolution of resistant populations of head lice.

5 Natural selection is sometimes described as 'survival of the fittest'. How good a description of natural selection do you think this is?

Find out about

- how Darwin explained evolution
- the argument his explanation caused
- how explanations get accepted

C The story of Charles Darwin

Today most scientists agree that evolution happens. But evolution wasn't always as well accepted. A very important person in the story of evolution was Charles Darwin. His ideas were a breakthrough in persuading people that evolution happens.

Darwin's big idea

Darwin worked out how evolution could happen. He explained how natural selection could produce evolution. But he didn't come up with this idea overnight. It took many years.

Charles Darwin was born in 1809. He was interested in plants and animals from a young age. When he was 22, Darwin was given the chance to sail on HMS *Beagle*. The ship was on a five-year, round-the-world trip to make maps.

Darwin on HMS *Beagle*

Journey of the *Beagle*

The *Beagle* stopped at lots of places along the way. At each stop Darwin looked at different types of animals and plants. He collected many specimens and made lots of observations about what he saw. He recorded these data in notes and pictures.

One place the *Beagle* stopped at was the Galápagos Islands, near South America. As he travelled between the different islands, Darwin noticed variation in the wildlife. One thing Darwin wrote about after his trip was the different species of finches living on the Galápagos Islands.

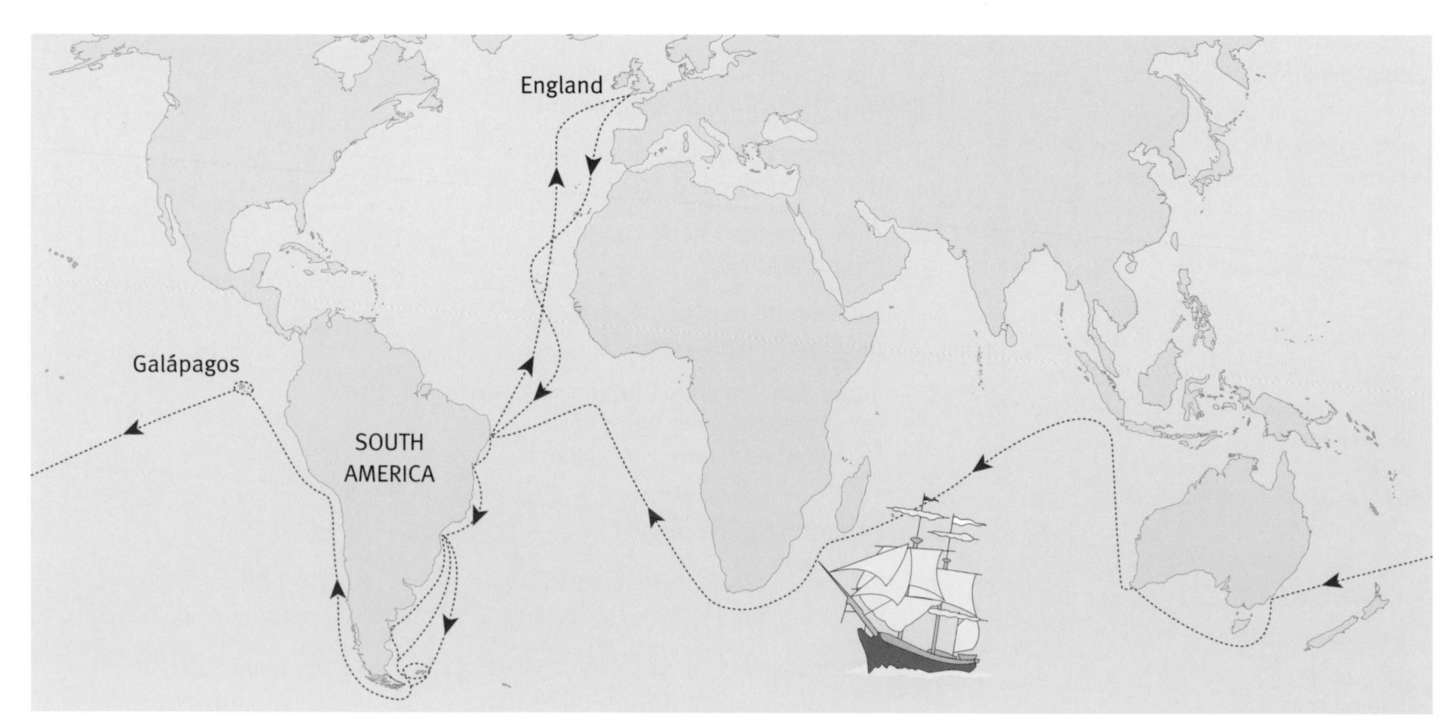

The *Beagle* stopped at different places around the world.

The famous Galápagos finches

Each species of finch seemed to have a beak designed for eating different things. For example, one had a beak like a parrot for cracking nuts. Another had a very tiny beak for eating seeds. It was as if the beaks were adapted to eating the food on each different island.

Different species of finch

In his notes, Darwin started to ask himself a question. He wondered if all the different finches could have evolved from just one species.

> *One might really fancy that from an original paucity of birds in this archipelago one species had been taken and modified for different ends.*

Charles Darwin, *The Voyage of the Beagle*, 1839

What was special about Darwin?

Darwin wasn't the first scientist to think that evolution happens. His own grandfather was one of several people who had written about it earlier. But most people at the time didn't agree with evolution. Darwin was the first person to make a strong enough argument to change their minds.

He started by looking at lots of living things. He made many observations which he would use as evidence for his argument. Then:

- He thought about the evidence in a way that no-one had done before. He was more creative and imaginative.
- He came up with an idea to explain *how* evolution could happen – natural selection.

> *I look to the future to young and rising naturalists who will be able to view both sides of the question with impartiality.*

Charles Darwin, *On the Origin of Species*, 1859

Darwin showed his notes to a friend, Thomas Huxley. Huxley was also a scientist. When he read them, Huxley said: "How stupid of me not to have thought of this first!"

Questions

1 Darwin made many observations about different species. How did he record his data?

2 What personal qualities did Darwin show that helped him develop his explanation of natural selection?

Darwin found more evidence for natural selection at home.

More evidence back home

Back in England, Darwin moved to a new home in Kent. For 20 years he worked on his idea of natural selection. He exchanged letters with other scientists in different parts of the world. All the time, Darwin was looking for more evidence to support his ideas.

His new home, Down House, had some pet pigeons. They had many different shapes and colours. But Darwin knew they all belonged to the same species. So he realized that:

- animals or plants from the same species are all different – there is **variation**

Too many to survive

Next, Darwin realized that:

- there are always too many of any species to survive

He came to this conclusion after reading the work of a famous economist, Thomas Malthus. At the end of the 18th century, Britain's population was growing very fast. Malthus pointed out that the numbers of any species had the potential to increase faster than any increase in their food supply. He predicted that the human population would grow too large for its food supply, and that poverty, starvation, and war would follow.

All the plants or animals of one species are in **competition** for food and space. A lot of them don't survive.

Darwin put these ideas together. He saw that some animals in a population were better suited to survive than others. They would breed and pass on their features to the next generation. This natural selection could make a species change over time. Darwin had explained how evolution could happen.

Elephants usually reproduce from age 30–90. Darwin worked out that after 750 years there would be nearly 19 million elephants from just one pair!

Owing to this struggle for life, any variation, however slight, if it be in any way profitable to an individual of any species, will tend to the preservation of that individual, and will generally be inherited by its offspring. I have called this principle, by which each variation, if useful, is preserved, by the term of Natural Selection.

Charles Darwin, *On the Origin of Species*, 1859

Key words
variation
competition

Same data, different explanations

Other scientists also saw that living things were different. They also saw fossils that showed changes in species. Before Darwin published his ideas, a French scientist called Lamarck had written a different explanation to Darwin's. He said that the history of life was like a ladder, with simple animals at the bottom and more complex ones at the top. He explained that animals changed during their lifetime. Then they passed these changes on to their young. He used the example of a giraffe.

Giraffe evolution explained by Lamarck

Why was Darwin's explanation better?

A good explanation does two things:

- it accounts for all the observations
- it explains a link between things that people hadn't thought of before

Lamarck's explanation said that 'nature' had started with simple living things. At each generation, these got more complicated. If this kept happening, simple living things, like single-celled animals, should disappear. So his idea didn't account for some observations, for example, why simple living things still existed on Earth.

Darwin's idea could account for these observations. It also linked together variation and competition, which hadn't been done before.

Was Lamarck a bad scientist?

Lamarck's ideas may sound a bit daft now, but he was a good scientist. He was trying to explain changes in species, but he wanted his explanation to be accepted by other people. He knew that people would be against natural selection.

Darwin was also worried about how people would react. He wrote his idea of natural selection into a book. Then he wrapped the manuscript in brown paper and stuffed it in a cupboard under the stairs. He wrote a note for his wife explaining how to publish the manuscript when he died. It stayed in the cupboard for almost 15 years.

Questions

3 How did Darwin try to get more evidence to support his ideas?

4 What two things make a good explanation?

5 What two things did Darwin link together to work out his explanation of natural selection?

I never saw a more striking coincidence. If Wallace had my manuscript sketch written out he could not have made a better abstract!

Charles Darwin, in a letter to the geologist Charles Lyell

People agreed with Darwin's observations. But they didn't agree with his explanation.

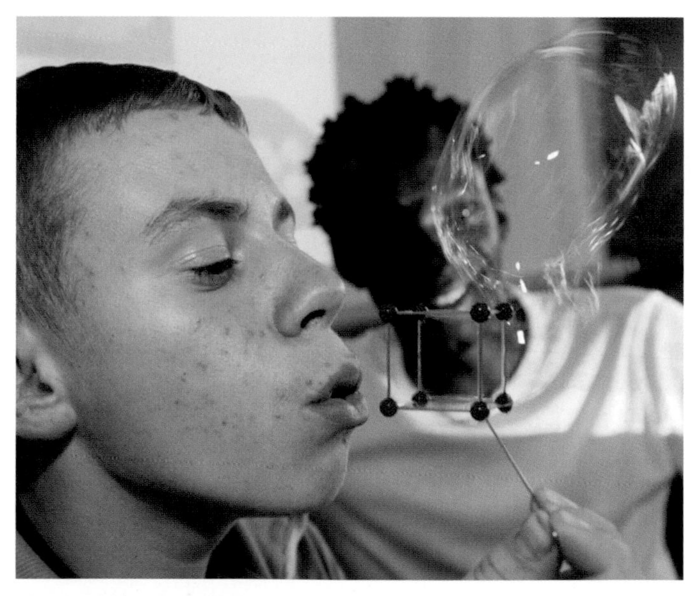

The British Association for the Advancement of Science (BA) meets every year.

On the Origin of Species

Then, in 1856, Darwin received a letter from another scientist, Alfred Russell Wallace. In it Wallace wrote about the idea of natural selection. Darwin was stunned. He gave Wallace credit for what he had done, and the two of them published a short report of some of their ideas. But now Darwin wanted to publish his full book before Wallace, or anyone else, beat him to it.

The now famous *On the Origin of Species* was published in November 1859. This book caused one of the biggest arguments in the history of science.

Why were people against natural selection?

Almost everyone in Victorian society disagreed with the idea of natural selection.

Most people thought that everything in the Bible should be believed just as it was written. The Bible said that all life on Earth was created in six days. There was no natural selection, and no evolution.

What changed people's minds?

The British Association for the Advancement of Science meets every year. Scientists meet to share their ideas. In 1860, many scientists argued against Darwin's idea.

But his two friends, Thomas Huxley and Joseph Hooker, defended it. They were very good scientists. They were also very good at speaking in public. So they helped to change many people's minds about natural selection.

Huxley and Hooker argued in favour of Darwin's theory.

The end of the story?

Natural selection was a good explanation. However, there were three big problems with it. But it wasn't Darwin's opponents who spotted these. It was Darwin himself.

Firstly, he knew that the record of fossils in the rocks was incomplete. At that time it was even more difficult to trace changes from one species to another than it is today. New fossil evidence has been found since then to support the idea of natural selection.

Secondly, the age of the Earth had not been worked out accurately enough. In Darwin's time it was thought to be about 6000 years old. This had been worked out using evidence in the Bible. So there didn't seem to have been enough time for evolution to have taken place. Scientists now have evidence to show that the Earth is much older.

The last problem was in two parts:

Questions

6 Most people in the 1800s disagreed with natural selection. What evidence did they have against this explanation?

7 Do you agree that evolution happens? Explain why you think this.

(8) Why are scientists sometimes reluctant to give up an accepted explanation, even when new data seem to show it is wrong?

Darwin could not explain why all the living things in one species were not all the same. Where did variation come from?

Also, he could not explain how living things passed features on from one generation to the next.

Both of these puzzles would have been easier for Darwin to answer if he had known about genes. Scientific discoveries since his time have allowed other scientists to do this.

The debate goes on

In 1996, the late Pope John Paul II, head of the Roman Catholic Church, acknowledged Darwin's ideas with the words: "... new scientific knowledge leads us to recognize more in the theory of evolution than hypothesis."

People continue to debate evolution. Because many of them have strong personal beliefs that are affected by this idea, it is unlikely to stop anytime soon.

Pope John Paul II

Solving the puzzle of inheritance

Gregor Mendel was born in 1822. He was a bright child, but very poor. So his teachers arranged for him to join a monastery. Here he learnt about plant breeding. He also got the chance to go to the University of Vienna. There he learnt how to plan and carry out scientific experiments.

One of Mendel's jobs at the monastery was to breed plants to produce better varieties. He used what he had learnt at university to investigate how features were passed on from one generation to the next. One of his experiments involved breeding different pea plants together.

The birth of genetics

Mendel crossed red-flowered plants with white-flowered plants. The new plants weren't pink – they were all red.

He took the new red-flowered plants and bred them together. This time he got mostly red flowered plants, with some white ones.

Mendel described the red colour as **dominant** and the white colour as **recessive**.

The parents' flower colours are not mixed together in the new plants.

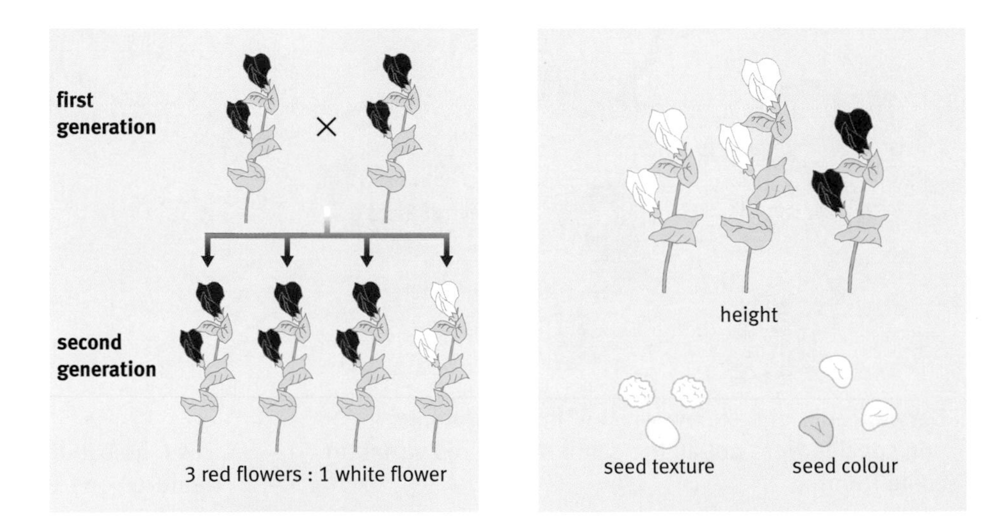

The next generation had a ratio of 3:1 red- to white-flowered plants.

Mendel investigated the inheritance of other pea features.

Mendel's evidence and natural selection

Mendel had discovered dominant and recessive alleles – different versions of the same genes. At the same time, Darwin was writing *On the Origin of Species*. He assumed that features were passed on from one generation to the next. Without this, natural selection could not work.

Mendel's work explained how features were passed on. He sent a copy of his work to Darwin. But Darwin didn't realize how important it was. Mendel's work was largely ignored until 16 years after his own death.

The double helix

On 28 February 1953, a young scientist called Francis Crick walked into a pub in Cambridge. He announced that he and James Watson had found the secret of life – the structure of DNA. Their idea was published later that year in the science journal *Nature*.

DNA carries the information on how an organism should develop. It is copied and passed on when new cells are made.

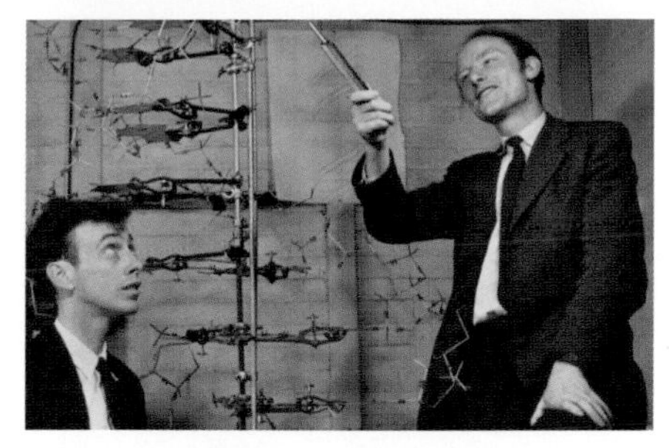

Crick and Watson with their first model of DNA.

Mutations

Suppose that, when DNA is being copied, a mistake is made. This **mutation** could result in a different coloured flower, or spots on an animal's fur. Mutations happen naturally, and they can also be caused by some chemicals or ionizing radiation.

A mutation in a gene controlling fur colour produced tigers with white fur.

Mutations cause variation

Mutations produce differences in a species. They are a cause of variation. This is very important for natural selection. Without variation, natural selection could not take place.

Most mutations have no effect on the plant or animal. They don't harm them or help them survive. Mutations that do have an effect are usually harmful. Only very, very rarely does a mutation cause a change that makes an organism better at surviving. If the mutation is in the organism's sex cells, it can be passed on to its offspring.

Evolution of a new species

Understanding more about DNA has helped scientists explain how a new species can evolve.

- Mutations produce variation in a population of the same species.
- A change happens in the environment.
- Natural selection means that only some of the population survive.
- Over many generations, these individuals form a new species.

Key words
dominant
mutation
recessive

Questions

9 Explain what a mutation is, and how they can happen.

10 What three processes combine to produce a new species?

11 Explain how the work of these scientists overcame problems with Darwin's explanation of natural selection:

a Mendel

b Crick and Watson

Find out about

- what the first life on Earth was like
- how scientists think life on Earth began

Key words

multicellular
specialized

D Where did life come from?

Life on Earth began about 3 500 million years ago. There are lots of clues to how it started. But scientists don't all agree about what the evidence means.

Living means reproducing

Living things can all reproduce. The first living things were molecules that could copy themselves – like DNA.

Where did it start?

Scientists have two main ideas about where life on Earth came from.

Scientists have found simple cells buried in wet rocks in Antarctica.

- Life started somewhere else in the Solar System. It was brought to Earth on a comet or a meteorite. Early Earth was too hostile for life to have started here. Life began in water-soaked rocks beneath the surface of another planet.
- Life started at the bottom of the oceans. Hot water springs on the ocean floor contain dissolved minerals. When the hot water from the springs meets cold sea water, minute bubbles of iron sulfide, filled with a solution of different chemicals, are formed. These bubbles could have acted like tiny cooking pots. The chemicals may have made a thin layer of fatty protein on the inside of the bubbles, making the first cell membranes.

The right conditions

The conditions on Earth 3 500 million years ago were very different to now. But they must have been just right for life to grow.

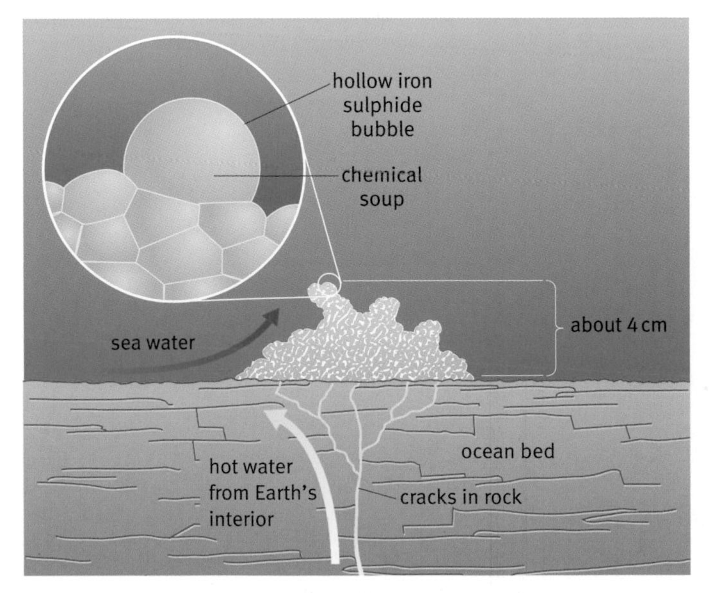

Iron sulfide bubbles on the sea floor.

When life began on Earth, the planet was very different. It may have looked like this.

Living things on Earth are suited to survive where they live. If they're not well suited, they die out. This is natural selection.

What if conditions on Earth had been different at any time in the last 3500 million years? Life as we know it might not exist. Very different living things might have evolved to suit living on Earth.

Iona marble

The oldest evidence for life in Britain

The oldest evidence for life in Britain comes from the Scottish island of Iona. The piece of marble on the right is nearly 3000 million years old. Marble is formed from limestone, which is formed from the remains of living things. It is thought that this Iona marble was formed from the remains of billions of single-celled organisms. They once lived along the edges of an ancient ocean.

This sea anemone is multicellular. It has hundreds of thousands of cells working together.

Living things get bigger

The first living things were only one cell big. **Multicellular** – many-celled – living things appeared hundreds of millions of years later.

Why did organisms get bigger?

Becoming multicellular had lots of advantages. For example, living things could get bigger. Also, cells could become **specialized**. Different cells changed so they could do one job better. Working together like this is more efficient than each cell trying to do every job.

The sea urchin has special cells to detect food. Different cells move the urchin to the food.

Was it all good news?

Becoming multicellular also caused one problem. Bigger organisms, like the sea urchin in the photo, need ways for cells to communicate.

Questions

1. How long ago did life begin on Earth?
2. What were the very first living things?
3. Explain *two* ideas scientists have for where life began.
4. What could have caused life on Earth to evolve differently?
5. What is meant by multicellular?
6. Write down two advantages of being multicellular.
7. What problem did living things have to solve when they became multicellular?

Becoming specialized means that you are very good at doing one particular job.

Find out about

- your body's communications systems
- how conditions inside your body are kept at the right levels

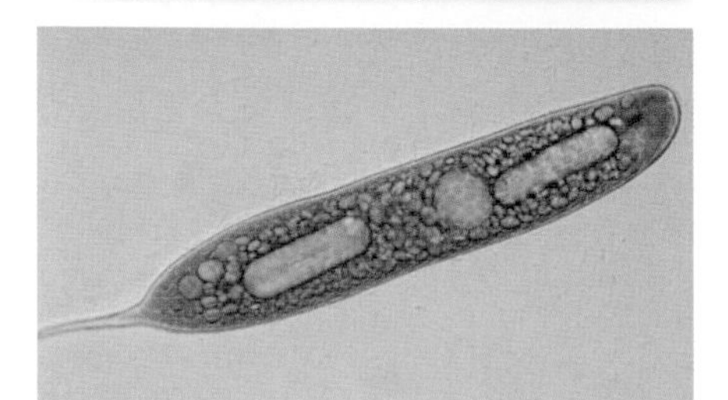

Euglena is a single-celled organism that lives in water. It will swim towards light. It uses light energy for photosynthesis to make food. (Mag: × 600 approx)

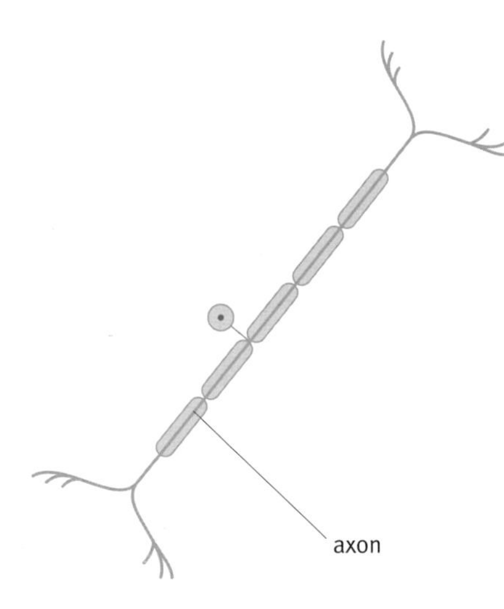

This diagram shows a nerve cell (neuron).

Key words
response
homeostasis
nerve cells
neurons
electrical impulses
hormones

E Keep in touch

Sound, sight, cold, and wet. All these things make the woman in the photo jump back from the car. One day this **response** could save her life.

Humans aren't the only animals that can sense something and react to it. All living things must do this to survive.

How does this work? Different parts of the body must be able to communicate with each other.

Changes happen inside too

Changes like those in the photos happen outside the organism's body. But many changes happen inside the body. The organism must respond to these as well, in order to survive. For example:

- Imagine you have just eaten a meal. Some of the food contained glucose – a sugar.
- The sugar is absorbed into your blood.
- Your blood sugar level rises above normal levels.
- If your body does not respond to this change, you will become unwell.

Your body's communication systems respond to changes inside the body. They keep your internal environment steady. This is called **homeostasis**.

What are the body's communication systems?

Parts of your body communicate with each other in two ways.

- **Nerve cells** (**neurons**) are very long, thin cells. They link up cells in different parts of the body. They carry **electrical impulses** around the body.
- Chemicals called **hormones** are carried in the blood. They are made in one part of the body. They make something happen in a different part of the body.

Why does the body need two communication systems?

Sometimes you need a fast response. Nerve cells carry electrical impulses very quickly. But their effect only lasts a very short time.

Sometimes you need a response that lasts for a longer time. For example, to control changes that take a long time, like growing. Hormones travel much more slowly. But their effects last much longer.

How does your nervous system work?

You touch a very hot plate – you move your hand away. This response protects your body from damage.

Let's look at another example.

- You walk from a dark cinema into a light room.
- Light receptors in the eyes detect the light.
- Muscles around your pupils contract.
- Your pupils get smaller.

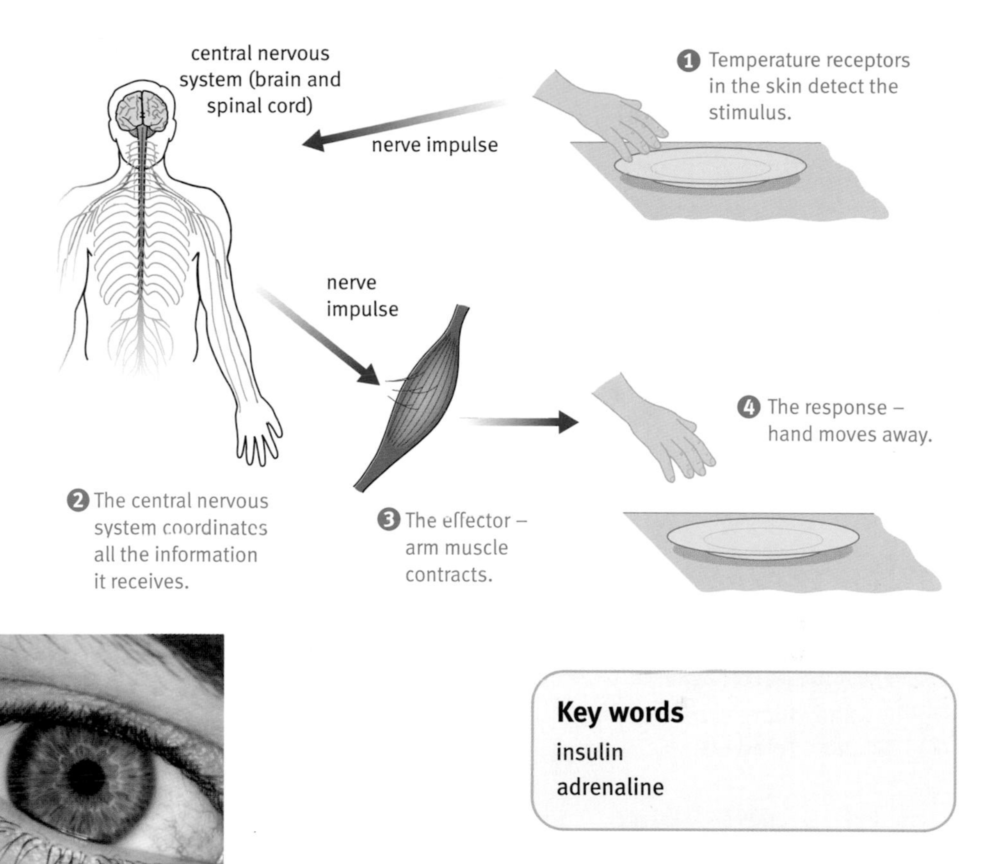

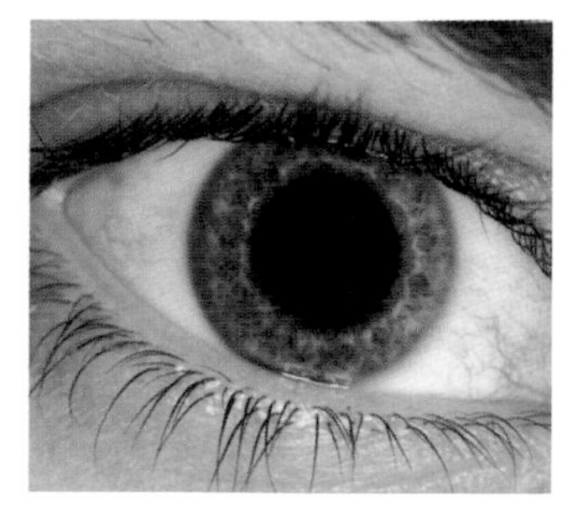

Hormone responses

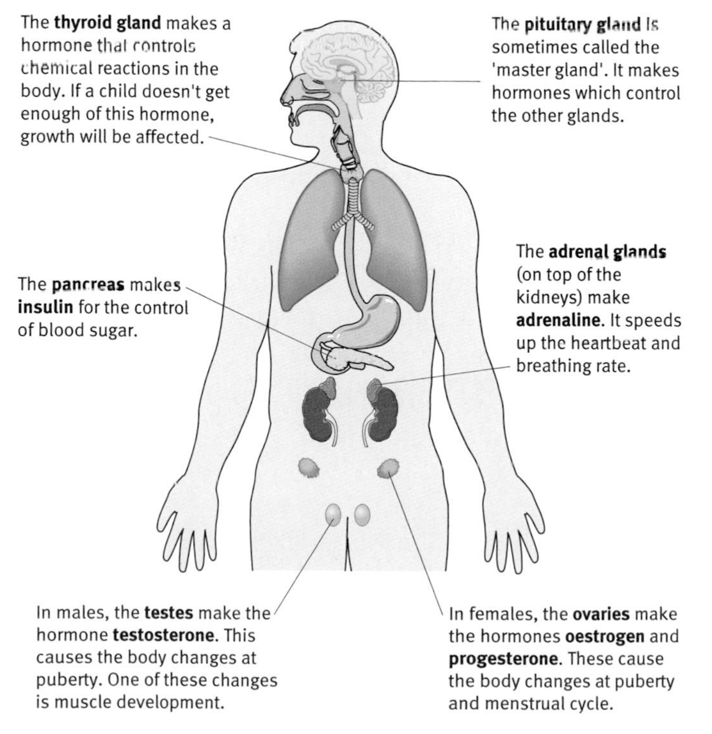

Hormones are made by parts of the body called glands.

Key words
insulin
adrenaline

Questions

1 How is information carried by the:
   a nervous system
   b hormonal system
2 What is the function of the body's brain and spinal cord?
3 Copy and complete the table to show examples of nervous communication.

Stimulus	Receptor	Effector	Response
Heat	temperature receptor in skin	muscle in arm	move hand away

4 Describe two examples of hormone communication.
5 Explain why your body needs two communication systems
6 Explain what is meant by homeostasis.

Find out about

- what we know about human evolution
- how new observations may make scientists change an explanation

Gorillas and chimpanzees are our closest living relatives.

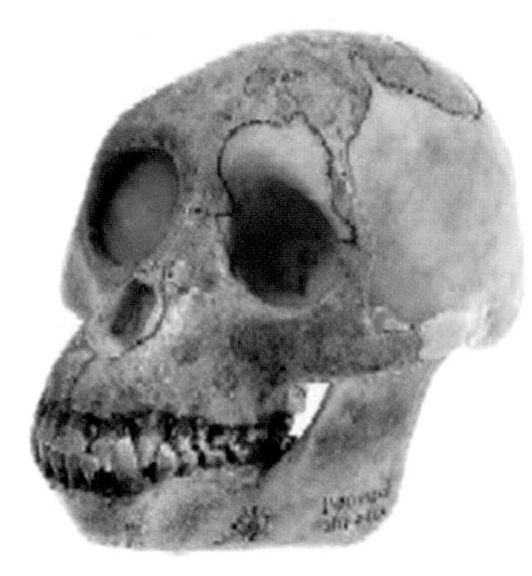

Scientists have dated these ape fossils to over 20 million years old.

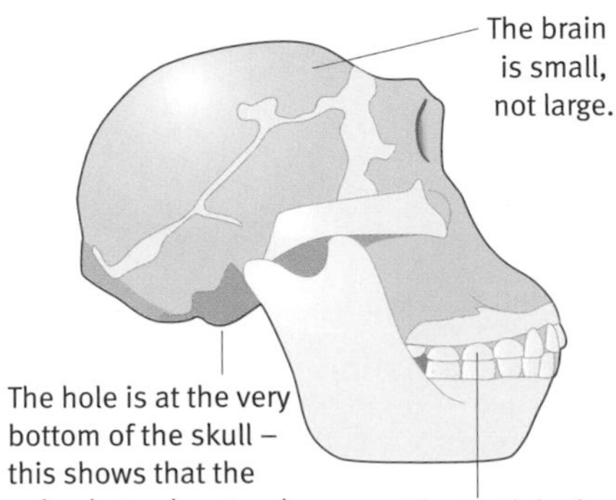

F Human evolution

Gorillas and chimpanzees are apes. Apes and human beings share many features. For example, human DNA is less than 2% different from chimp DNA.

So does this mean that human beings evolved from apes? No. But apes and humans do share an ancestor.

Where did the first humans come from?

The photo below is of a fossil skull of an ape-like animal which lived in Africa over 20 million years ago.

Modern apes and human beings evolved from an ancestor like this. At some point, they started to develop differently.

Human beings have bigger brains

Human beings have two big differences from apes:

- bigger brains
- walk upright

At first, scientists explained that apes which had developed big brains were able to stand up. So they predicted that big brains came before walking upright. Any observation they found that agreed with this prediction would increase the scientists' confidence in their explanation.

Just one observation would not be enough to prove that the explanation was correct. The scientists hoped to make several new observations which would do this. But the new evidence they found didn't agree with their prediction at all.

Hominids

In 1924, a skull was dug up in South Africa. It was the first skull found of a **hominid**. Hominids are animals that are more like humans than apes. They lived in Africa between 1.5 and 4 million years ago.

The skull really surprised scientists because the animal:

- had a small brain, not much different to apes
- walked upright

These observations disagreed with the scientists' prediction. Either the observations were wrong or the prediction was wrong. So scientists started to doubt their explanation of why hominids started to walk upright.

A new explanation

Around 7 million years ago, Africa was getting drier. Areas of trees were becoming grass. Apes that could find food in the grasslands wouldn't have to compete with other apes in the trees. An ape that walked upright would be able to see over the tall grass. This would have helped them survive.

Key words
hominid
common ancestor

More evidence for walking hominids

The most complete early hominid skeleton known was found in 1974. She's known as Lucy. Her skeleton also seemed to show that she walked upright. Then, in 1978, a set of footprints was found preserved in mud. These footprints showed that early hominids like Lucy really did walk upright.

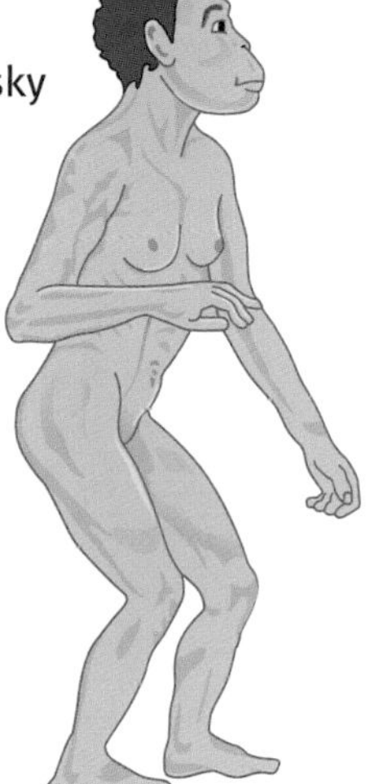

Lucy was named after the Beatles song 'Lucy in the sky with diamonds', which the scientists were playing in their camp at the time of the discovery.

These footprints were found preserved in volcanic mud.

Early humans

There were several different species of hominid. They shared a **common ancestor**. Over time, most of these hominids died out. But one species had the largest brains. This helped them to survive. They were early humans. By 150 000 years ago, a small group of them had evolved into modern humans *Homo sapiens*. They started to leave Africa and explore the rest of the world.

Big brains helped early humans learn to use tools, hunt, and make fire.

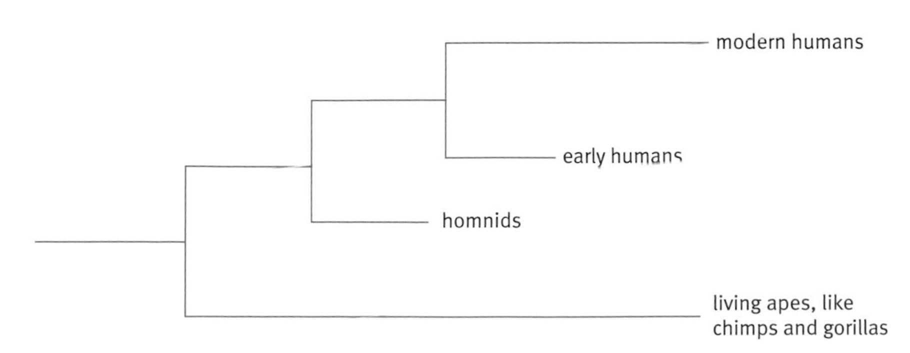

Hominids shared a common ancestor. Only one species survived and evolved into *Homo sapiens*.

Questions

1 What is a hominid?

2 Draw a diagram to show how all hominids had a common ancestor.

3 Give *two* ways in which big brains helped some early humans to survive.

(4) Scientists predicted that hominids had big brains before they started to walk upright. This was proved wrong. Explain how.

(5) Observations that disagree with a prediction may decrease our confidence in an explanation. Give an example from these pages of this happening.

(6) Explanations of human evolution are constantly changing. Why do you think this is?

Find out about

- why some species are under threat
- whether it matters if species become extinct

Royal Bengal tigers are already endangered. Rising sea levels from global warming may flood their last habitat.

A large international study says that up to a quarter of the species on Earth face extinction from global warming.

G Extinction!

Over the last few million years many species of plants and animals have lived on Earth. Most of these species have died out. They are **extinct**.

There is fossil evidence of at least five mass extinctions on Earth. Now we are at the beginning of another.

Endangered species

Where an animal or plant lives is called its **habitat**. Any quick changes in their habitat can put them at risk of extinction.

Around the world over 12 000 species of plants and animals are at risk of extinction. They are **endangered**.

Changes in the environment

All living things need factors like water and the right temperature to survive. Rising temperatures are changing many habitats. This global warming is putting many species at risk.

New species

New species moving into the habitat can put another one at risk.

- Animals and plants compete with each other for the things they need. Two different species that need exactly the same things cannot live together.
- The new species could be a **predator** of the species already living there.
- If the new species causes **disease**, it could wipe out the native population.

Red squirrels used to live all over the UK. Now the larger American grey squirrels have taken over most of their habitats.

In the 1960s, the virus that causes Dutch Elm disease came to the UK. It destroyed most of the UK elm population.

Going hungry

Plants and animals need other species in their habitat. For example, in this food chain spiders eat caterpillars.

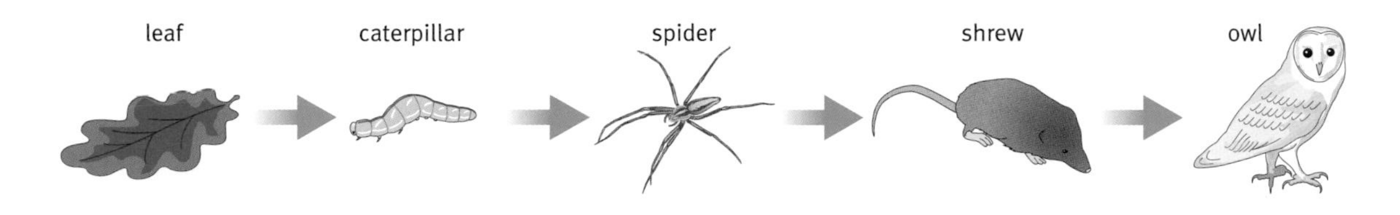

So if the caterpillars all died, the spiders could be at risk. That could also endanger the shrew and the owl.

The food web

Most animals eat more than one thing. Many different food chains contain the same animals. They can be joined together into a **food web**.

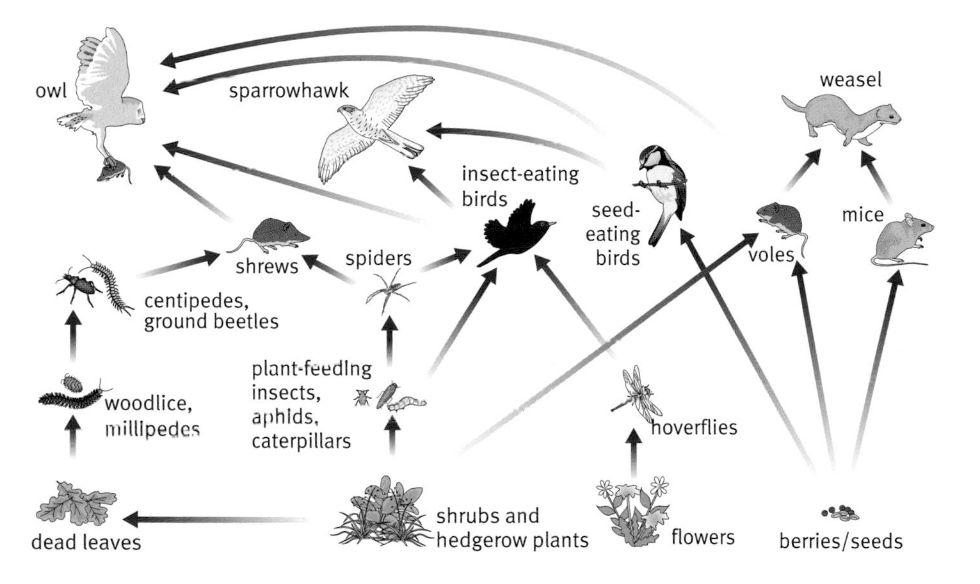

A new animal coming into a food web can affect plants and animals already living there.

Key words
- extinct
- habitat
- endangered
- predator
- disease
- food web

Questions

1 Look at the food web on this page.

- **a** A disease kills all the flowering plants. Explain what happens to the number of hoverflies.
- **b** Mink move into the habitat. They eat voles.
  - **i** The number of mice decreases. Explain why.
  - **ii** Explain what would happen to the number of caterpillars.

2 Explain what is meant by

- **a** extinct
- **b** endangered.

③ Name two things that

- **a** plant species may compete for
- **b** animal species may compete for

Dodos were not able to survive the changes in their environment. This is a disaster for any species.

Are humans to blame for some extinctions?

In 1598, Dutch sailors arrived on the island of Mauritius in the Indian Ocean. In the wooded areas along the coast they found fat, flightless birds that they called dodos. By 1700, all the dodos were dead. The species had become extinct. The popular belief is that sailors ate them all. But this explanation appears too simple. Written reports from the time suggest that they were not very nice to eat.

What killed the dodos?

Humans may not have eaten dodos. But did they cause their extinction without meaning to? When the sailors arrived, they brought with them rats, cats, and dogs. These may have attacked the dodos' chicks or eaten their eggs. The sailors also cut down trees to make space for their houses. Maybe this took away the dodos' habitat.

So human beings can cause other species to become extinct:

- directly, for example, by hunting
- **indirectly**, for example, by taking away their habitat, or bringing other species into the habitat

Pandas are endangered. They eat bamboo but there are only small areas of this left in China.

Isn't extinction just part of life?

Twenty First Century Science put this question to Georgina Mace of the UK Zoological Society.

Georgina Mace

"It is true that species have always gone extinct. This is a natural process. But the pattern of extinction today is different from what has been recorded in the past.

- The rate of species extinction today is thousands of times higher than in the past.
- Current extinctions are almost all due to humans."

Does extinction matter?

If many species become extinct, there will be less variety on Earth. This variety is very important. For example:

- People depend on other species for many things. Food, fuel, and natural fibres (such as cotton and wool) all come from other species.
- Many medicines have come from wild plants and animals. There are probably many other medicines in plants that haven't been found yet.

The variety of life on Earth is called **biodiversity**.

Foxgloves are very poisonous. But they have given us a powerful medicine to treat heart disease.

Biodiversity and sustainability

The Earth is 4500 million years old. Human beings have been here for about 160 000 years. If Earth is going to be a good home for future generations, then people today must take care of the planet.

Keeping biodiversity is part of using Earth in a sustainable way. **Sustainability** means meeting the needs of people today without damaging Earth for people of the future.

Questions

4 Explain how humans can cause extinction of other species:

a directly **b** indirectly

(5) Find out how human beings have caused the extinction of

a *Didus ineptus* (the dodo)

b *Equus quagga*

c *Ectopistes migratorius*

d *Achatinella mustelina*

Key words

indirectly
biodiversity
sustainability

B3 Life on Earth

Science explanations

In this chapter you have learnt how life on Earth has evolved. You have also seen how scientists work out explanations for things they see happening on Earth.

You should know:

- all life on Earth has evolved from the first very simple living things
- evidence for evolution comes from fossils and by comparing the DNA of different organisms
- the first living things appeared on Earth 3500 million years ago and were molecules that could copy themselves
- these living things may have developed on Earth, or they may have come to Earth from somewhere else
- if conditions on Earth had been different at any time since life first began, then evolution may have happened differently
- members of a species are not identical, there is variation between them
- variation is caused by the environment or genes, but most features are affected by both
- evolution happens by natural selection:
  - members of a species are all different from each other (variation)
  - they compete with each other for different resources
  - some have features that given them a better chance of surviving and reproducing
  - they pass on features through their genes to the next generation
- more of the next generation have these useful features
- the difference between natural selection and selective breeding
- genetic variation is caused by mutations in an organism's genes
- the main parts of the human nervous system
- how nervous and hormonal systems communicate information around the body
- that the evolution of a larger brain gave some early human a better chance of survival
- that many hominid species evolved from a common ancestor, but only one survived and became modern humans
- living organisms depend on their environment and each other for survival
- animals and plants in the same habitat compete for different resources
- how a change in a food web can affect all the species there
- species may become extinct if:
  - their environment changes
  - a new species arrives that is a competitor a predator, or causes disease
- another plant or animal in the food web becomes extinct
- two examples of modern extinctions caused:
  - directly by humans, for example, by hunting
  - indirectly by humans, for example, destroying their habitat
- why keeping biodiversity is important for us and for future generations

Ideas about science

Working out how something happens – an explanation – takes imagination and creativity. Scientists don't always agree about what the correct explanation for something is.

This Module looks at several explanations, including natural selection and where life on Earth began. From these you should be able to identify:

- statements that are data
- statements that are all or part of an explanation
- data or observations that are an explanation can account for
- data or observations that don't agree with an explanation

Scientists don't always come to the same conclusion about what some data means. The debate about Darwin's idea of natural selection is one example of this. You should know:

- working out an explanation takes creativity and imagination
- why Darwin's explanation was a good one
- why other scientists disagreed with his ideas at the time

New observations about human evolution are being found. Sometimes scientists use an explanation to predict an observation which hasn't been made yet. For example, scientists predicted that human evolved a big brain before they began to walk upright. Then they found fossils which did not agree with this prediction. You should know:

- how observations that agree or disagree with a prediction can make scientists more or less confidence about an explanation

Some scientific questions have not been answered yet. You should know:

- scientists have two different explanations for how life on Earth began, but there is not enough evidence to decide between them

Why study homeostasis?

Every moment of your life your body is reacting to changes. Some of these changes happen outside your body. Others happen inside your body, for example your body's water level dropping. Your body responds to these changes to make sure that conditions inside your body stay steady. This is vital for survival.

The science

Keeping a steady state inside your body is called homeostasis. Automatic systems in the body control water balance and body temperature. Water molecules are constantly moving in and out of cells. The amount of heat you lose to your environment depends on several factors. For example, is it a warm or a cold day? Enzymes speed up chemical reactions in cells. They need a particular temperature to work at their fastest rate.

Biology in action

Some diseases damage the body's ability to keep conditions inside steady. Extreme environments and sports can put too much strain on the body, and homeostasis starts to fail. Understanding homeostasis is crucial to help many people, from a baby in an incubator, to a person with kidney disease.

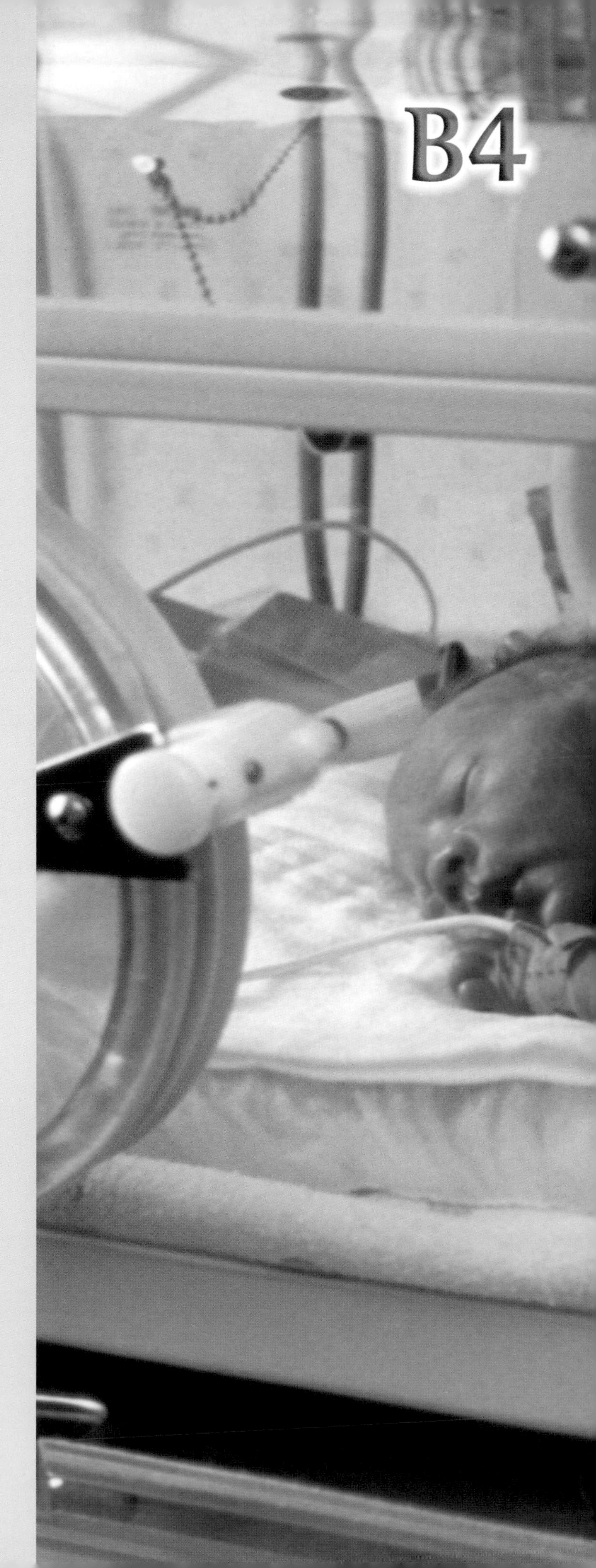

Homeostasis

Find out about:

- why homeostasis is important for your cells
- how temperature affects your enzymes
- how your body temperature is kept constant
- how different chemicals move in and out of your cells
- how your kidneys control your body's water level

Find out about:

- homeostasis
- why it is important

A Changing to stay the same . . .

Inside your cells thousands of chemical reactions are happening every second. These reactions are keeping you alive. But for your cells to work properly they need certain conditions. Keeping conditions inside your body the same is called **homeostasis.**

Homeostasis is not easy – lots of things have to happen for your body to 'stay the same'. Look at just a few of the changes happening every second:

Running has made this athlete hot. His body is sweating more to cool back down. This is an example of homeostasis.

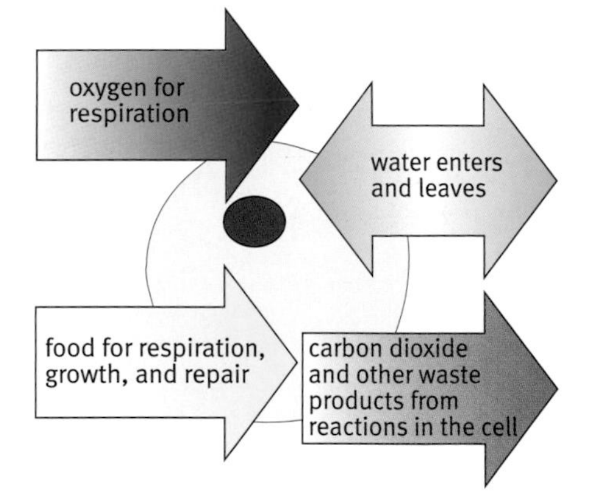

Cells must take in materials for respiration, growth, and repair. They must also get rid of waste.

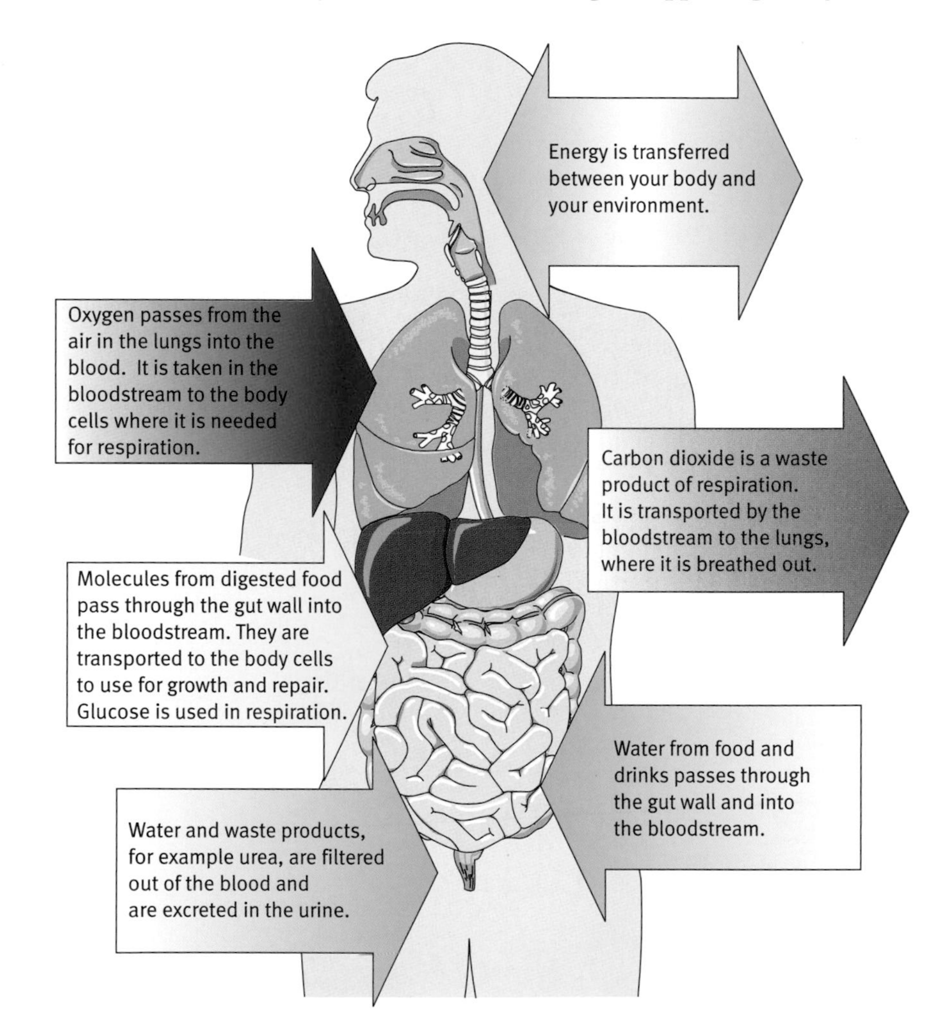

Some of the inputs and outputs that are going on all the time in your body. Materials enter and leave. Energy is also transferred in or out along a temperature gradient. The direction of transfer depends on the temperature of the environment.

So your body must work hard to:

- keep a constant body temperature
- keep the correct levels of water and salt
- control the amounts of nutrients, for example glucose
- take in enough oxygen for respiration
- get rid of toxic waste products, for example carbon dioxide and urea

Control systems

Premature babies cannot control their temperature, so they are put in incubators. The incubator is an artificial control system.

The control systems keeping a steady state in your body work in a similar way to artificial control systems.

All control systems have:

- a **receptor**, which detects the stimuli (the change)
- a **processing centre**, which receives the information
- an **effector**, which produces an automatic response

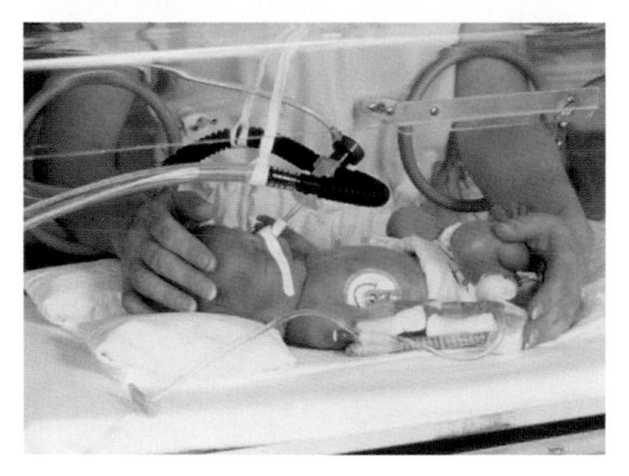

An artificial control system is keeping this baby's temperature steady.

How does the incubator work?

An incubator has a temperature sensor, a thermostat with a switch, and a heater.

The temperature inside the incubator is detected by the sensor. If the temperature falls, the thermostat switches on the heater. When the temperature gets back up to its normal setting of 32 °C, the thermostat switches off the heater.

What about your body?

Some of the temperature control in your body is automatic too. For example, you do not consciously decide to sweat when you are hot. But you can also consciously control your temperature by doing things like putting a coat on or taking it off, or moving between a shady and a sunny place.

There are temperature receptors in your brain and skin. If you are cold, you can warm yourself by putting on a jacket.

Questions

1 Write down a definition for homeostasis.

2 In an incubator, name:

   a a receptor

   b a processing centre

   c an effector

3 You can cool yourself down by taking a coat off. In this control system, name:

   a the receptors

   b the processing centre

   c the effectors

4 Explain why the temperature in an incubator is set below 37 °C even though the baby's body must stay at this temperature. (*Hint:* Respiration in the baby's cells will be releasing energy.)

Key words

homeostasis
receptor
processing centre
effector

Find out about:

- negative feedback
- how some effectors work in opposite pairs

H

B Feedback in control systems

If the temperature in an incubator falls too low, the heater is switched on. The temperature goes up. When the temperature is high enough, the heater is switched off. This type of control is called **negative feedback**:

- Any change in the system results in an action that reverses the change.

The diagram below explains how this works.

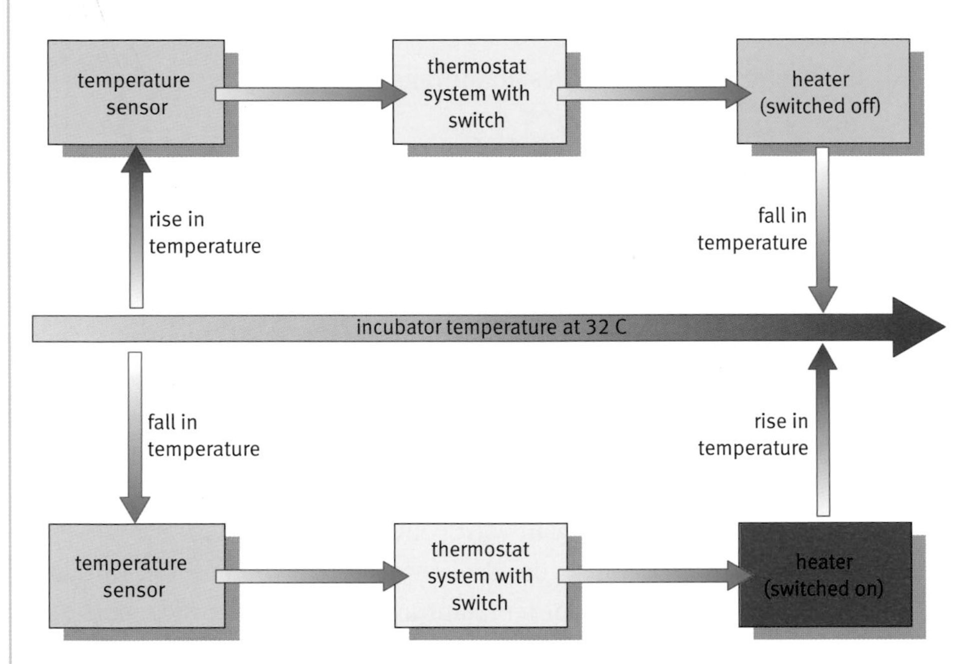

Negative feedback in an incubator to control temperature

Negative feedback systems are all around you. For example, if the temperature inside your fridge goes up, the motor switches on to cool it down. When it is cool enough, the motor switches off.

Your body uses negative feedback systems too. However, your body systems are more complicated than those in an incubator or a fridge. You have effectors to cool you down ***and*** to warm you up.

Control systems with more than one effector

Imagine you are driving a car in a 30 mph zone. You need to make sure that you do not speed, but you should not drive too slowly either. In this system, your brain is the processing centre. There are two effectors: an accelerator and a brake.

Now imagine you could have only one of those effectors, and you are going down a hill.

- If you only have a brake, you will be fine – until you get to the bottom.
- If you only have an accelerator, you have problems! Taking your foot off the accelerator is not going to be enough to control your speed.

You need both of these effectors to have proper control of the car. The brake and accelerator work together to maintain a constant speed. When the car speeds up, you brake. If the speed falls too low, you use the accelerator to speed it up again.

Because they have opposite effects, the brake and accelerator are called **antagonistic effectors**. You can use them to adjust the system as soon as there is a change – either too fast or too slow. This means that the response is much more sensitive. Your body uses antagonistic effectors too.

You need both an accelerator and a brake to control a car's speed.

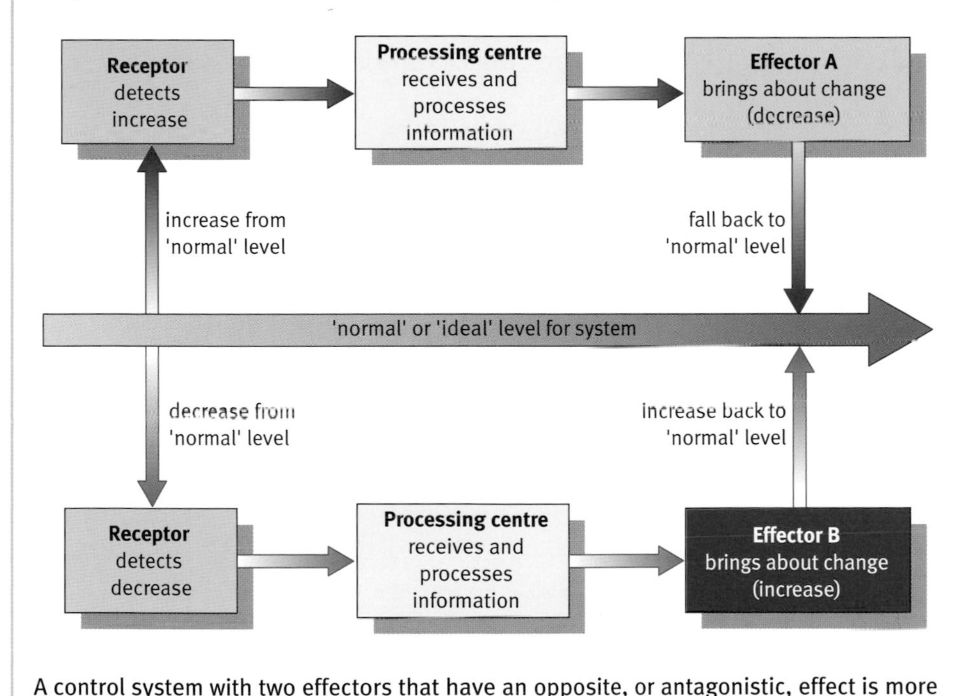

A control system with two effectors that have an opposite, or antagonistic, effect is more sensitive than a system with only one effector.

Key words

negative feedback
antagonistic effectors

Questions

1 Write down a definition for negative feedback.

2 If the diagram above represents a car, what is:

a effector A? **b** effector B?

3 Why are the brake and accelerator of a car called antagonistic effectors?

4 Why is it an advantage to have antagonistic effectors in a car?

5 Central heating systems use negative feedback to control the temperature of a house. Explain simply how this works.

Find out about:
- why you cannot live without enzymes
- how enzymes work
- how temperature and pH effect enzymes

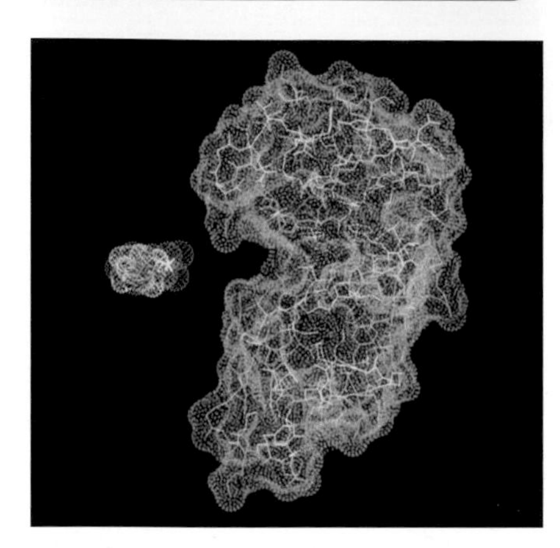

A computer graphic of an enzyme, its active site, and the product of a reaction.

C Enzymes

The chemical reactions that take place in cells rely on **enzymes**. Enzymes work best in certain conditions, for example at a certain temperature. This is an important reason why you need a steady state inside your cells.

What are enzymes?

Enzymes are the **catalysts** that speed up chemical reactions in living organisms. They are proteins – large molecules made up of long chains of amino acids. The amino acid chains are different in each protein, so they fold up into different shapes. An enzyme's shape is very important to how it works.

How do enzymes work?

Some enzymes break down large molecules into smaller ones. Others join small molecules together.

In all cases, the molecules must fit exactly into a part of the enzyme called the **active site**. It is a bit like fitting the right key in a lock. So scientists call the explanation of how an enzyme works the **lock-and-key model**.

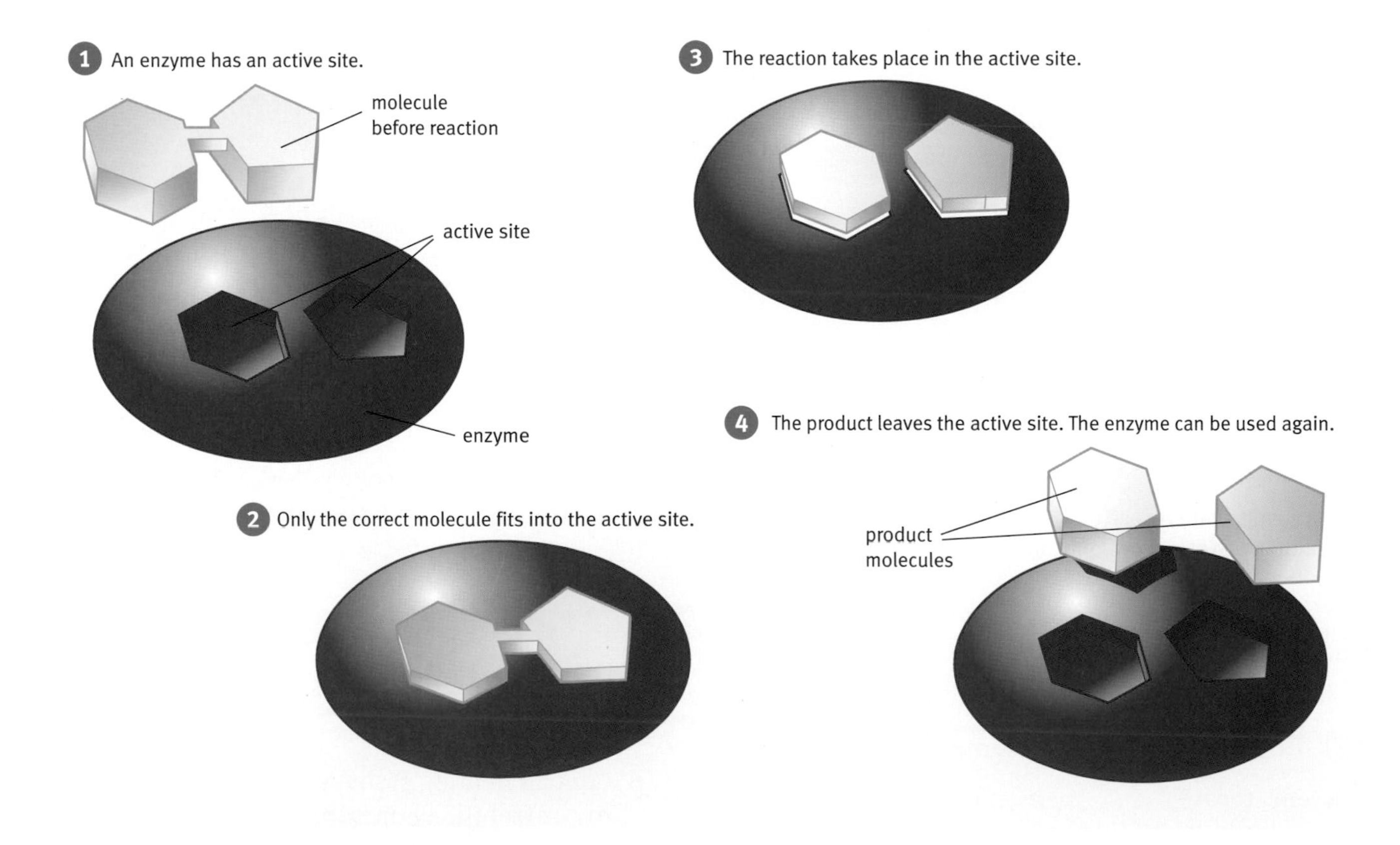

The lock-and-key model of enzyme function. (*Note:* This diagram is schematic. This means that if you were able to see the molecules and enzymes, they would not look like they do here. Enzymes consist of many molecules – see the photograph above left for an idea of scale. However, this is a useful way of picturing, or modelling, how an enzyme works.)

Why do we need enzymes?

At 37 °C, chemical reactions in your body would happen too slowly to keep you alive.

One way of speeding up a reaction is to increase the temperature. As the temperature rises, molecules

- have more energy
- move around faster and collide more often
- react more easily when they do collide

A higher body temperature could speed up the chemical reactions in your body. But higher temperatures damage human cells. Also, to keep your body warm, you have to release energy from respiration. For a higher body temperature, you would need a lot more food to fuel respiration.

So, we rely on enzymes to give us the rates of reaction that we need. They can increase rates of reaction by up to 10 000 000 000 times. It is not possible to live without enzymes.

If you had to maintain a higher body temperature, you would have to spend a lot more time eating to provide the calories to heat your body. Already 80% of the energy from your food is used for keeping warm.

Questions

1 Write down:

a what enzymes are made of **b** what enzymes do

2 Explain how an enzyme works. Use the key words on this page in your answer.

3 The enzyme amylase breaks down starch to sugar (maltose); catalase breaks down hydrogen peroxide to water and oxygen. Explain why catalase does not break down starch.

(4) a Calculate from your own body mass how much food you would need to consume each day if you had to eat the same proportion of your body mass as a shrew.

b How many 500 g loaves of bread does this mass of food represent?

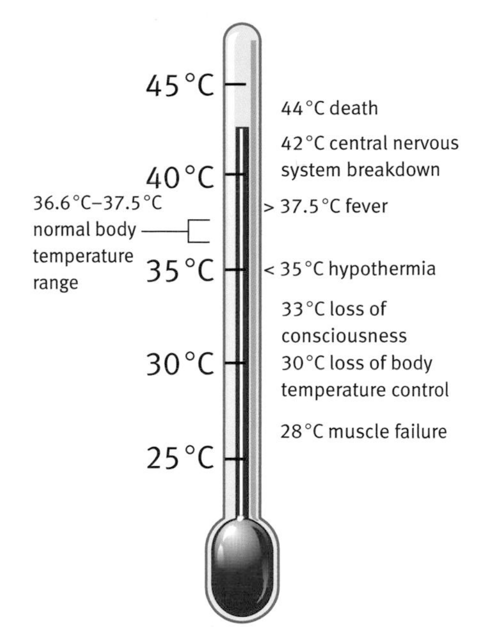

Your core body temperature is about 37 °C, but a small variation either side of this is normal. A core temperature over 42 °C or under 28 °C usually results in death.

Shrews have a large surface area for their volume. They lose heat to the environment over their whole body surface. To release enough energy to maintain their body temperature, they have to eat 75% of their body mass in food each day. If not, they die within 2–3 hours.

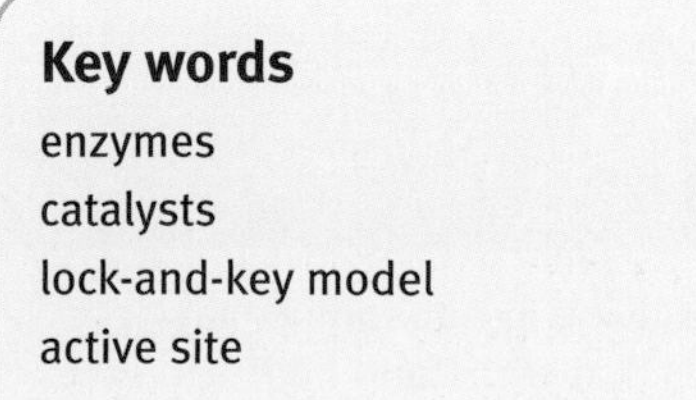

Key words

enzymes
catalysts
lock-and-key model
active site

Why do higher temperatures speed up reactions?

For any reaction to happen, the molecules must bump into each other. At low temperatures, molecules move slowly, so they are less likely to collide. If the temperature increases, the molecules have more energy. They move around faster. The molecules collide more often and with more energy. So the rate of reaction increases.

How does temperature affect enzymes?

At low temperatures, enzyme reactions get faster if the temperature is increased. The enzymes and other molecules collide more often, so they react more frequently.

But there is a difference with enzyme reactions. Above a certain temperature the reaction stops. This is because enzymes are proteins. Higher temperatures change an enzyme's shape so that it no longer works. The diagram below explains what happens using the lock-and-key model.

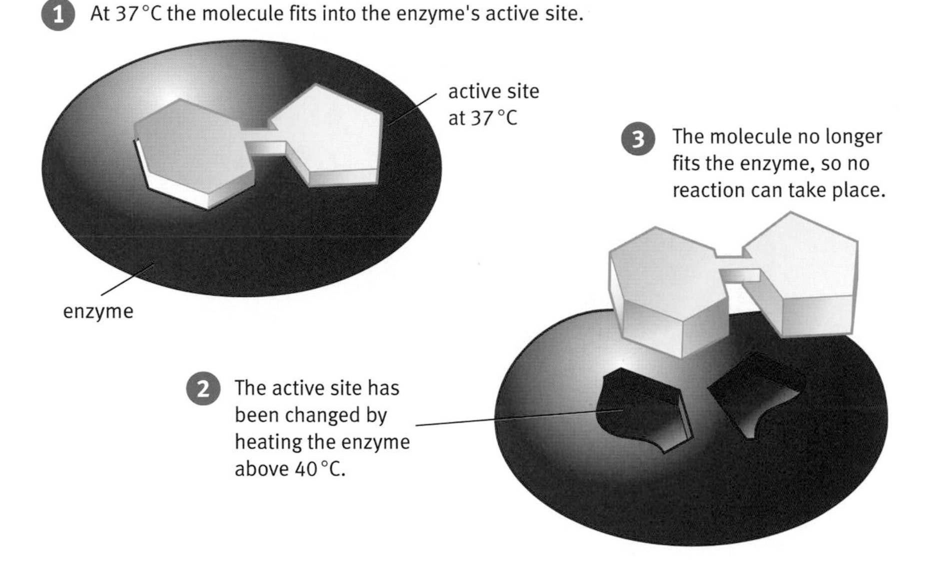

How an enzyme reaction can be stopped by a rise in temperature.

'Ice fish' such as Antarctic cod are active at 2 °C because their enzymes work best at low temperatures. Most organisms would be dead or very sluggish at 2 °C.

Bacteria living in hot springs have enzymes that withstand high temperatures.

Enzymes are denatured at high temperatures

High temperatures change the shape of an enzyme. They do not destroy it completely. But even when an enzyme cools down, it does not go back to its original shape. Like cooked egg white, the protein cannot be changed back. The enzyme is said to be **denatured**.

Why 37 °C?

The temperature at which an enzyme works best is called its **optimum temperature**. It is a temperature too low to denature the enzyme. But it is high enough for collisions between the enzyme and other molecules to be frequent and energetic. Enzymes in humans work best at around 37 °C. Some organisms have cells and enzymes adapted to different temperatures.

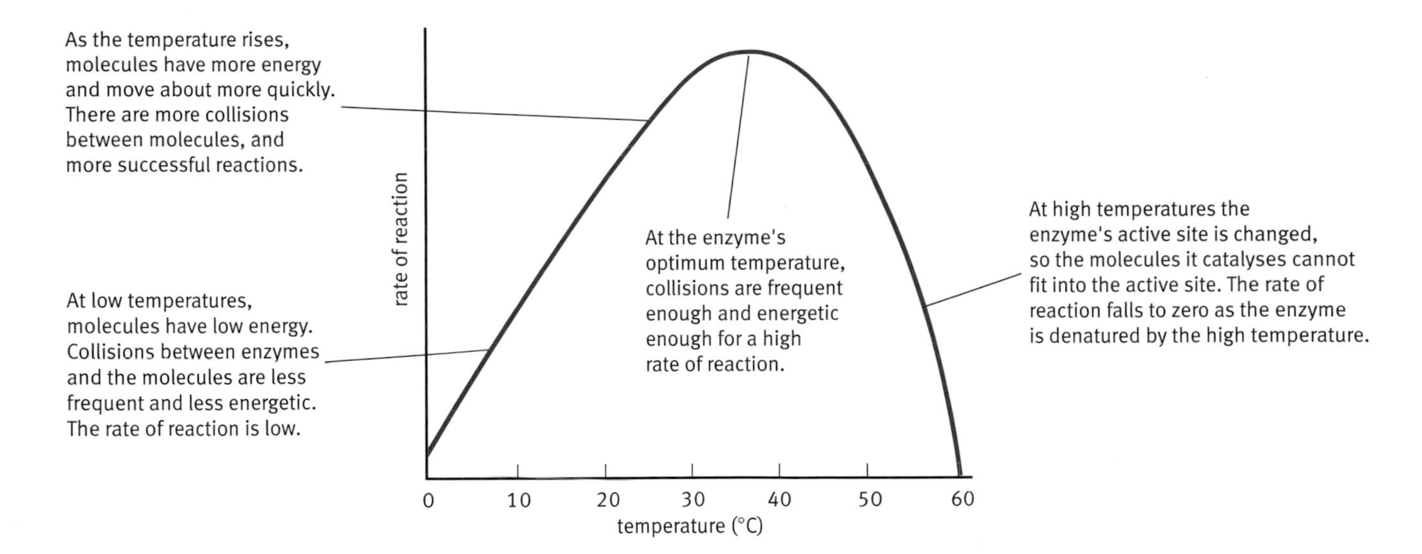

The optimum temperature of an enzyme

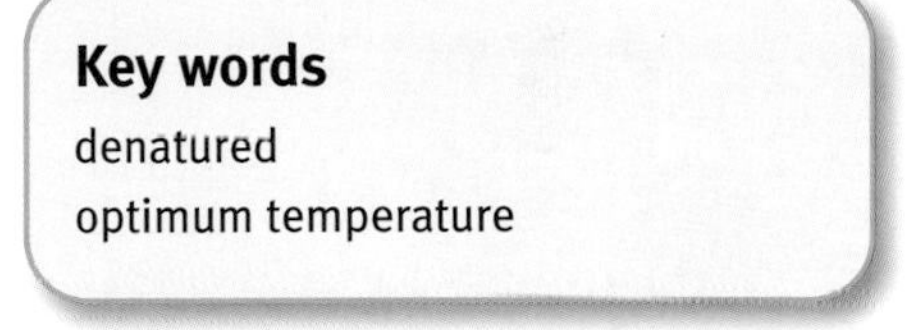

Key words

denatured

optimum temperature

pH also affects enzymes

The shape of an enzyme's active site is also affected by pH. Every enzyme has an optimum pH at which it works best.

Enzyme	What it does	Optimum pH
salivary amylase	breaks down starch to sugar (maltose)	4.8
pepsin	breaks down proteins into short chains of amino acids	2.0
catalase	breaks down hydrogen peroxide into water and oxygen	7.6

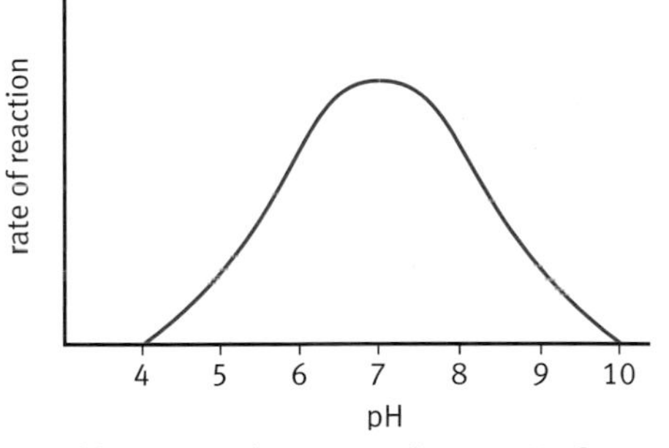

This enzyme has an optimum pH of 7.0.

Questions

5 Explain why increasing the temperature causes enzyme reactions to:

a get faster at low temperatures

b stop at higher temperatures

6 What is meant by an enzyme's optimum temperature?

7 Amylase is an enzyme that changes starch to maltose. Some students compared the rate of reaction at 17 °C, 37 °C, and 57 °C.

a Say what would happen if the students:

i raised the temperature of the tube at 17 °C to 37 °C?

ii lowered the temperature of the tube at 57 °C to 37 °C?

b Explain both of these changes.

8 Enzymes in food and enzymes from microorganisms such as bacteria and fungi make food decay. Why does food stay fresh in a refrigerator for a few days but for months in a freezer?

9 Amylase is in your saliva and your small intestines. Suggest why amylase cannot work in your stomach.

Find out about:
- how you warm up and cool down

D Getting hot, getting cold!

Imagine you are on a tropical island in summer. You are lying on the beach. The sand is baking hot. You slowly drift off to sleep. . . but inside, your body is working hard to keep a constant body temperature.

Gaining heat

If your environment is hotter than you are, energy will be transferred to your body. Your body also gains heat from **respiration**. During respiration, glucose is broken down to release energy for cells. This energy is used by muscles for movement. And some of the energy warms up the body.

Luckily, these swimmers' bodies respond quickly to their fall in temperature to bring it back to normal.

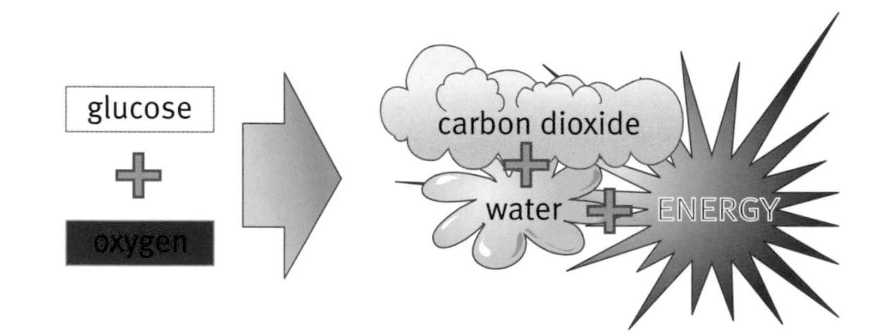

Respiration releases energy from the breakdown of glucose.

Losing heat

If your environment is cooler than you are, energy will be transferred away from your body. The bigger the temperature difference, the greater the rate of cooling. The swimmers on the left are losing energy to the sea and air.

Getting the balance right

For your temperature to remain constant, energy gain must be balanced by energy loss.

In other words, if heat gained = heat lost, your body temperature stays the same.

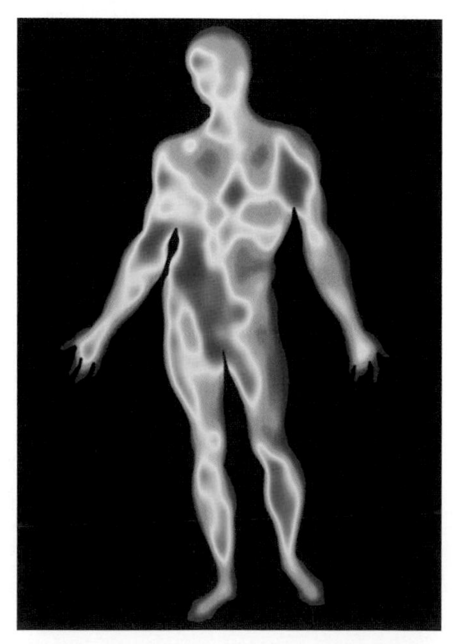

On this thermal image, the hottest parts of the body are red and the coolest, blue.

Different parts, different temperatures

Not all of your body is at the same temperature. Your **extremities** (hands and feet) are cooler than your **core** (deeper parts). They have a larger surface area compared with their size. So they lose energy to the environment faster than the main parts of your body.

The temperature of your hands, feet, and skin falls, but your core temperature hardly changes. Your core body temperature is the temperature inside your trunk and your brain. It should be between 36 °C and 37.5 °C for your body to work properly. 'Normal' body temperature will vary from one person o the next.

The amount of energy released in respiration and other reactions is greatest in your muscles and liver. The circulation of your blood transfers this energy to other parts. When you are cold, the circulation to your extremities is reduced. This means that your core stays warm.

Investigating temperature control

A physiologist called Sir Charles Blagden was Secretary of the Royal Society towards the end of the eighteenth century. Like many physiologists, he experimented on himself. He went into a very hot room to see how his body would react. The account below is adapted from Harry Houdini's book, *The Miracle Mongers: An Exposé*.

Another account describes Blagden taking a dog into the room with him, along with steak and eggs. The dog was unharmed but had to stay in its basket so it did not burn its feet on the hot floor.

Key words
respiration
extremities
core

Blagden's experiment

Sir Charles Blagden went into a room where the temperature was 1 degree or 2 degrees above 127 °C, and remained eight minutes in this situation, frequently walking about to all the different parts of the room, but standing still most of the time in the coolest spot, where the temperature was above 116 °C. The air, though very hot, gave no pain, and Sir Charles and all the other gentlemen were of the opinion that they could support a much greater heat.

During seven minutes Sir Charles' breathing continued perfectly good, but after that time he felt an oppression in his lungs, with a sense of anxiety, which induced him to leave the room. His pulse was then 144, double its ordinary quickness. In order to prove that the thermometer was not faulty, they placed some eggs and a beef-steak upon a tin frame near the thermometer, but more distant from the furnace than from the wall of the room. In twenty minutes the eggs were roasted quite hard, and in forty-seven minutes the steak was not only cooked, but almost dry.

Blagden's dog takes part in the experiment.

Questions

1 What two things must be balanced to keep your body temperature balanced?

2 What part of your body is warmest?

3 What temperature does this warmest part need to be kept at?

4 Why is your blood important in keeping extremities warm?

5 What effects did the very high temperatures have on Sir Charles Blagden?

⑥ What happened to the proteins in the steak and eggs as they cooked?

⑦ Why did the same thing not happen to Sir Charles' proteins?

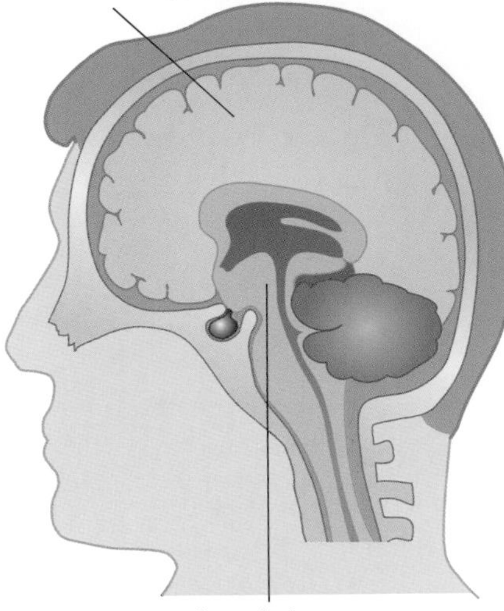

The **hypothalamus** is the processing centre in the brain for sleep, water balance, body temperature, appetite, and other functions. The cerebral hemispheres are where you make conscious decisions to warm or cool yourself.

Goose pimples are the result of contraction of the muscles in your skin that make your hair stand on end. The insulating layer of air trapped by hair is effective in furry mammals.

Sensing and control

Changes in your body temperature can have serious effects on health. So temperature receptors in your skin can detect a change in air temperature of as little as 0.5 °C. Your brain is particularly sensitive to temperature changes. So there are receptors here that detect blood temperature. Your brain also contains the processing centre for temperature control. It receives information from the temperature receptors. When the temperature in your brain is above or below 37 °C, it automatically triggers effectors to bring your body temperature back to normal. These effectors include muscles and sweat glands.

Warming up

Shivering is one way your body keeps warm. When you shiver, muscle cells contract quickly. They must respire faster to release the energy for this movement. So there is also more energy for keeping warm.

Shivering is an automatic response. You may also take a conscious decision to do something that will warm you up, for example drinking a warm drink, putting on more clothes, or going inside.

Cooling down

When you are too hot, nerve impulses from the brain stimulate your **sweat glands**. They make sweat, which passes out of small pores onto the skin surface. When sweat **evaporates** from your skin, it cools you down. Even when it is cool and you are not very active, you can lose nearly a litre of water a day in sweat. When you are hot and active, you can lose up to three litres of water an hour. That amount of water is hard to replace, so you could become dehydrated.

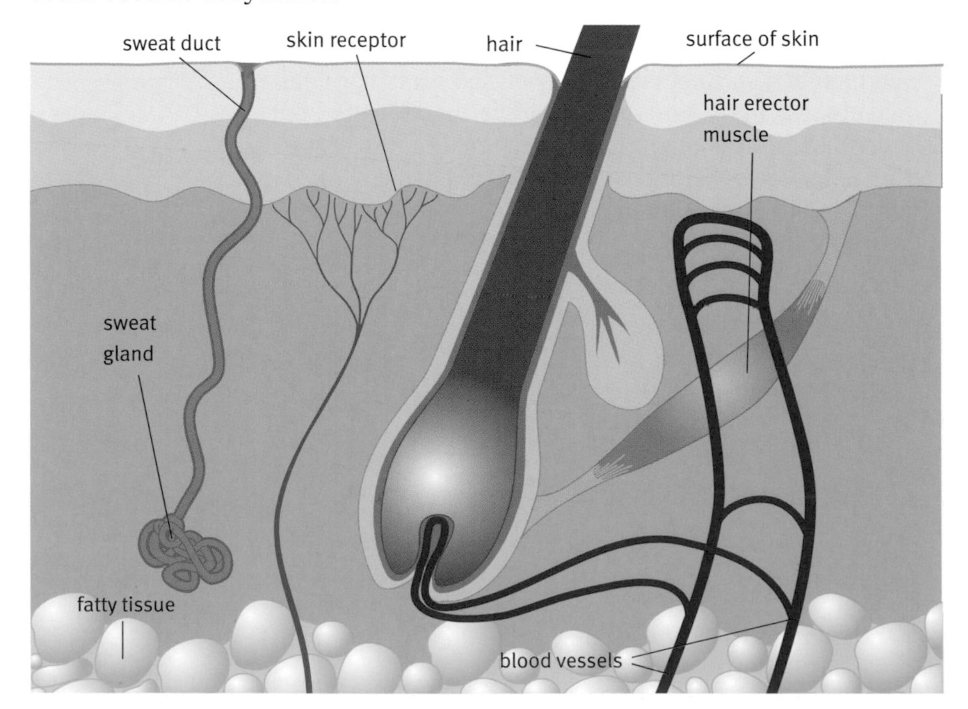

The structure of skin

How does sweat cool you down?

Water molecules in sweat gain energy from your skin. Soon they move quickly enough to evaporate. This cools you down.

Sweating only works to cool you down if the sweat can evaporate quickly. In a hot, humid climate sweat drips off you and you feel uncomfortably hot. Air currents increase the rate of evaporation and so increase the cooling effect.

Is your body temperature the same all day?

All these responses keep your body temperature within a narrow range. Average core body temperature is about 36.9 °C. But this varies from person to person and it varies throughout the day. For example, when you are sleeping, you move around less and respire more slowly. So your body temperature drops to its lowest point at night.

In hot, humid conditions, sweat cannot evaporate easily. Your clothes may become soaked in sweat.

The fans in this Sikh temple provide a welcome breeze to help to keep the people cool.

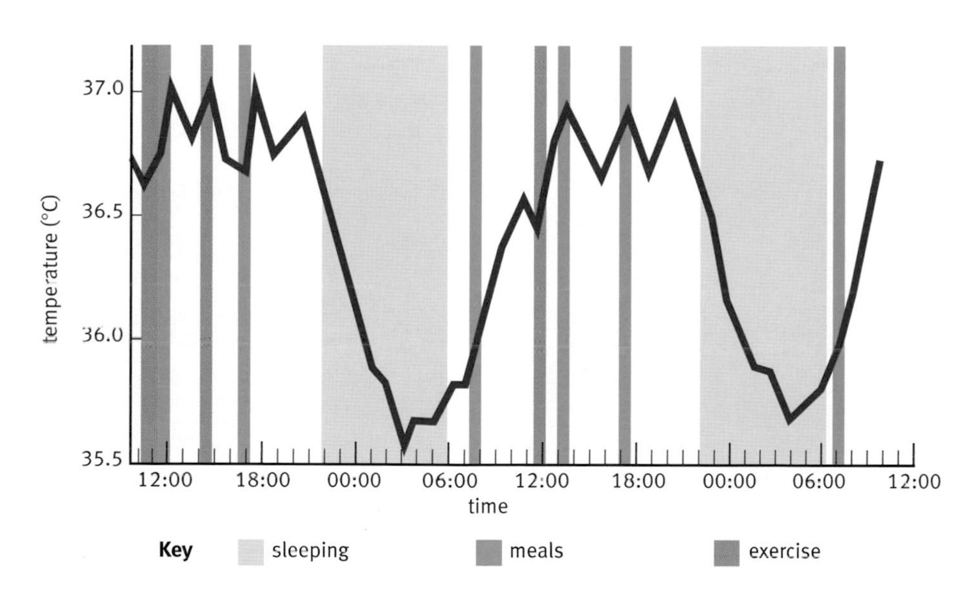

The daily cycle of variation in body temperature. You can see that eating, sleeping and activity affect body temperature. But these fluctuations happen when you are at rest too – they are controlled by our 'biological clocks'.

Questions

8 Where in your body would you find:
  a temperature receptors?
  b the temperature processing centre?

9 Name two effectors for controlling body temperature.

10 Explain how shivering warms you up.

11 Explain how sweating cools you down.

Key words

hypothalamus
shivering
sweat glands
evaporates

Too hot or too cold?

The table below lists ways ways of controlling body temperature.

Too cold?	Too hot?
Shivering Energy warms the body tissues when muscles contract. This also happens when you exercise.	**Sweating (or getting wet)** Energy is lost from the skin molecules when the water in sweat evaporates.
Warm food or drinks Energy is transferred from the warm food or drink to your body.	**Cold food or drinks** You are cooled as energy from your body heats the cold food or drink.
Clothes and hair-raising Clothes and hair trap an insulating layer of air. This slows the loss of energy from your body to the environment.	**Protective clothing** Hats, sun umbrellas, and protective clothing can reduce the heating effect of direct sunlight. An insulating layer can keep you cool in hot weather.
Heater Go into an area that is warmer than your body, for example near a heater or in the sun. Energy is transferred to your body from the environment.	**Fan** Moving air near the skin can increase the rate of evaporation of sweat. This increases the cooling effect.
Vasoconstriction H Loss of energy from your body's surface is reduced.	**Vasodilation** H Energy loss from your body's surface is increased.

A

B

C

D

E

F

Questions

12 For each picture on this page, say why the action cools the body.

Vasoconstriction and vasodilation

The boy on the right is flushed as a result of **vasodilation**. 'Vaso' is from the Latin for vessel, so vasodilation means widening of the blood vessels. More blood flows into the capillaries in the skin, so there is more energy transfer to the environment. The opposite is **vasoconstriction**. Less blood reaches the capillaries in the skin, so energy loss is reduced.

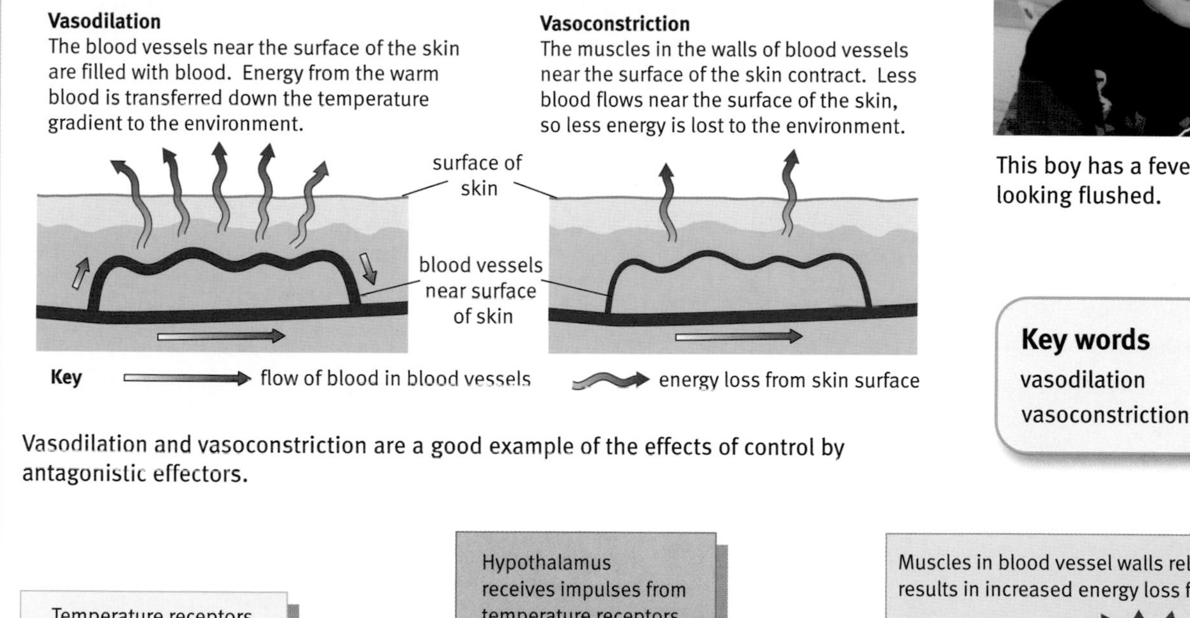

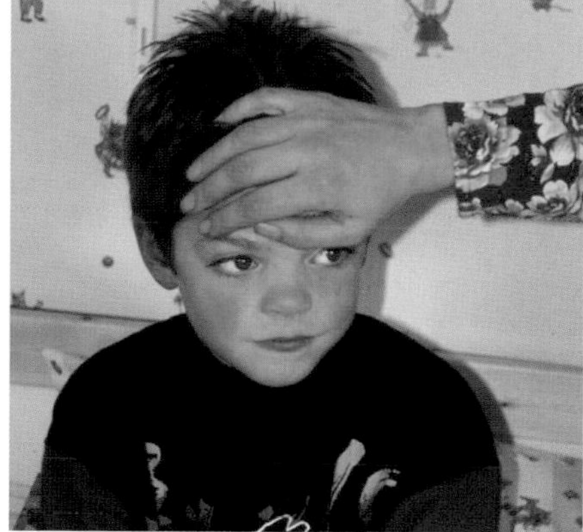

This boy has a fever. He is sweating and looking flushed.

Key words
vasodilation
vasoconstriction

Vasodilation and vasoconstriction are a good example of the effects of control by antagonistic effectors.

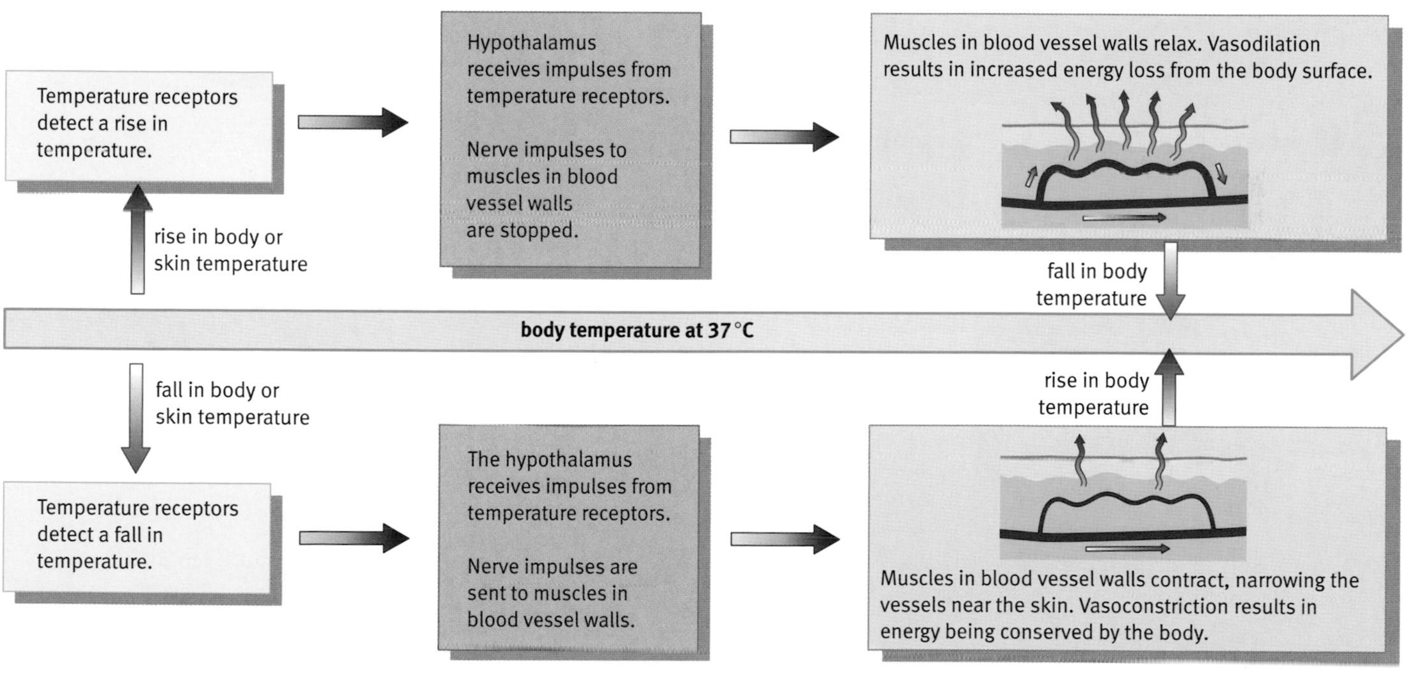

Temperature control is another example of negative feedback.

Questions

13 What are the effectors that cause vasodilation and vasoconstriction?

14 Explain why vasodilation and vasoconstriction are said to be controlled by antagonistic effectors.

(15) Why does vasodilation not cool you when the temperature of the environment is higher than your body temperature?

Find out about:
- how chemicals move in and out of cells
- why cells need a steady water balance

E Diffusion and osmosis

Your cells need a constant supply of raw materials for chemical reactions. Waste products must also be removed from cells. So molecules move in and out of cells all the time.

Diffusion

Molecules in gases and liquids move about randomly. They collide with each other and change direction. This makes them spread out.

Overall, more molecules move away from where they are concentrated than move the other way. We say the molecules diffuse from areas of their high concentration to areas of low concentration.

For example, molecules diffuse out of a tea-bag when you make a cup of tea. Diffusion is a **passive** process. It does not need any energy.

1 Water just poured onto tea-bag

2 About 30 seconds later

3 About 2 minutes later

Molecules diffusing out of a tea-bag.

This swimmer's cells need oxygen and glucose for respiration. They must get rid of carbon dioxide. These molecules move in and out of cells by diffusion.

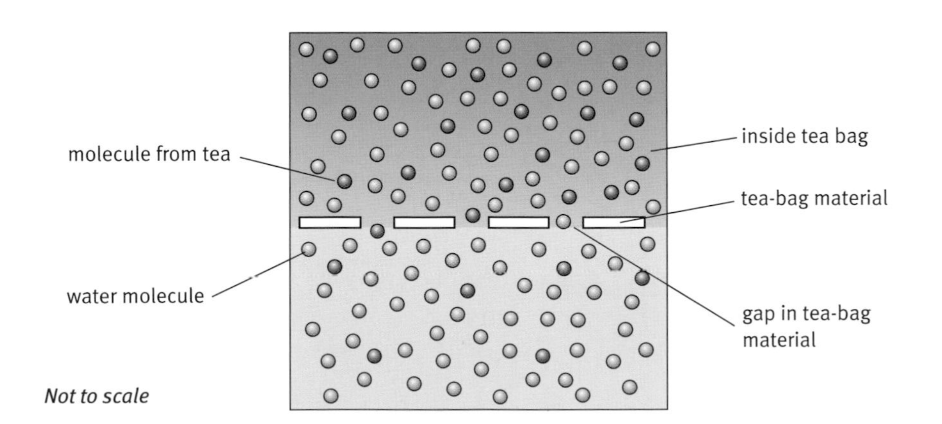

Dissolved molecules from the tea are more concentrated in the tea-bag than in the surrounding water. As the tea brews, more of the dissolved molecules move out of the tea-bag than in. More water molecules will move into the tea bag than out. Eventually, the tea molecules will be evenly spread. (*Note:* In this diagram the circles represent molecules, not individual atoms. The tea-bag material is also made of molecules.)

Cell membranes are partially permeable

The gaps in a tea bag allow water and dissolved molecules to diffuse in and out freely. The bag is permeable. Cell membranes let some molecules through, but not others. We call them **partially permeable membranes**.

The diagrams below show partially permeable membranes. These membranes let water through but not glucose.

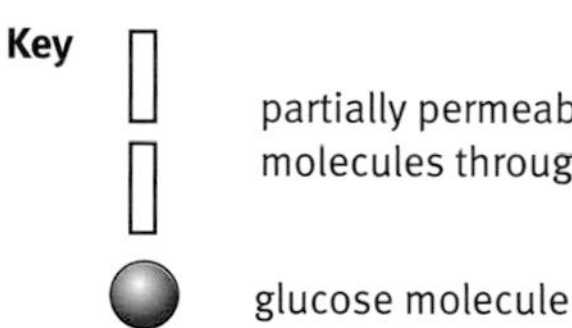

partially permeable membrane allows some molecules through and acts as a barrier to others

glucose molecule

water molecule

water molecules associated with glucose molecule

(*Note:* In these diagrams, the circles represent molecules, not individual atoms. Cell membranes are also made of molecules.)

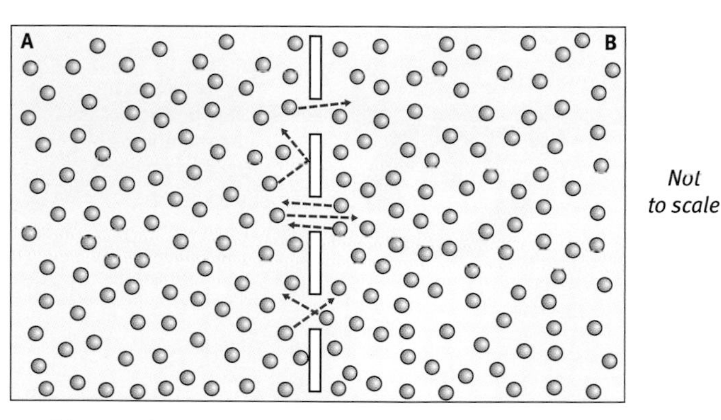

Not to scale

a This membrane is separating water molecules. Water molecules move at random. So as many pass from left (A) to right (B) as pass from right (B) to left (A).

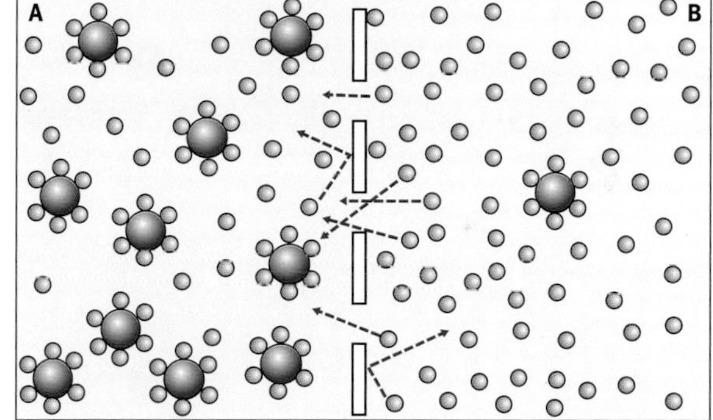

b This membrane is separating two glucose solutions. There are more water molecules and fewer glucose molecules on the right (B). In other words, water is in higher concentration on the right. So there is overall movement of water from right (B) to left (A).

Osmosis

A solution that has a high concentration of water molecules is a dilute solution. For example, if you make a dilute drink of squash, it has lots of water in it. A more concentrated drink of squash would have less water.

More water molecules move away from an area of higher concentration of water molecules. Think of it as diffusion of water. This overall flow of water from a dilute to a more concentrated solution across a partially permeable membrane is called **osmosis**.

Key words
diffusion
partially permeable membrane
osmosis

Questions

1. Write down a definition for diffusion.
2. Name three chemicals that move in and out of cells by diffusion.
3. Explain how diffusion lets you detect a fish and chip shop when you are still around the corner.
4. Explain what is meant by a partially permeable membrane.
5. Write down a definition of osmosis.
6. Draw a diagram like the ones above showing an overall movement of water molecules from solution A to solution B.

H Another way of getting molecules into cells

Remember that cell membranes are partially permeable. Some molecules that the cells need cannot diffuse through membranes into cells. Others diffuse through very slowly.

Sometimes a cell needs to take in molecules that are in higher concentration inside the cell than outside. The molecules cannot move by diffusion.

So cells have another way of moving molecules in or out. It is called **active transport**. Cells use the energy from respiration to transport molecules across the membrane. Glucose is one chemical that is moved into cells by active transport.

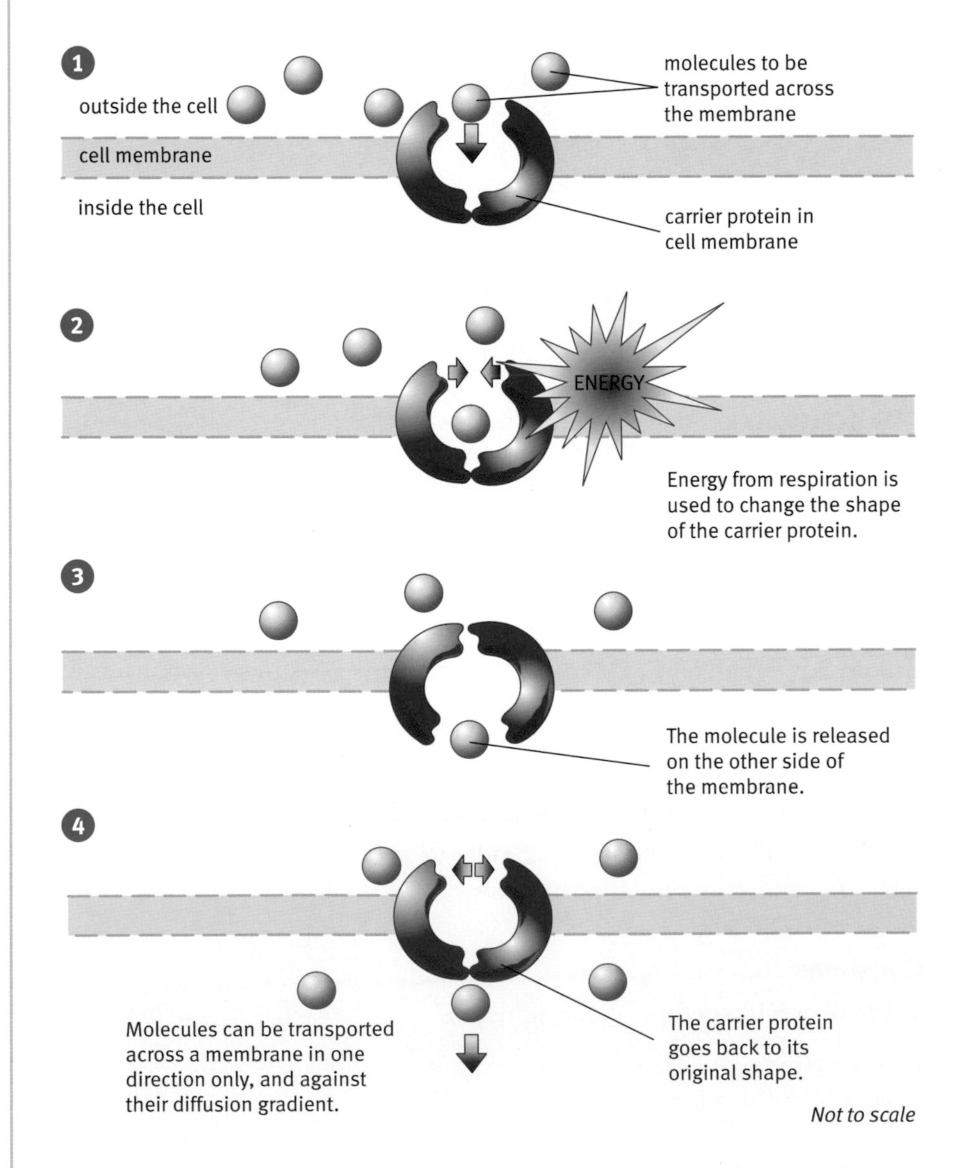

Movement of molecules across a cell membrane by active transport (schematic).

Questions

7 Write down two main differences between diffusion and active transport.

8 Name one chemical which is moved into cells by active transport.

9 Explain why cells sometimes need to use active transport.

Keeping the balance

In most animals that live in the sea, the concentration of dissolved chemicals in their cells and body fluids is about the same as that of seawater. So they do not have a problem with water balance. Water molecules enter and leave their bodies at the same rate.

Freshwater animals do have a problem. The concentration of dissolved chemicals in their bodies is much higher than in freshwater. Water constantly enters their bodies by osmosis. Freshwater animals, such as *Paramecium* (see the diagram on the right), have to use energy to pump water out of their body.

Animals that live on land have the opposite problem. They lose water to the environment all the time, for example in sweating, panting, and urination. So land animals must replace water all the time. If they do not, their body fluids become too concentrated and their cells do not work properly.

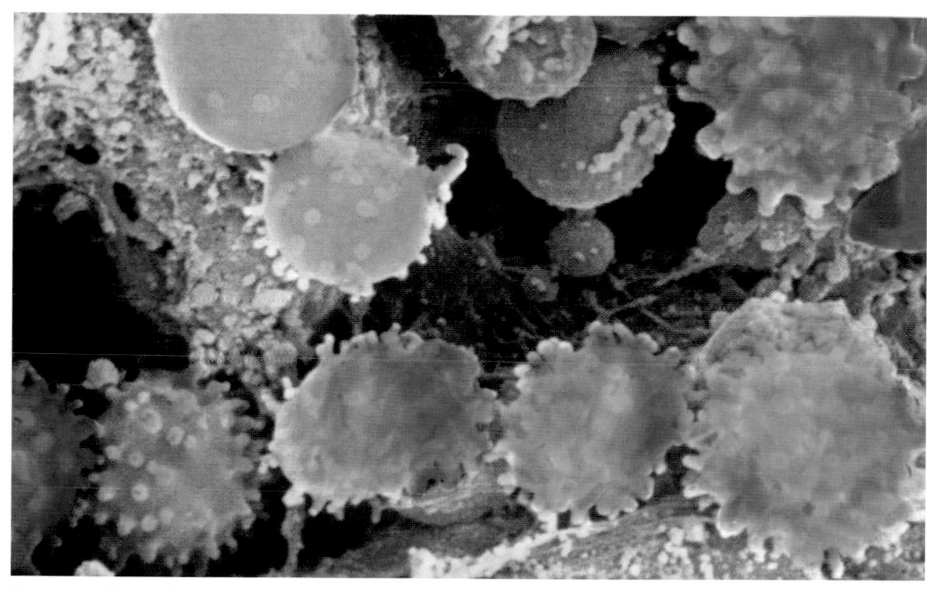

These red blood cells appear wrinkled because they have lost water by osmosis.

About two-thirds of your body is water – in cells, tissue fluid, and blood. Tissue fluid and blood must have a steady water level, or your cells may gain or lose water by osmosis. Your kidneys keep water levels in the body balanced.

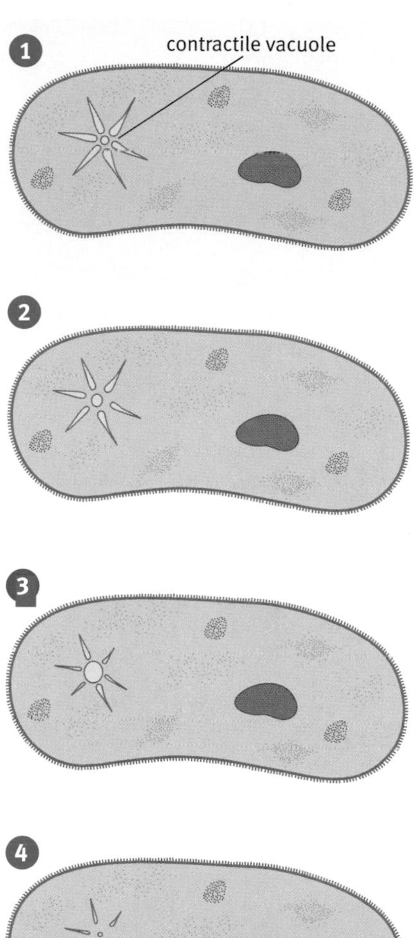

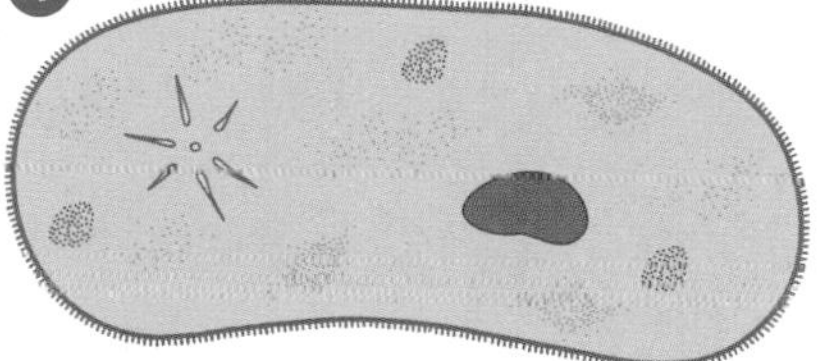

Paramecium is a single-celled organism that lives in fresh water. It has a contractile vacuole that fills and bursts over and over again to get rid of water. If this did not happen, the whole cell would swell and burst.

Questions

10 The red blood cells in the photo on this page are an unusual shape. They have lost water by osmosis. This can happen when the environment of the cell changes. To produce this effect, were there more dissolved molecules in the blood cells or in the surrounding plasma?

11 What would happen if red blood cells were placed in freshwater?

Key words

active transport

Find out about:
- how your body gains and loses water
- how your kidneys get rid of wastes
- how kidneys balance your water level

F Water homeostasis

Experiments carried out with students showed that being able to drink water in the classroom
- increased their concentration time
- improved test results

Keeping a steady water level is done by balancing your body's water inputs and water outputs. The diagram below shows how you gain and lose water.

Water intake and water loss must balance for your body to work well.

Your **kidneys** control the water balance in your body. They do this by changing the amount of urine that you make. On a hot day, or when you have been running, you lose a lot of water in sweat. So your kidneys make a smaller volume of urine. Your urine will be more concentrated that day.

What else do your kidneys do?

Your kidneys have two jobs: water homeostasis and **excretion**. Excretion is getting rid of toxic waste products from chemical reactions in your cells. These two jobs are linked because you use water to flush out waste products such as **urea**.

Liver cells make urea when they break down amino acids your body cannot use. Urea diffuses into your blood and is carried around your body. Except in very low concentrations, urea is poisonous. So as blood passes through your kidneys, urea is filtered out.

How do your kidneys work?

Kidneys work like sieves. Small molecules are filtered out of the blood as it passes through your kidneys. These small molecules are water, sugar (glucose), and urea, and ions of salt. Blood cells and large molecules, such as proteins, are too big, so they stay in the blood.

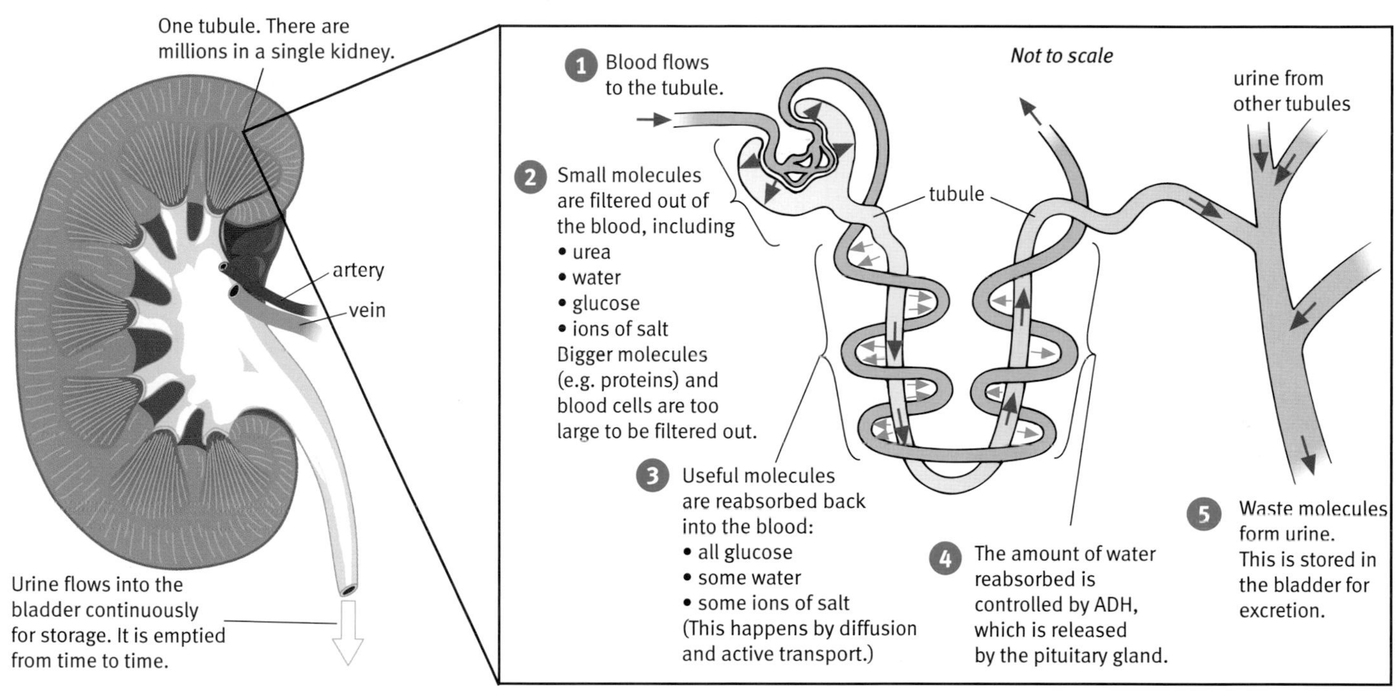

a Cross-section through a kidney

b How a kidney tubule filters out waste molecules. (*Note:* You do not have to remember the structure of the tubule.)

Getting the balance

Some of the small molecules are useful to the body. You do not want to lose them. So the kidneys **reabsorb** what the body needs. These useful chemicals go back into the blood:

- all of the sugar (glucose) for respiration
- as much salt as the body needs
- as much water as the body needs

The rest of the filtered chemicals go to your bladder. They make up urine.

Key words
kidneys
excretion
urea
reabsorb
urine

Questions

1 List three ways in which your body gains water and three ways in which it loses water.

2 Suggest *two* different things that could cause you to make:

a a small amount of concentrated urine

b lots of dilute urine

3 Which molecules do kidneys:

a filter out of your blood?

b reabsorb into your blood?

(4) Your body is two-thirds water. If your body mass were 60 kg, how many kilogrammes of that would be water?

(5) Explain why blood cells or protein in a person's urine is a sign of kidney damage.

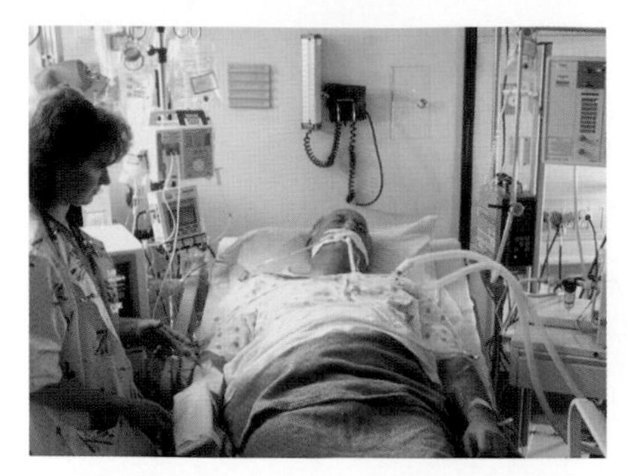

After serious illness or surgery, homeostatic systems in the body may not work reliably. Doctors and nurses monitor patients carefully to keep their bodies in the ideal state to encourage healing and recovery. They may use intravenous drips to prevent dehydration.

Dry air in aeroplanes can cause dehydration. Drinking alcohol and coffee can make the dehydration worse.

More about water balance

Remember that the concentration and volume of your urine varies. On cold days you probably make lots of pale-coloured urine. On hot days you make a smaller volume of darker, more concentrated urine.

The salt concentration of your blood determines how much water your kidneys reabsorb, and how much you excrete in urine. The salt concentration of your blood can become higher than normal because of:

- excess sweating
- not drinking enough water
- eating salty food

Drugs and urine

Some drugs affect the amount of urine a person makes. **Caffeine** in tea and coffee causes a greater volume of dilute urine to be produced. **Alcohol** has an even greater effect and can make people very dehydrated.

The drug **Ecstasy** has the opposite effect. It reduces the volume of urine a person makes. Overheating may also lead to the person drinking too much water. The amount of water in the body can become dangerously high.

H

Controlling water balance

The control system for water balance is a negative feedback system:

- Receptors in the hypothalamus in your brain detect any changes in salt concentration in the blood.
- The hypothalamus is also the processing centre. When the salt concentration is too high, it triggers the release of a hormone called **ADH** from the **pituitary gland.** This gland is also in the brain, just below the hypothalamus. When the salt concentration is low, no ADH is released.
- The ADH travels in the blood to the kidney tubules. These are the effectors. ADH affects the amount of water that can be reabsorbed back into the blood. The more ADH, the more water is reabsorbed.

Key words

caffeine
alcohol
Ecstasy
ADH
pituitary gland

Questions

6 Describe the affect that these drugs have on urine production:

a caffeine

b alcohol

c Ecstasy

H

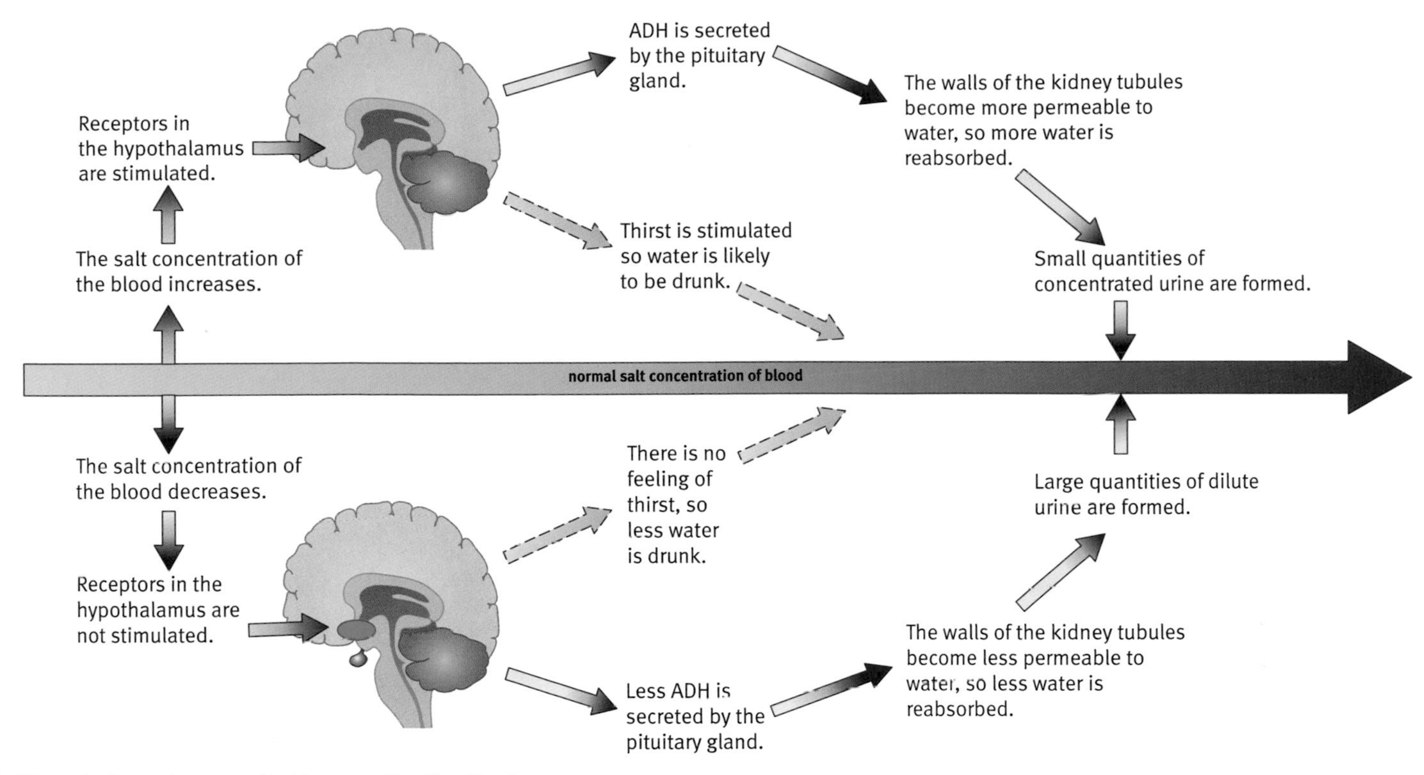

Water balance is controlled by negative feedback.

Drugs affect ADH control

Caffeine, alcohol, and Ecstasy change the volume of urine a person makes because they affect ADH production. For example, alcohol suppresses ADH production. Less water is reabsorbed in the kidneys, so a larger volume of urine is made.

Questions

7 ADH is a hormone.

a Where is ADH made?

b What affect does ADH have on the body?

8 Copy and complete the table below to show what happens when the salt concentration of the blood changes.

	Salt concentration of blood falls	**Salt concentration of blood rises**
Pituitary gland secretes	less ADH	
Kidney tubules reabsorb		
Urine volume		
Urine concentration		increases

9 Ecstasy triggers release of ADH. Explain the effect this will have on water balance.

⑩ In a few people the cells that produce ADH are destroyed. Suggest what happens to them and what treatment they need.

Find out about:
- how extreme conditions can upset homeostasis

G When it all goes wrong

Sometimes a person's control systems cannot cope with the conditions. For example, strenuous exercise, very hot or cold climates, or sports like scuba-diving and mountain climbing can all affect homeostasis. In extreme conditions, the person's life is at risk.

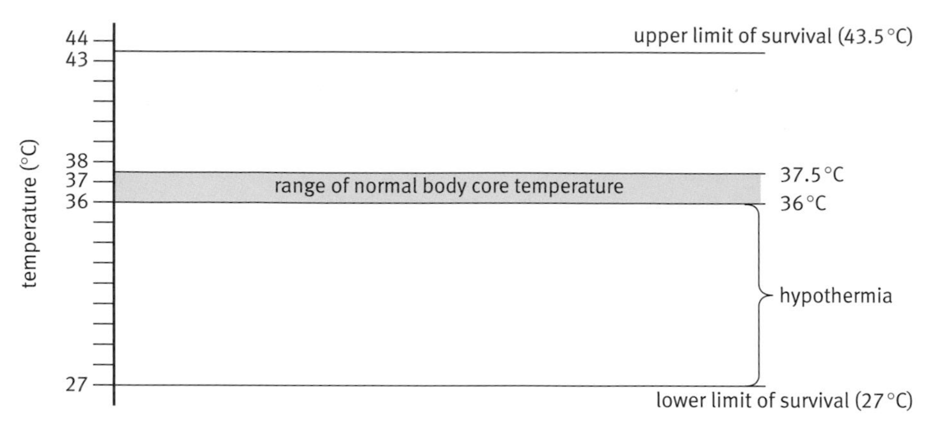

Survival limits for body temperature

Ecstasy can cause dehydration because it increases sweat production. Dehydration can lead to heat stroke. (*Note:* Because it affects water control by your kidneys, Ecstasy can also cause over-hydration.)

Heat stroke

Heat stroke happens when your body cannot lose heat fast enough. For example, in humid conditions, you sweat, but the sweat cannot evaporate. It may be that you have become dehydrated and produce too little sweat. Fever, prolonged exercise, and over-exposure to the sun and drugs such as Ecstasy can all lead to heat stroke.

Core body temperatures of 42 °C and over affect the temperature processing centre in your brain (the hypothalamus). Your temperature control system fails, and you stop sweating. Your body temperature rises out of control.

Symptom of heat stroke	Cause
hot, dry skin	sweating stops
rapid pulse rate	dehydration, stress, increased metabolic rate
dizziness and confusion	nerve cell damage in the brain

Treatment for heat stroke

Rapid cooling of the patient is essential, for example by:

- sponging them with water
- wrapping them in wet towels
- use of a fan
- putting ice in their armpits and groin

Doctors may use cooled intravenous drips.

Hypothermia

Hypothermia happens when your core body temperature falls below 35 °C. It can happen to anyone exposed to low temperatures for long enough. A person's body heat cannot be replaced as fast as it is being lost. Young babies and elderly people are at greatest risk, because their temperature control systems work least well. Babies have a large surface area compared with their volume, so they lose heat particularly quickly.

Hypothermia causes about 30 000 deaths a year in the UK.

Core body temperature	What happens
below 35 °C	shivering, confusion, drowsiness, slurred speech, loss of coordination
30 °C	coma
28 °C	breathing stops

Climbers are at risk of hypothermia in freezing external temperatures. They watch out for signs of hypothermia in each other.

Treatment for hypothermia

Increasing the patient's core temperature is essential. However, it is important not to heat their skin and limbs as this increases blood flow to them. Heat loss then increases, and core temperature falls even further. This sudden cooling increases the risk of heart failure.

Do

- insulate them, particularly their head, neck, armpits, and groin
- handle them gently to keep blood flow to the limbs low
- warm them gently with warm towels
- give them warm drinks (not alcohol)

Do not

- give them food, as digesting food lowers metabolic rate
- use hot water bottles

Key words

heat stroke
hypothermia

Questions

1 For heat stroke ***and*** hypothermia list:

a causes

b symptoms

c treatments

2 Explain why a person with heat stroke stops sweating, even though they need to lose body heat.

3 Why are babies most at risk of hypothermia?

4 Why do people not recognize signs of hypothermia in themselves?

Summary

In this module you have found out how your body temperature and water levels are kept at a steady state. Homeostasis is vital for cells to be able to work properly.

Keeping a steady state

- Automatic control systems in the body keep factors steady, including body temperature and water level.
- Artificial control systems are similar to body control systems; they both have receptors, processing centres, and effectors.
- Strenuous exercise and extreme climates can affect homeostasis (body temperature, water balance, blood oxygen level H, salt level H).
- Negative feedback is used in many control systems, so that any change from a normal level causes a response that returns the system back to normal. H
- Some effectors work antagonistically, which gives a more sensitive response. H

Moving in and out of cells

- Chemicals move in and out of cells by diffusion, e.g. O_2, CO_2, dissolved food.
- Diffusion is the passive movement of molecules from where they are in high concentration to where they are in low concentration.
- Osmosis is the overall movement of water molecules from a dilute to more concentrated solution through a partially permeable membrane.
- Keeping the body's water level balanced is important so that cell contents are at the correct concentration for the cell to work properly.
- Some chemicals, e.g. glucose, are moved in or out of cells by active transport, which requires energy. H

Enzymes

- Enzymes are proteins that speed up chemical reactions in cells.
- A small part of an enzyme (the active site) is shaped so that only molecules with the correct shape can fit with the enzyme; this is the 'lock-and-key' model.
- The 'lock-and-key' model explains why each type of enzyme only speeds up one particular chemical reaction.
- Enzymes work best at a certain temperature, which is why the body's temperature must be kept at a steady level.
- At low temperatures, increasing the temperature will increase the rate of a reaction because molecules collide more often and with more energy.
- At high temperatures an enzyme stops working (it is denatured) because its active site shape is changed.

Body temperature

- Energy gain and loss must be balanced to keep a steady body temperature.
- Receptors in the skin and brain detect changes in temperature of the air and the blood.
- The brain processes information about temperature control in the hypothalamus, and effectors (sweat glands and muscle) produces responses to keep body temperature steady.
- Vasodilation and vasoconstriction control the amount of blood flowing near to the skin surface. H
- You should be able to describe the cause, symptoms, and first aid treatment for heatstroke and hypothermia.

Body water balance

- Water is gained from food, drinks, and respiration, and is lost through sweating, breathing, faeces, and urine.
- The amount and concentration of urine made by the body depend on several factors.
- Kidneys remove waste urea from the body and balance the level of other chemicals in the blood.
- Concentration of urine is controlled by the hormone ADH, released by the pituitary gland in the brain. ADH production is controlled by negative feedback.
- Alcohol and Ecstasy both effect the amount and concentration of urine made by the body (by affecting ADH production H).

Questions

1 a Write definitions for these key words:

i homeostasis **ii** stimulus **iii** response **iv** receptor **v** effector

b Explain why it is important that these factors are kept at steady levels in the body:

i temperature **ii** water

c Draw a flowchart to explain how the body reacts when body temperature increases above normal.

2 A person sprays air freshener at the other side of a room. After a few minutes you can smell the air freshener molecules.

a Name the process by which the air freshener molecules have moved towards you.

b Write down a definition to describe this process.

c Say whether this process is passive (no energy needed) or active (requires energy).

d Name two chemicals that move in and out of body cells by this process.

3 Water moves in and out of cells.

a Name the process by which this happens.

b Write bullet point notes to explain why a single-celled organism in fresh water will gain water. Start with:

- The solution around the cell is more dilute than the cell contents.

4 a Draw a graph showing how temperature effects the rate of an enzyme reaction.

b Label your graph with these notes:

A: At low temperatures, increasing the temperature increases the frequency and energy of collisions. The rate of reaction increases.

B: The enzyme has an optimum temperature. It works at its fastest rate.

C: At higher temperature, the active site changes shape. The enzyme is denatured and stops working.

c The enzyme catalase speeds up the breakdown of hydrogen peroxide to water and oxygen. Explain why catalase cannot speed up any other reactions. Use the 'lock-and-key' model in your answer.

5 Some chemicals are filtered out of the blood when it goes through a kidney. The table below shows the concentration of some chemicals at different places.

A: blood entering the kidney

B: liquid just after it is filtered from the blood

C: filtered liquid just before it leaves the kidney

a Name one chemical that is not filtered out by the kidney.

Where	Concentration (g in 100 cm^3)		
	Protein	**Glucose**	**Urea**
A	7.5	0.1	0.03
B	0.0	0.1	0.03
C	0.0	0.0	0.15

b Explain why this is not filtered out.

c What happens to the glucose between **B** and **C**?

d Why does the concentration of urea increase between **B** and **C**?

6 Alcohol causes more dilute urine to be produced.

a Explain why this can lead to dehydration.

b Describe how alcohol interferes with ADH control of water balance. H

Why study growth and development?

How does a human embryo develop?
What makes cells with the same genes develop differently? Exploring questions like these is part of the fast-moving world of modern biology.

The science

Sex cells carry the genetic information to make a new individual. After fertilization, the egg cell divides many times to form an embryo. Cells become specialized because of the different proteins they make. DNA is the chemical that genes are made of. It has a unique structure which determines the proteins a cell makes. There are differences in how plants and animals grow, and we use these differences to grow cloned plants.

Biology in action

Genetic technologies are at the cutting edge of modern science. Our expanding knowledge of how human beings grow and develop has the potential to offer great benefits for this and future generations. For example, research into how cell growth is controlled could be crucial in combating cancer.

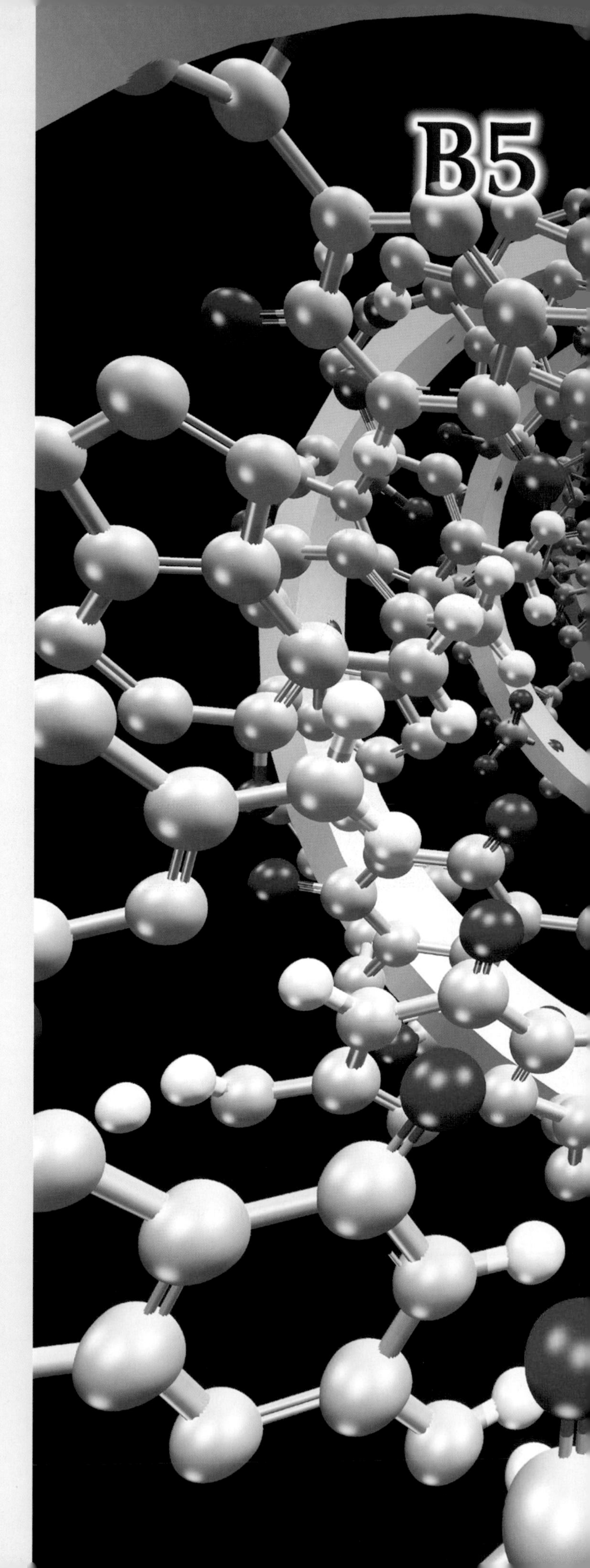

Growth and development

Find out about:

- the structure of DNA, and how it controls the proteins a cell makes
- how cells divide to make sex cells and to make new body cells
- how cells become specialized
- the differences between plant and animal growth

Find out about:
- different cells, tissues, and organs
- growing up

A Growing and changing

You began life as a single cell. By the time you were born you were made of millions of cells. You probably weighed about 3 or 4 kilogrammes. Now you probably weigh over 50 kilogrammes. Not only have you grown, but you have changed in many ways. In other words, you have developed.

Since you were born, your **development** has been gradual. In some plants and animals, development involves big changes. For example, the young and the adults in these photographs look very different:

Key words
development

Questions

1 Match the young and the adults in pictures A to F.

(2) Which animal in the pictures has a life cycle most like a fly?

Life cycles

Until 1668 people thought that maggots just appeared in decaying meat. The scientist Francesco Redi observed flies on the meat with maggots. He did experiments to find out if they were linked. He put pieces of meat in pots and covered some of them with gauze. He left the others uncovered. Maggots appeared only on the meat in the uncovered pots. Where flies could not get to the meat, there were no maggots.

Redi eventually worked out the whole life cycle of the fly:

- Flies lay eggs.
- Eggs hatch into maggots.
- Maggots change into pupae. Inside the pupa, the tissues reorganize into a fly.

Building blocks

Like you, the plants and animals in the pictures opposite are multicellular. Your body has more than 300 different kinds of cell. Each cell is **specialized** to do a particular job.

Tissues and organs

All newly formed human cells look much the same. Then they develop into groups of specialized cells called **tissues**.

Plant cells are different from animal cells, but they are specialized too. Plant cells have walls, and some have spaces called vacuoles.

As an animal embryo or plant grows, groups of tissues arrange themselves into **organs**, for example the heart and brain in humans, and roots, leaves, and flowers in plants.

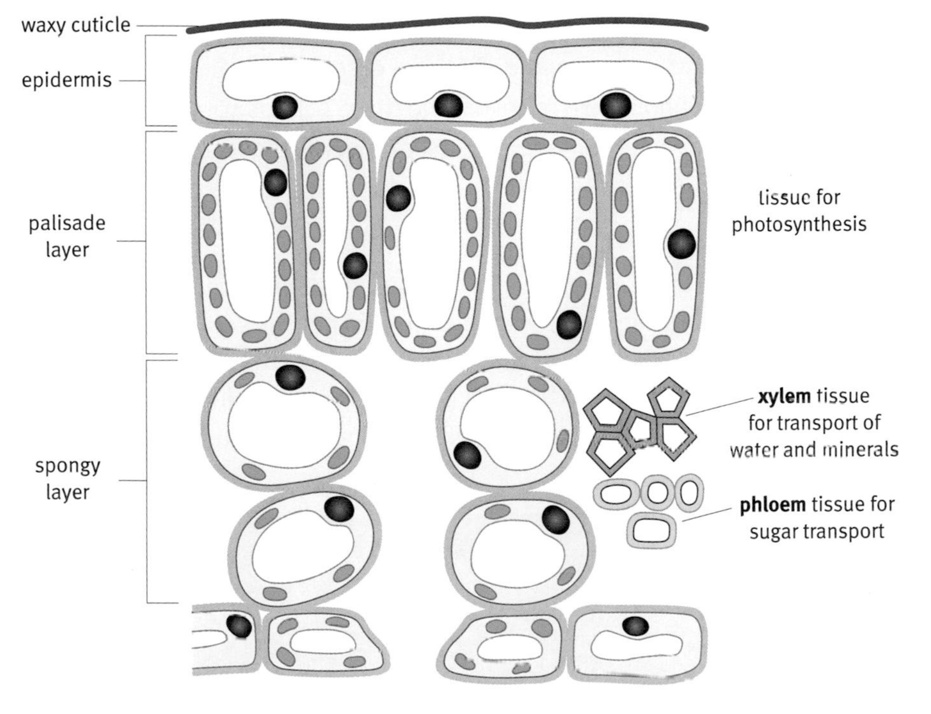

Different tissues in a leaf – a plant organ.

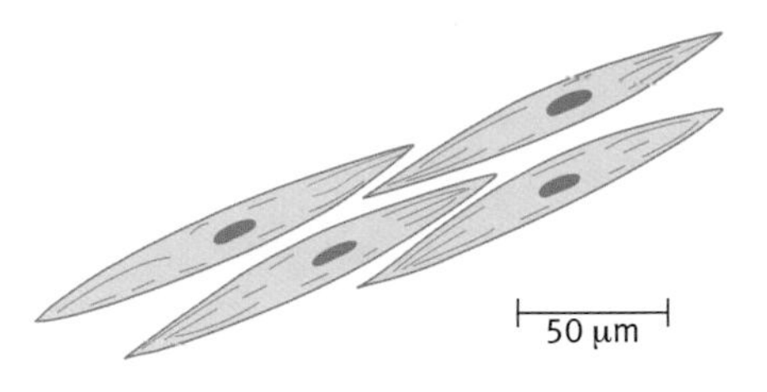

Muscle cells contract and relax to cause movement.

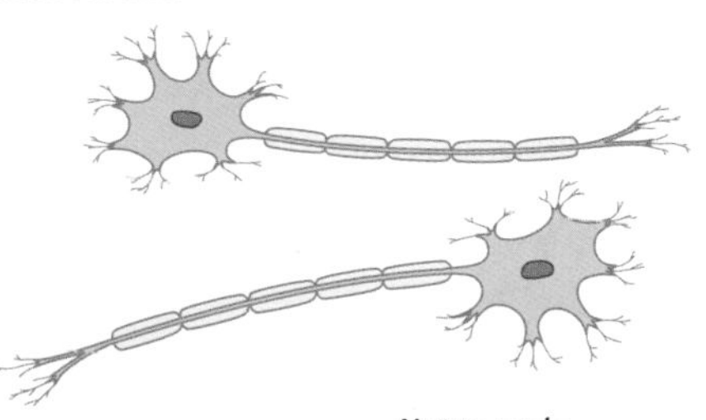

Not to scale

Nerve cells carry nerve impulses.

Questions

3 What does *specialized* mean?

4 What is the job of these tissues:
  a muscle?
  b xylem?
  c phloem?

5 Name three plant organs.

6 Explain the difference between a tissue and an organ.

Key words

development
specialized
tissues
organs
xylem
phloem

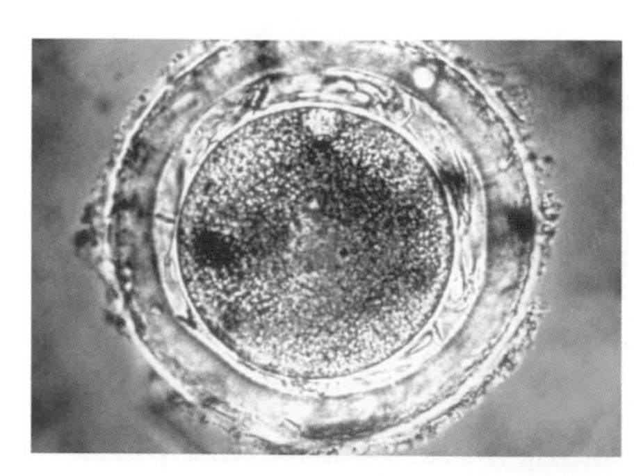

A fertilized human egg cell

From single cell to adult

All the cells in your body come from just one original cell – a fertilized egg cell or **zygote**.

So the zygote must contain instructions for making all the different types of cells in your body, for example muscle cells, bone cells, and blood cells. It also has the information to make sure that each type of cell develops in the right place and at the right time. This information is in your DNA – the chemical that your genes are made of.

In humans:

sperm + egg cell —fertilization→ zygote (fertilized egg cell)

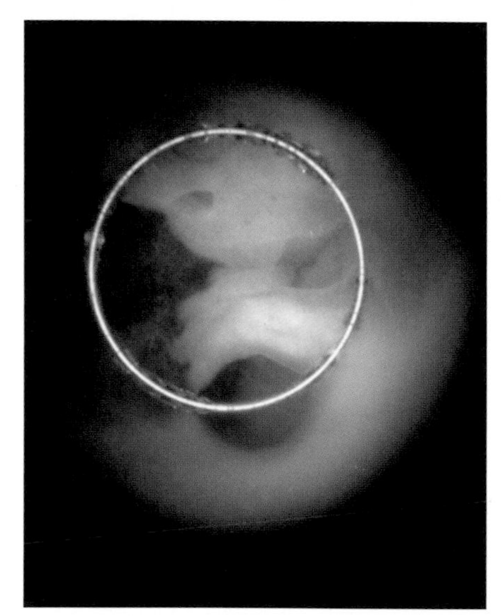

A human fetus at eight weeks. This photograph was taken from inside the mother's uterus. The fetus is about 2.5 cm long.

The growing baby

During the first week of growth, the zygote develops into a ball of about 100 cells. The nucleus of each cell contains an exact copy of the original DNA. As the embryo grows, some of the new cells become specialized and form tissues. After about two months, the main organs have formed and the developing baby is called a **fetus**. A six-day embryo is made of about 50 cells. Adults contain about 10^{14} cells, each with the DNA faithfully copied.

When the embryo is a ball of cells, it occasionally splits into two. A separate embryo develops from each section. When this happens, identical twins are produced. They are clones of each other. This shows that there are cells in the early embryo that can develop into complete individuals. These are **embryonic stem cells.** You can read more about these cells in Section H.

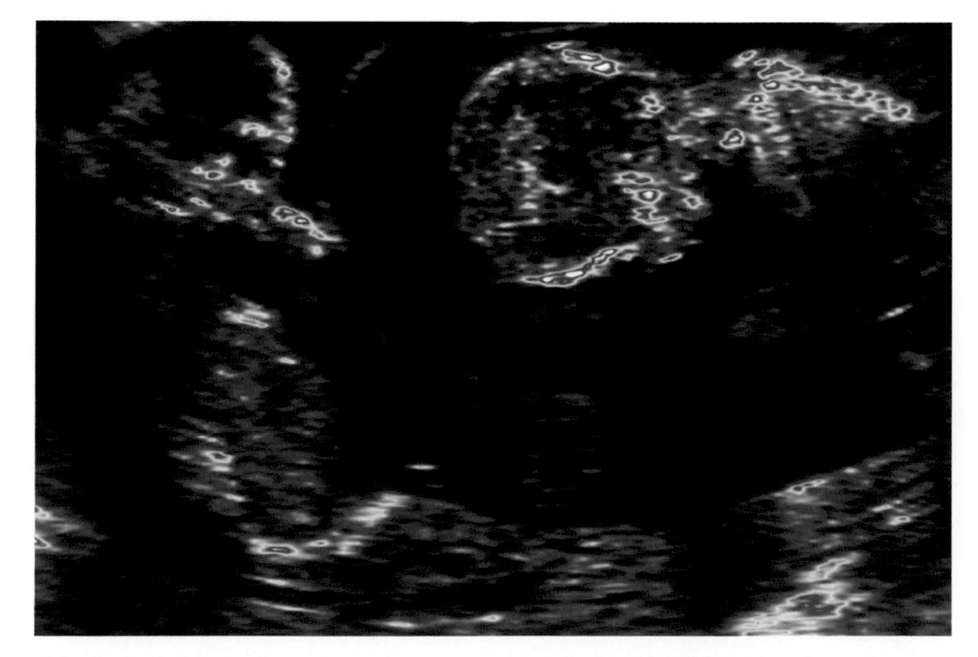

This ultrasound scan shows that twins are expected. Both of the babies' heads can be seen.

Key words
- zygote
- fetus
- embryonic stem cells
- meristem cells

Questions

7 What is a zygote?

8 When does a human embryo become a fetus?

Growth patterns

For living things to grow bigger, some of their cells must divide to make new cells. You will probably stop growing taller by the time you are about 18–20 years old.

Flowering plants continue to grow throughout their lives.

- Their stems grow taller.
- Their roots grow longer.
- To hold themselves upright, most increase in girth or have some other means of support.

Plants increase in length by making new cells at the tips of both shoots and roots. They also have rings of dividing cells in their stems and roots to increase their girth. These dividing cells are called **meristem cells**.

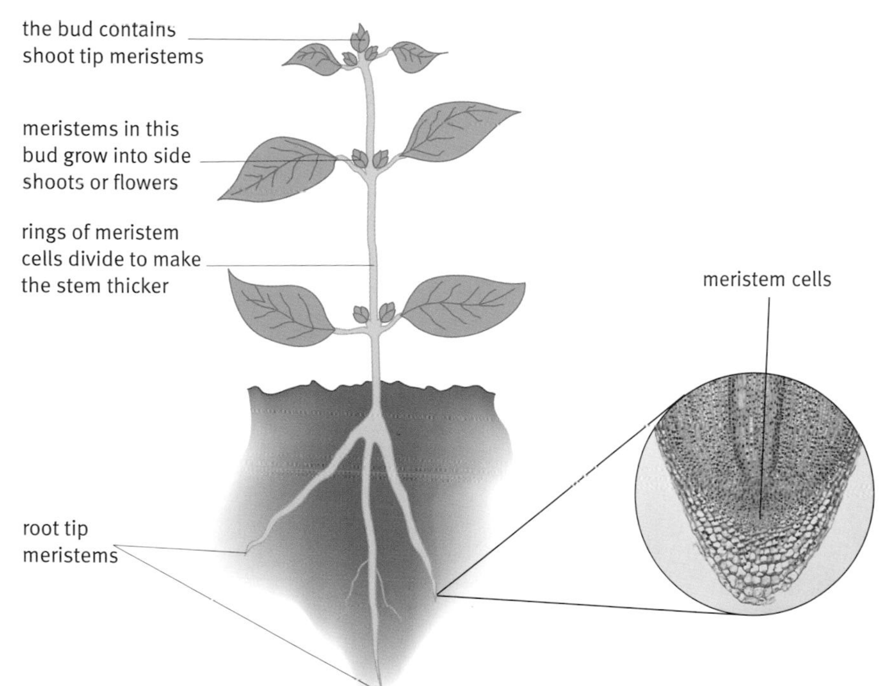

Meristem cells divide to make stems and roots longer and make the stem thicker. On the right is a root tip meristem (photographed through a microscope).

This giant sequoia tree is over 2000 years old, 83 m tall, and 26 m in girth (circumference). It is known as 'General Sherman' and is officially the largest giant sequoia tree and the largest living thing on Earth.

Honeysuckle twines its stem around other plants for support.

Questions

9 Why is it important that all living things have cells that can divide?

10 Name the type of cell in a plant that can divide.

11 Explain how plants:
- **a** grow taller
- **b** grow longer roots
- **c** grow thicker in girth

Find out about:
- why plants are so good at repairing damage

B Growing plants

Cells in your body divide when you are growing. If you cut yourself, cells can also divide to repair your body. But your body can make only small repairs. Many plants and some animals can replace whole organs.

Why can plants grow back?

Plant meristem cells are **unspecialized**. Plants keep some meristem cells all through their lives. These are spare back-up cells that can divide to make any kind of cell the body needs. So plants can regrow whole organs, such as leaves, if they are damaged.

How do newts grow?

Animals also have spare back-up cells called **stem cells**. These cells divide, grow, and develop into any kind of cell the body needs.

Animal growth

Newts' stem cells stay unspecialized throughout their life. So newts can grow new legs if they need to – or even an eye.

The stem cells in adult humans are not as useful, because they are already specialized. For example, the stem cells in your skin can only develop into skin cells.

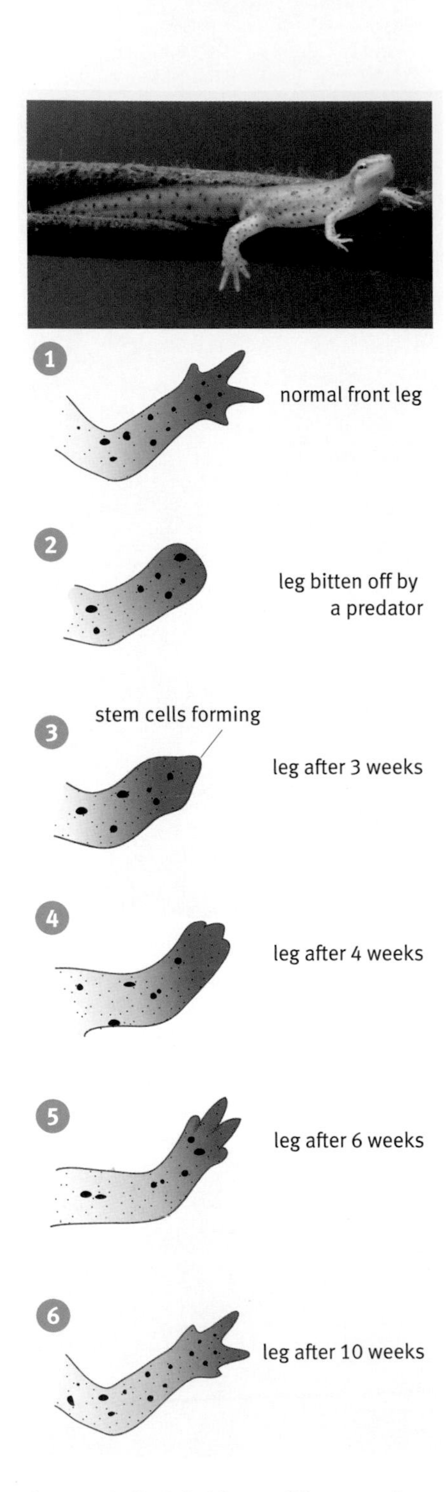

If a newt's limb is bitten off by a predator, it can grow a replacement. Most animals can only make small repairs to their body.

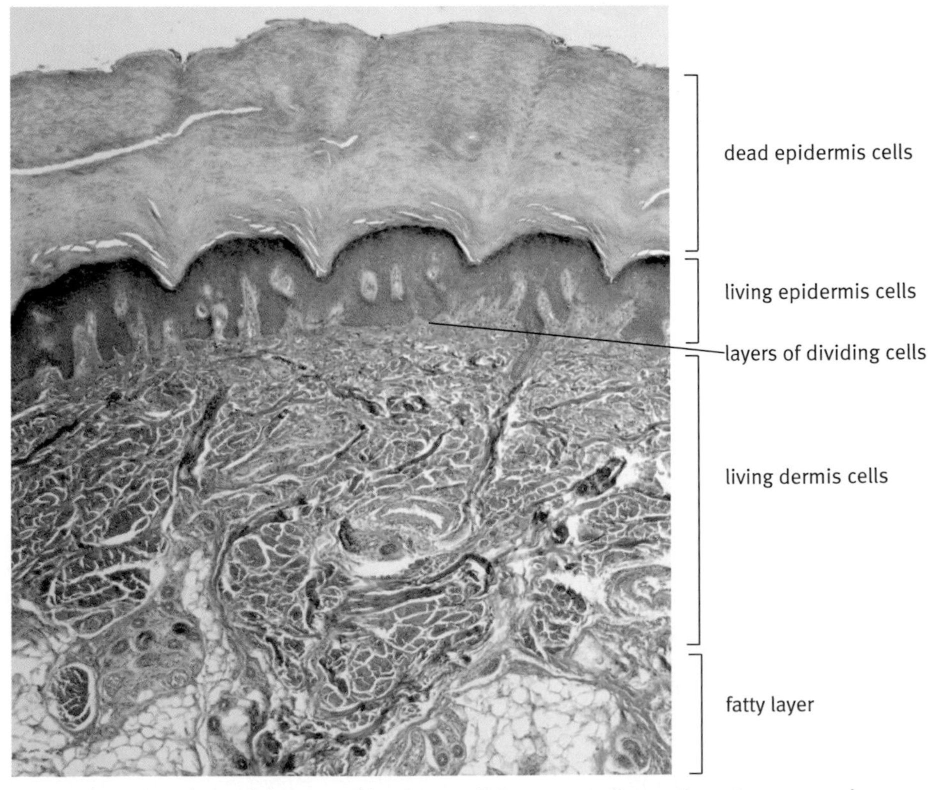

A cross-section through human skin. Some of the stem cells continue to grow and divide. Others replace skin cells at a wound or those that wear off at the surface. (×36)

You replace millions of skin cells every day. Most of the dust in your home is worn-off skin cells. Other tissues in your body need a constant supply of new cells. For example, bone marrow contains stem cells to make new blood cells.

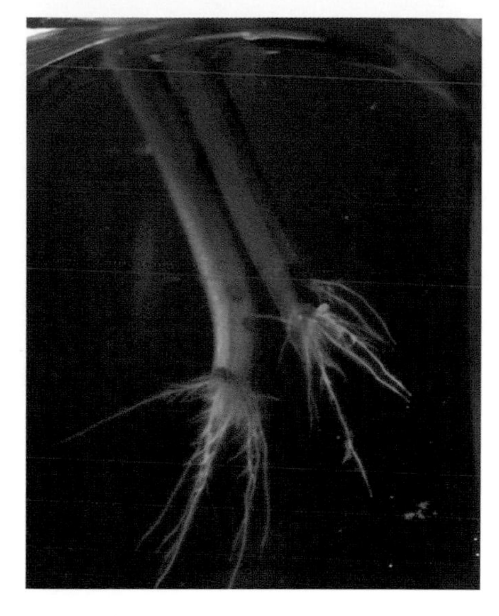

This willow cutting needed only water to grow.

Using meristems to make more plants

Gardeners use meristems when they grow new plants by taking **cuttings.** Cuttings are just shoots or leaves cut from a plant. In the right conditions they develop roots and grow into new plants.

Some cuttings grow new roots when you put them in water or compost. Others grow better when you dip the cut ends in **rooting powder** before you plant them. Rooting powder contains plant hormones called **auxins**. Auxins make the new cells produced by the shoot meristem develop into roots.

By taking cuttings, gardeners can produce lots of new plants quickly and cheaply. But this is not the only reason that they do it. All the cuttings taken from one plant have identical DNA. They are genetically identical: we say that they are clones. So taking cuttings is a good way of reproducing a plant with exactly the features that you want. When flowering plants produce seeds, they are reproducing sexually. So new plants grown from seeds vary. They are not identical.

Rooting powder

Questions

1 Name two parts of your body where you have stem cells.

2 Explain why a newt can regrow a leg but a human cannot.

③ For each of these types of cell, say whether they are fully unspecialized or not:
  a meristem cells
  b embryonic human stem cells
  c adult human stem cells

4 Give two reasons for growing plants from cuttings.

5 Explain how rooting powder helps a plant cutting to grow.

⑥ A gardener wants to grow dahlias with a variety of colours and sizes. Should she grow them from cuttings or seeds? Explain your answer.

Key words

unspecialized
stem cells
cuttings
rooting powder
auxins

Find out about:
- where genes are kept inside your cells

C A look inside the nucleus

All cells start their lives with a nucleus. A few specialized cells lose their nuclei when they finish growing. Human red blood cells are one example. Their job is to carry oxygen attached to haemoglobin molecules.

Red blood cells develop from stem cells in your bone marrow. As they develop, they make more and more haemoglobin. By the time they leave the bone marrow, they are full of haemoglobin and their nuclei have broken down.

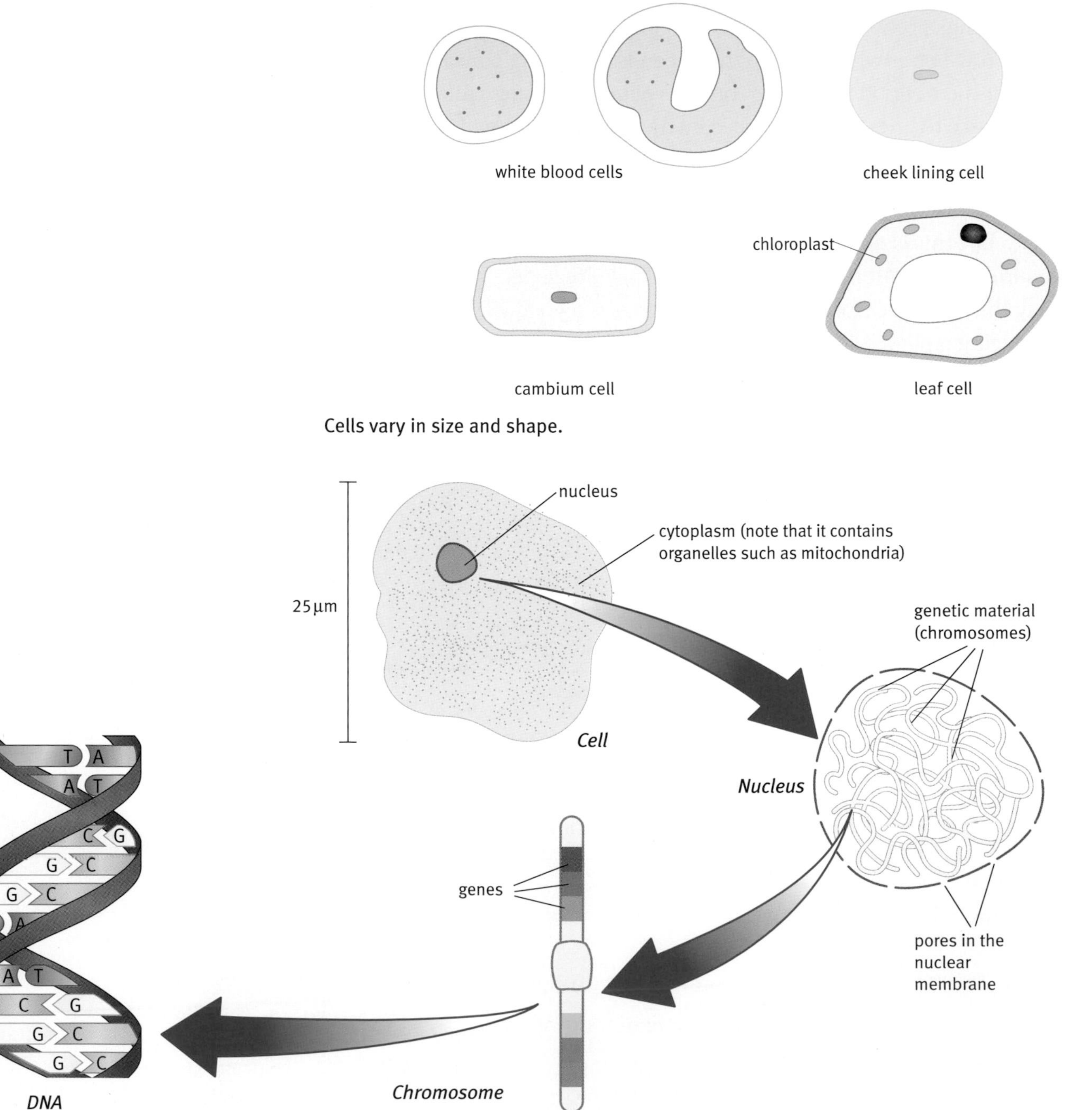

Cells vary in size and shape.

Each **chromosome** in the nucleus contains thousands of **genes**.

Chromosomes

Imagine a chromosome as a long molecule of DNA wound around a protein framework. You have about a metre of DNA in each of your nuclei. This is made up of about 30 000 genes. Each gene probably codes for a protein.

Different species have different numbers of chromosomes and different numbers of genes (see the table on the right).

You have 23 pairs of chromosomes in your nuclei. You got one set of 23 from your mother's egg cell nucleus and the other set from the nucleus of your father's sperm.

Organism	Estimated gene number	Chromosome number
human	~30 000	46
mouse	~30 000	40
fruit fly	13 600	8
Arabidopsis thaliana (plant)	25 500	5
roundworm	19 100	6
yeast	6 300	16
Escherichia coli (bacterium)	3 200	1

Key words

chromosomes
genes

These people are more alike than it appears. 99.9 % of their genes are the same.

What is special about DNA?

You will find out later in the module that the molecule of DNA has a particular structure that allows it to:

- make exact copies of itself
- provide instructions so that the cell can make the right proteins at the right time

Questions

1. Name two ways in which your red blood cells are different from the other cells in your body.
2. Suggest why red blood cells wear out and have to be replaced every 2–3 months.
3. How many different types of protein are there likely to be in yeast?
4. Which animal has half as many genes as yeast?
5. To work as your genetic material, what two properties does DNA have?

Find out about:
- how your cells divide for growth and repair your body

D Making new cells

It is not always easy to tell whether something is a living organism or not.

You could ask yourself:

- Can it grow and reproduce?
- Is it made of cells?

Life cannot exist without the growth, repair, and reproduction of cells.

Imagine a space probe bringing back objects like this from Mars. Scientists would need to find out whether they were alive or not. It might be a living organism able to colonize Earth!

Cell division

When new body cells are made, they contain the same number of chromosomes as each other and the parent cell. They also contain the same cell parts, called **organelles**. So, before a cell divides, it must grow and make copies of:

- other organelles – such as **ribosomes** and **mitochondria**
- its nucleus, including the chromosomes

Only then does the cell divide. This part of the process is called **mitosis**.

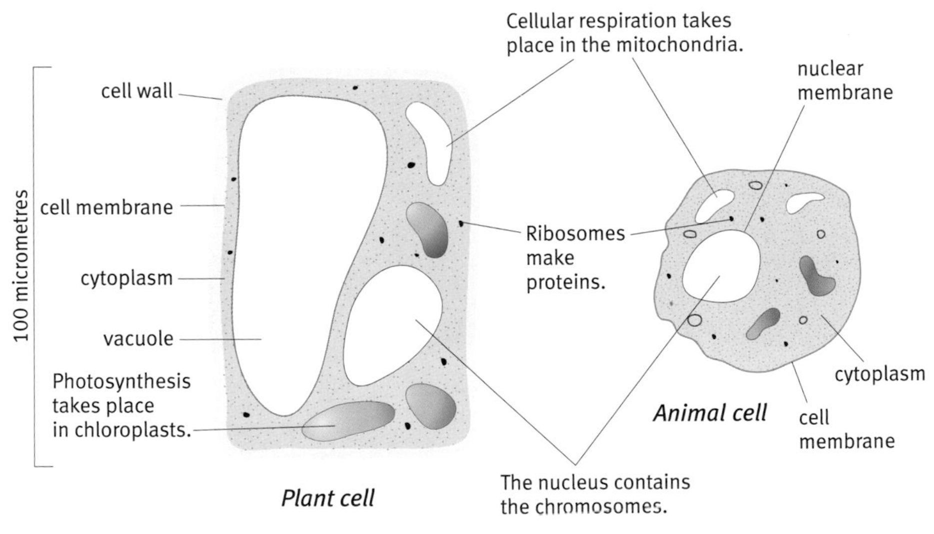

Cell organelles, such as mitochondria, are copied before a cell divides.

single-stranded chromosome in non-dividing cell

double chromosome in dividing cell

You can see chromosomes only in dividing cells.

Copying chromosomes

You can see chromosomes by using a light microscope, but only in dividing cells. The DNA is too spread out in other cells. After the chromosomes are copied, the DNA strands become shorter and fatter. Then you can see them. You can read more about how DNA is copied in Section F.

Mitosis

During mitosis, copies of chromosomes separate and the whole cell divides.

First, a complete set of chromosomes goes to each end of the dividing cell and forms two new nuclei. A complete set of organelles also goes to each end. Then the cytoplasm divides to form two identical cells. In plant cells, a new cell wall forms too.

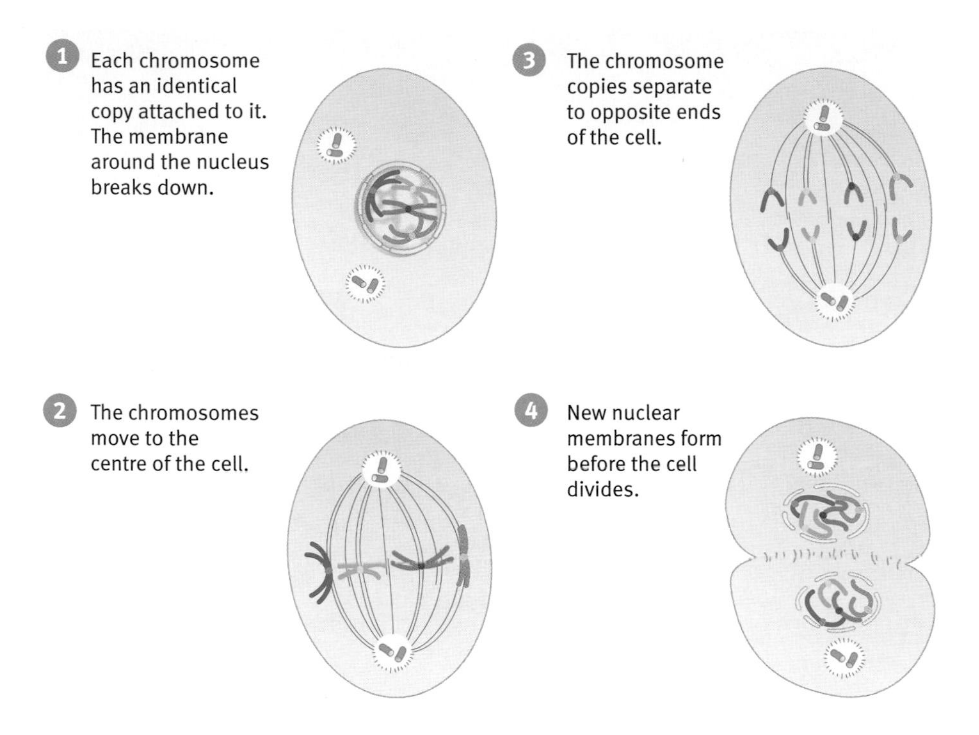

The pictures show what happens in mitosis in an animal cell. They show only 4 chromosomes.

Mitosis and asexual reproduction

Some plants and animals reproduce asexually. They use mitosis to produce cells for a new individual.

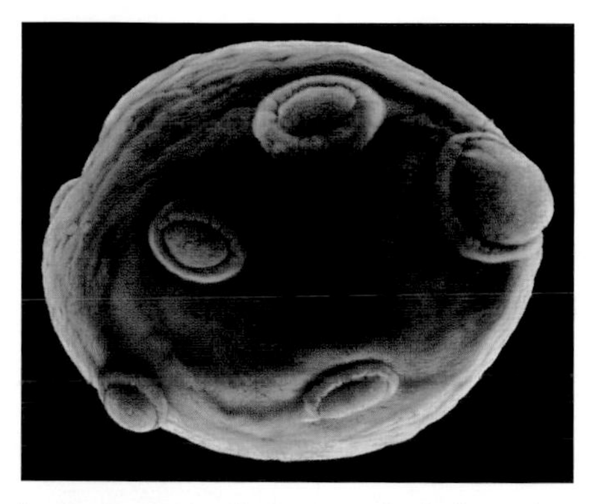

In yeast, new cells grow as buds from the parent.

Daffodil bulbs divide to form new ones.

This means that each of the individuals produced in asexual reproduction is genetically identical to the parent, so it is a clone of its parent.

Key words
organelles
ribosomes
mitochondria
mitosis

Questions

1 What might scientists do to find out whether the objects from Mars in the drawing opposite are living things or not?

2 Before a cell divides, it grows. What two steps happen during cell growth?

3 What two main steps happen during cell division by mitosis?

4 a How many cells are made by mitosis?

b What are these new cells like compared to their parent cell?

Find out about:
- cell division to make gametes

E Sexual reproduction

Most plants and animals reproduce sexually. Males and females make sex cells or **gametes**, which join up at fertilization. The fertilized egg, or zygote, develops into the new life.

It is not always obvious which plant or animal is male and which is female. Some plants and animals are both.

It is easy to tell which is male and which is female.

The holly on the right is female. You cannot be sure about the one on the left.

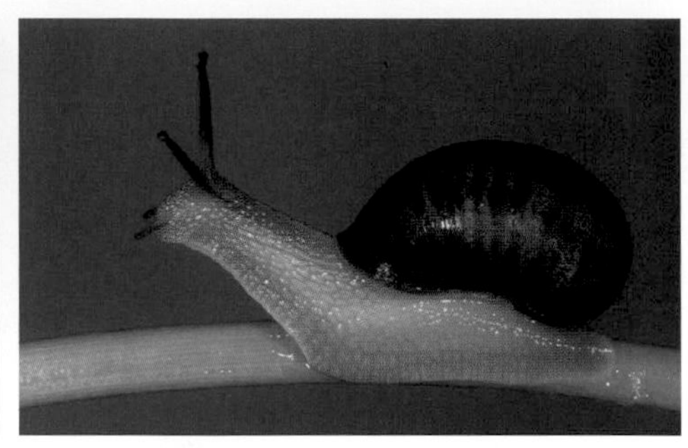

A snail has both male and female sex organs.

The only way to be sure about the sex of an organism is to look at its gametes. Males have small gametes that move. Females have large gametes that stay in one place.

Human males produce sperm in their testes. Females produce egg cells in their ovaries.

Sperm develop in tubules in the testes.

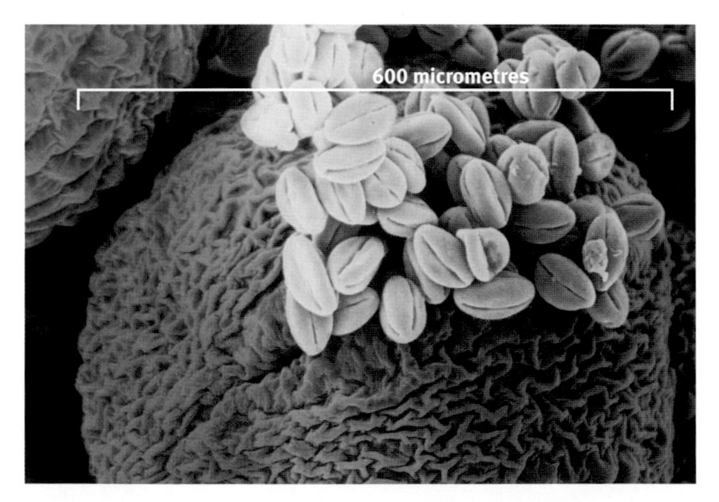

Pollen contains the male gametes of a flowering plant.

Male gametes are usually made in very large numbers. They move to the female gamete by swimming or being carried by the wind or an insect.

What is special about gametes?

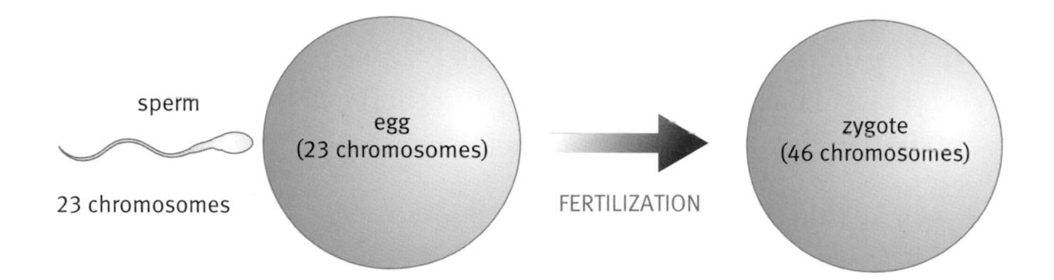

Meiosis halves the number of chromosomes in gametes. Fertilization restores the number in the zygote.

Human body cells have 23 pairs of chromosomes – 46 in total. Gametes have only 23 single chromosomes. This is important. When a sperm cell fertilizes an egg cell, their nuclei join up. So the fertilized egg cell (zygote) gets the correct number of chromosomes: 23 pairs – 46 in total. Gametes are made by a special kind of division called **meiosis**.

In humans, meiosis makes gametes that:

- have 23 single chromosomes (one from each pair)
- are all different – no two gametes have exactly the same genetic information

Offspring from sexual reproduction are different from each other and from their parents. We say that they show **genetic variation**.

Meiosis

Meiosis starts with normal body cells. It only happens in sex organs.

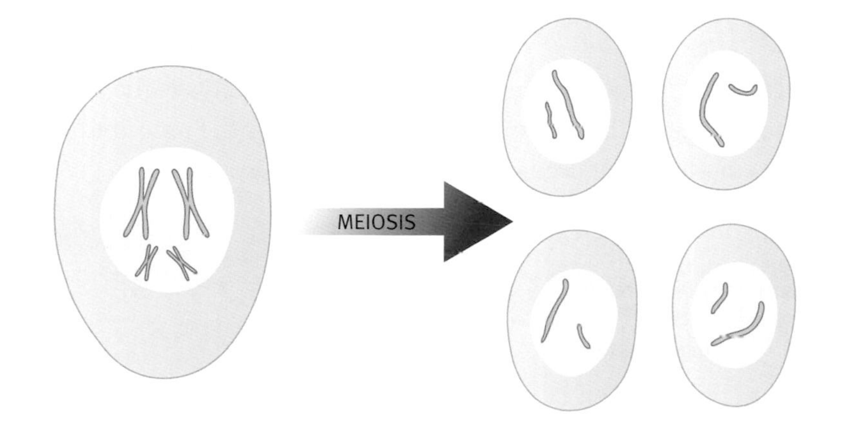

The chromosomes in this diagram have been copied. The parent cell divides twice, producing four cells. There are four cells after meiosis. They have half the number of chromosomes as the parent cell.

Key words
gametes
meiosis
genetic variation

Questions

1 Look again at the photos of holly at the top of the opposite page. You can be sure that the holly on the right is female. Why can you not be sure about the sex of the holly on the left?

(2) Why are male gametes made in such large numbers?

3 Why is it important that gametes have only one set of chromosomes?

Find out about:
- the structure of DNA
- how DNA is copied for cell division

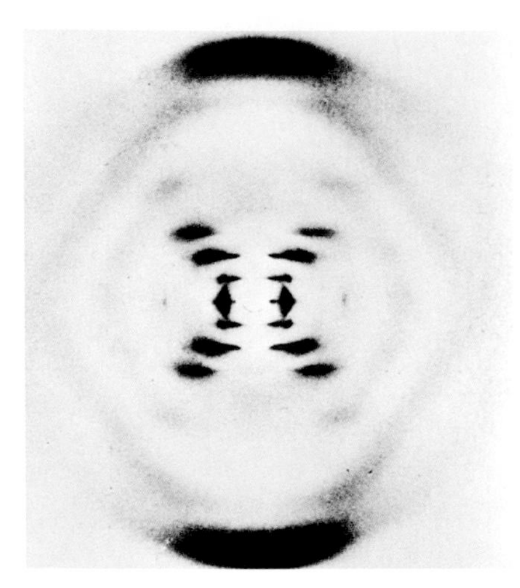

X-ray diffraction pictures of DNA, like this, show a repeating pattern.

Watson (left) and Crick

F The mystery of inheritance

In 1865 Gregor Mendel published his work on pea plants. You learnt about this in Module B3 *Life on Earth*. Mendel showed how features could be passed on from parents to their young. But there was still a mystery – how was the information passed from cell to cell? It took the work of many scientists to solve it.

1859	A chemical was extracted from nuclei and named 'nuclein'.
1944	'Nuclein' was recognized as genetic material.
Late 1940s	Erwin Chargaff discovered a pattern in the number of bases in DNA.
1951	Linus Pauling and Robert Corey showed that proteins have a helix structure.
1952–3	Rosalind Franklin and Maurice Wilkins produced X-ray diffraction pictures of DNA. They showed that the molecule had a regular, repeating structure.

Some of the discoveries that led up to the discovery of DNA structure.

Solving the mystery

In 1953, Francis Crick and James Watson published their now famous paper. 'A Structure for Deoxyribose Nucleic Acid', in the scientific journal *Nature* Their paper brought together all the work done on DNA. They used it work out the **double helix** structure of DNA.

Base pairing

There are four bases in DNA: adenine (A), thymine (T), guanine (G), and cytosine (C).

Erwin Chargaff had discovered that the amount of A is always the same as the amount of T, and the amount of G is the same as the amount of C. This is true no matter what organism the DNA comes from.

Crick and Watson concluded from this evidence that:

- A always pairs with T.
- G always pairs with C.

This is **base pairing**.

The double helix

Watson made cardboard models of the bases. He found that A+T was the same size as G+C. Suddenly, he realized what that meant. In a molecular model, he and Crick fitted the pairs of bases between two chains of the other chemicals in DNA, a sugar called deoxyribose and phosphates. It worked. The shape turned out to be a double helix – a bit like a twisted ladder. Better still, it matched the X-ray evidence.

How DNA passes on information

Base pairing means that it is possible to make exact copies of DNA:

- Weak bonds between the bases split, opening up the DNA from one end to form two strands.
- Immediately, new strands start to form from free bases in the cell.
- As A always pairs with T, and G always pairs with C, the two new chains are identical to the original.

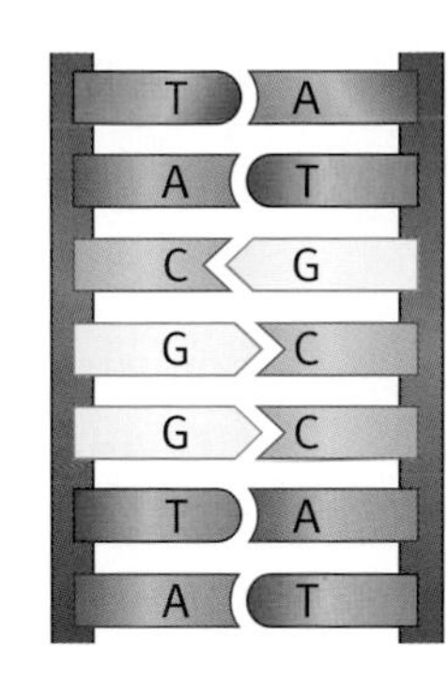

The rungs of the ladder are the pairs of bases held together by weak chemical bonds.

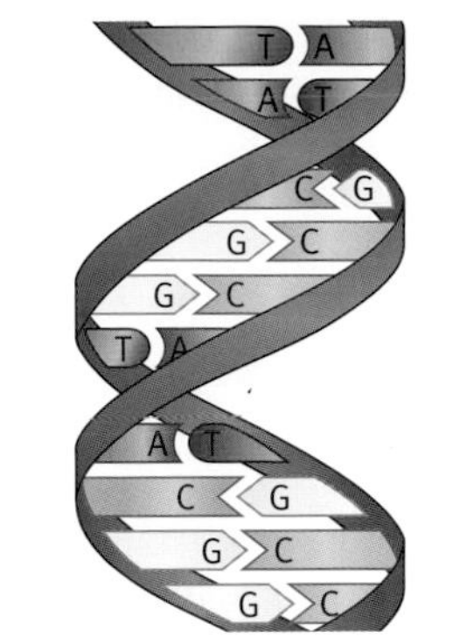

There are ten pairs of bases for each twist in the helix.

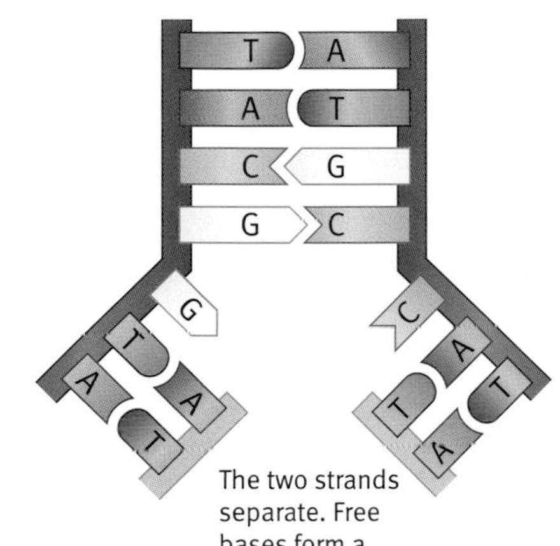

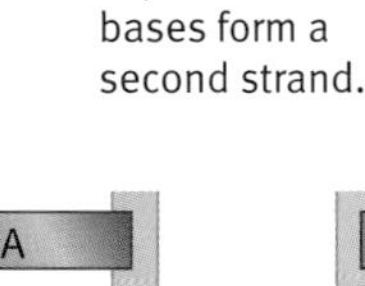

The two strands separate. Free bases form a second strand.

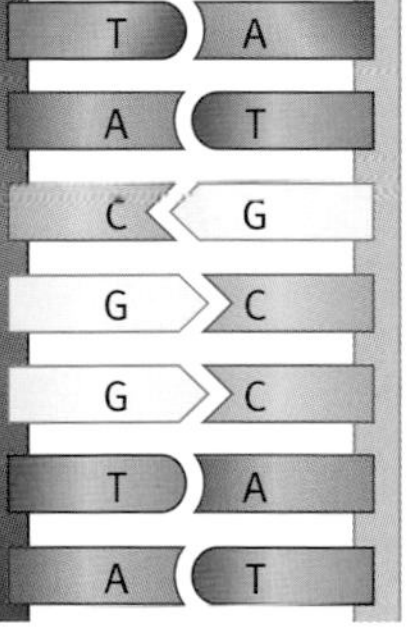

Each DNA molecule is made of half old DNA (black) and half new DNA (grey).

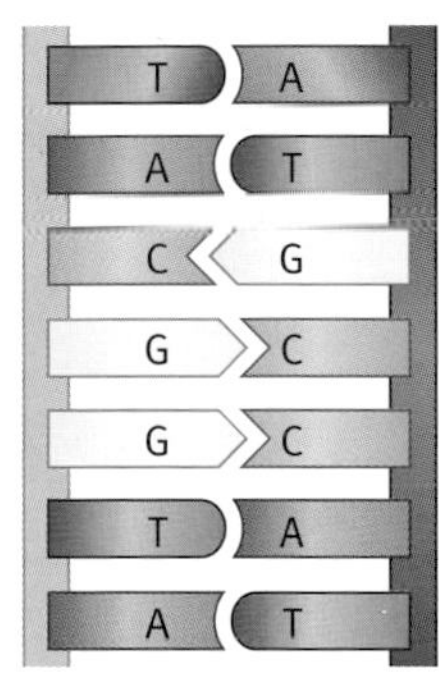

Each half of the split DNA molecule is complete, making two identical DNA molecules.

The structure of DNA (schematic)

Questions

1 The shape of a DNA molecule is sometimes described as a twisted ladder.

a What makes the sides of the ladder?

b What part of the ladder are the bases?

2 The bases in a DNA molecule always pair up the same way. Which base pairs with:

a A?

b C?

c G?

d T?

3 Describe what happens when a DNA molecule is copied.

4 Which observations made by other scientists did Crick and Watson's model account for?

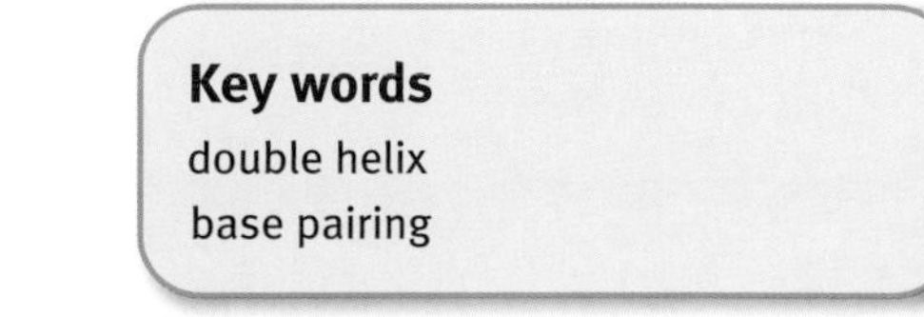

Key words
double helix
base pairing

Find out about:
- some different proteins in your body
- why cells become specialized

G Specialized cells – special proteins

An oak tree has about 30 different types of cell. Your body has more than 300 types of cell. Each cell type has its own set of proteins.

Some proteins make up the framework of cells and tissues. These are **structural proteins**. If we take away all the water in an animal cell, 90% of the rest is proteins.

Protein	Found in . . .	Property
keratin	hair, nails, skin	strong and insoluble
elastin	skin	springy
collagen	skin, bone, tendons, ligaments	tough and not very stretchy

Different structural proteins have different properties.

The flesh of meat and fish is the animals' muscles. They are mainly protein and provide the protein in many people's diets. Plant seeds such as soya and other beans are rich in proteins and are the basis of many vegetarian meals.

Other proteins are essential for the chemical reactions that keep our bodies working. For example, **enzymes** speed up the chemical reactions in a cell. **Antibodies** are the proteins that help to defend us against disease.

An oak tree has about 30 cell types.

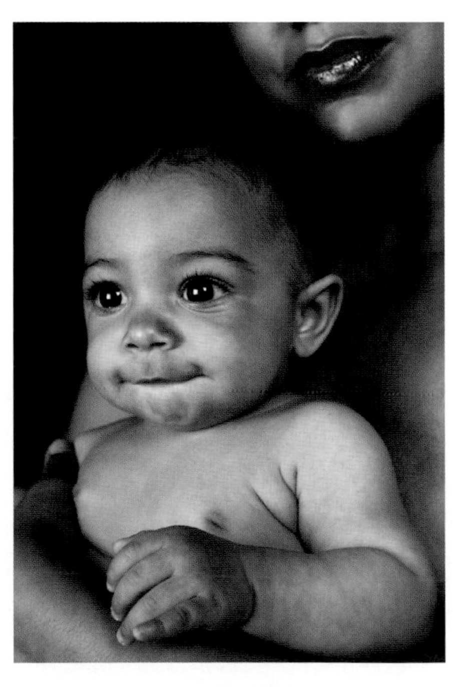

Humans have over 300 different types of cell.

Questions

1 Name three types of protein in your body.

2 Name one structural protein and say how it is suited to do its job.

What is the link between genes and proteins?

All the cells in your body come from just one original cell, the zygote. This divides to form a ball of cells. After about three weeks, cells start to specialize. They make the proteins needed to become a particular type of cell.

DNA is a cell's genetic code. Each gene is the instruction for a cell to make a different protein. By controlling what proteins a cell makes, genes control how a cell develops.

Each of your cells has a copy of all your genes. Something must make some cells turn into nerve cells and others into heart cells and all the other cell types. There must be **genetic switches**, but how they work is still a bit of a mystery.

Key words
structural proteins
enzymes
antibodies
genetic switches

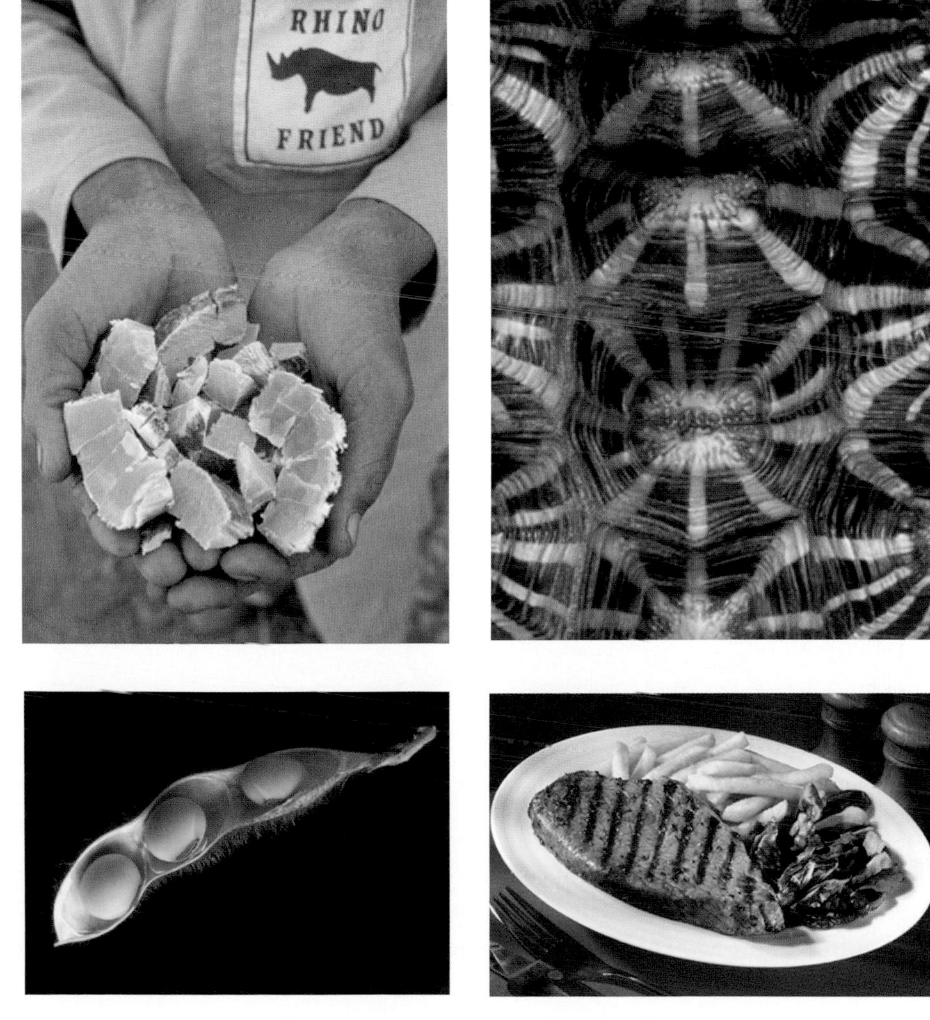

Rhino horn, tortoise shell, soya beans, steak . . . lots of different proteins.

Questions

3 DNA is a cell's genetic code. What does it do in a cell?

4 How do genes control how a cell develops?

(5) Name a cell that would make these proteins:

- **a** collagen
- **b** amylase
- **c** haemoglobin

Gene switching

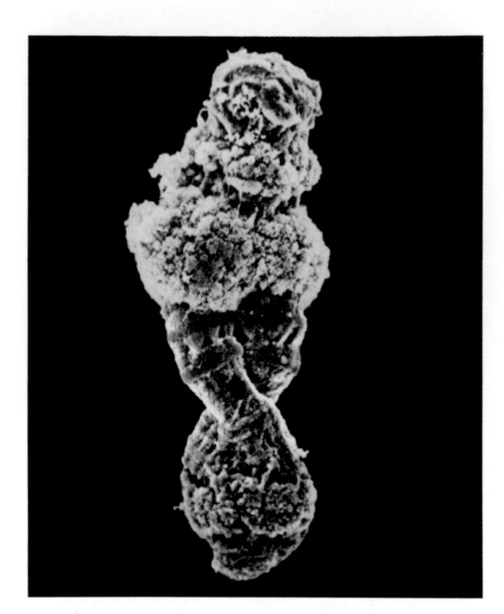

This is one chromosome from a salivary gland of a midge. The green areas are the active genes where DNA unravels whilst the protein is being made. One of these is instructing the cell to make a lot of amylase.

The **one-gene-one-protein theory** says that each gene controls the manufacture of one type of protein. So, in any organism, there are as many genes as there are different types of protein. In humans there are 20 000 to 25 000 genes.

Not all these genes are active in every cell. As cells grow and specialize, some genes switch off.

In a hair cell, the genes for the enzymes that make keratin will be switched on:

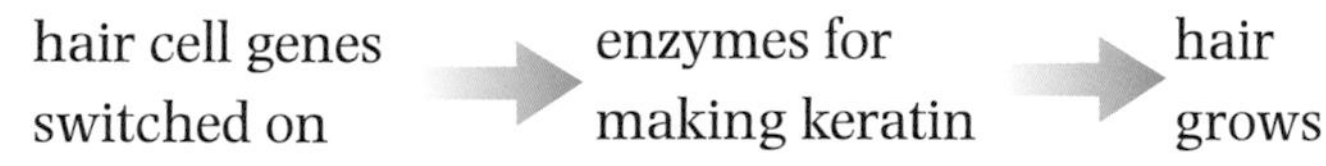

But the genes for those that make amylase will be switched off.

In a salivary gland cell:

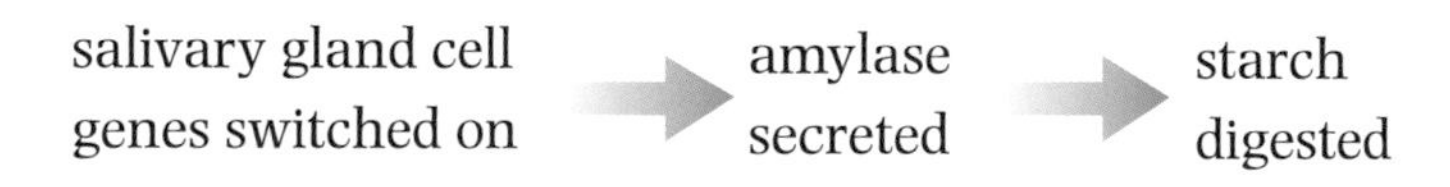

Gene switching in embryos

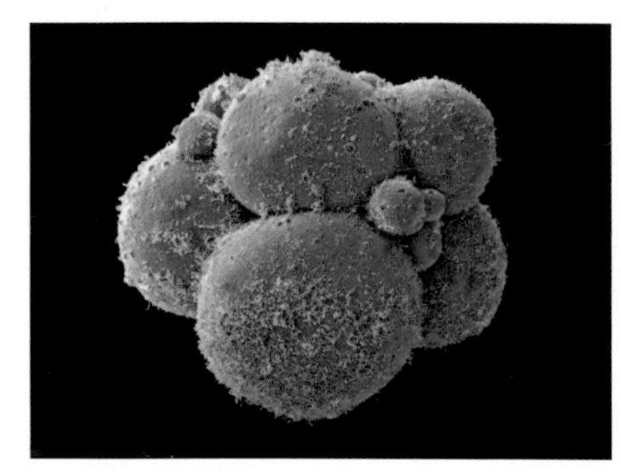

At the 8-cell stage, each cell of an embryo can develop into any kind of cell – or even into a whole organism.

An early embryo is made entirely of embryonic stem cells. These cells are unspecialized. All the genes in these cells are switched on. As the embryo develops, cells specialize. Different genes switch off in different cells.

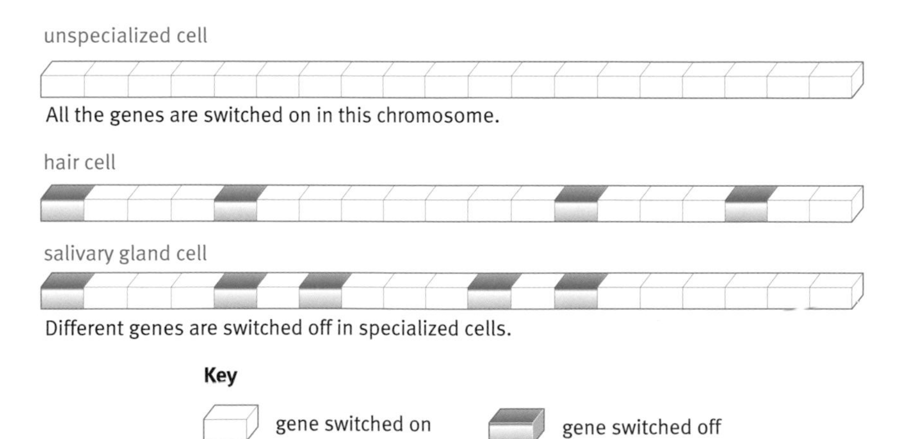

Some proteins are found in each type of cell, for example the enzymes needed for respiration. All cells respire, so the genes needed for respiration are switched on in all cells.

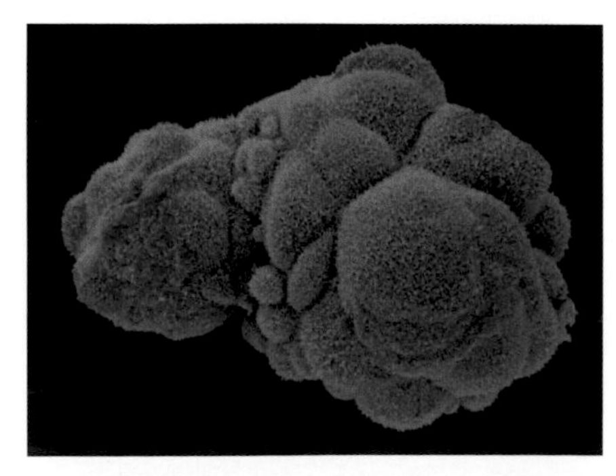

At six days and about 50 cells, a cell can develop into any kind of cell, but not into a whole organism.

In adults, there are stem cells in parts of the body where there is regular replacement of worn out cells. These can develop only into cells of the particular organ where they are. So some of their genes must be switched off.

Right cell, right place

Compare the fingers on your right hand with the same fingers on your left hand. They are probably almost mirror images of each other. As we grow, the position and type of cells must be controlled, so each tissue and organ develops in the right place.

In some animals, such as frogs, the place where the sperm enters the egg affects where the head and tail of the animal will be. In mammals, the head and tail end are probably already decided. There are differences in the amount of proteins in different parts of the egg cell. This will affect which genes are active. In simple terms, genes for the front end will become active in one half of the embryo. Genes for the back end will be active in the other half. Early development of embryos is an important area of research.

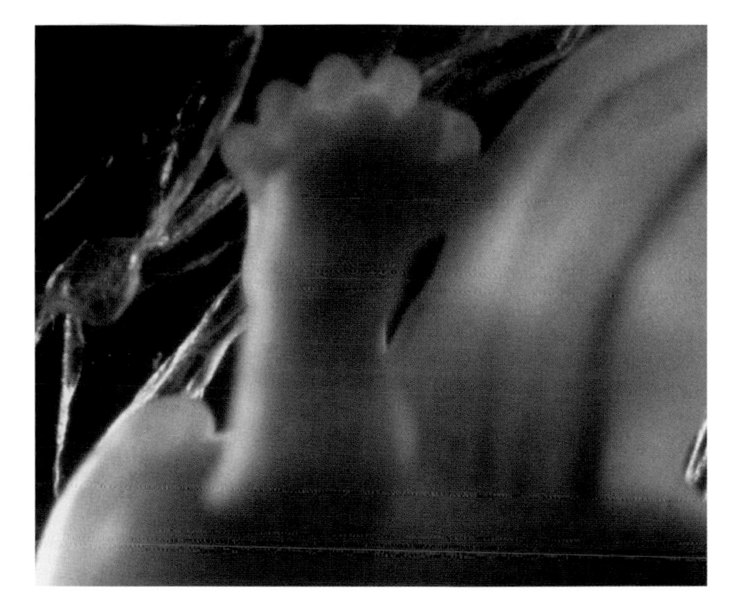

Cells near the end of a limb will make fingers. Cells nearer the body will make the arm. This happens because of the difference in the concentrations of chemical signals in each region of the embryo.

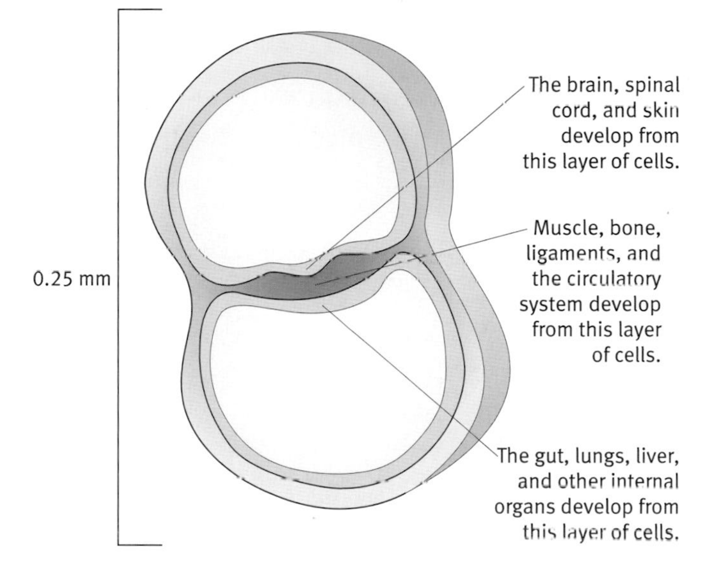

It is possible to map specialized parts of the body onto a diagram of a 14-day-old human embryo. Here you can see which groups of cells in the embryo will develop into future tissues and organs.

Questions

6. Suggest a function, other than respiration, that all cells carry out.
7. At the 8-cell stage of any embryo, how many genes are switched on?
8. What is the evidence that some genes are switched off at the 50–100-cell stage?

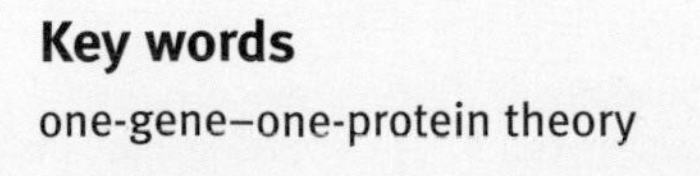

Key words

one-gene–one-protein theory

Find out about:

- scientific research to use stem cells for treating some diseases

Scientists grew this skin from skin stem cells in sterile conditions. Doctors use it for skin grafts.

H

H Stem cells

A lot of research is going on into stem cells. This is because many scientists see the possibility of using them for:

- the treatment of some diseases
- the replacement of damaged tissue

Imagine if scientists could produce . . .	They might use them to treat . . .
nerve cells	Parkinson's disease and spinal cord injuries
heart muscle cells	damage caused by a heart attack
insulin-secreting cells	diabetes
skin cells	burns and ulcers
retina cells	some kinds of blindness

The problem is to find stem cells of the correct type and then to grow them to produce enough cells. Stem cells can come from early embryos, umbilical cord blood, and adults. Embryonic stem cells are the most useful because the cells are not yet specialized. All their genes are still switched on.

There are problems with this new technology. For example:

- Tissues from the embryonic stem cells do not have the same genes as the person getting the transpant. Transplanted tissue is rejected if your body recognizes that the cells are not from your body.

Cloning from your own cells

Another possibility is to remove the nucleus of a zygote and replace it with the nucleus from a patient's own body cell. The new embryo would have the same genes as the patient. So the embryonic stem cells produced would match those of the patient. There would be no problem of rejecting a transplant of tissues grown from these cells. This is **therapeutic cloning**. Some people do not agree that cloned embryos should be produced. You can learn more about this in Module B1 *You and your genes*.

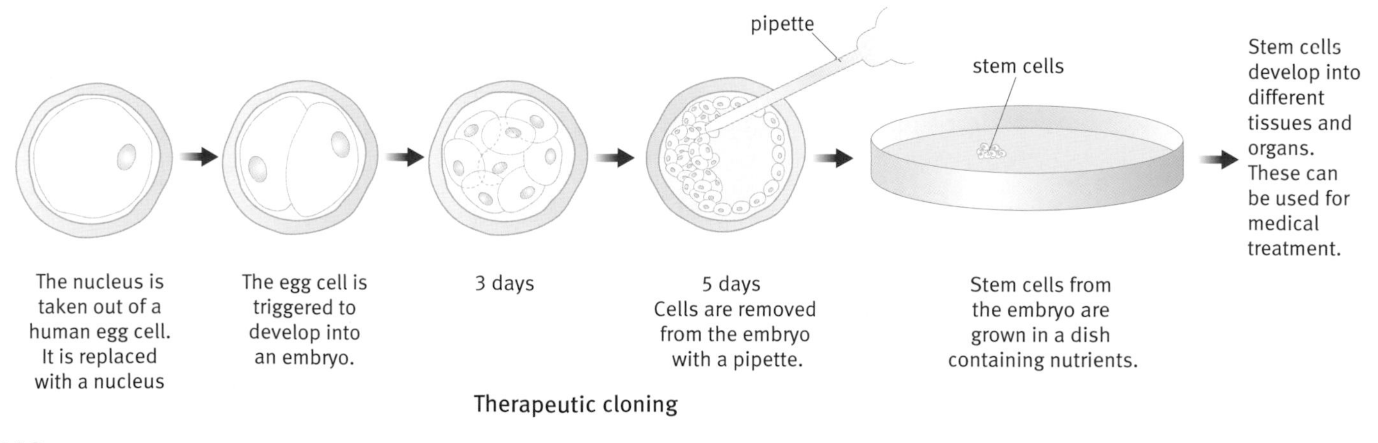

Therapeutic cloning

H

Could adult stem cells be used instead?

Most adult human stem cells grow into only a few cell types because many genes are switched off. Suppose scientists could find a way to switch the genes back on. Then they could use a person's own stem cells to produce any cells they needed. But it is not easy to reactivate genes, and there are other problems with this technology. For example:

- Bone marrow, where blood cells form, is particularly rich in stem cells. But only 1 in 10 000 cells in bone marrow is a stem cell. So it is hard to separate them from the millions of other cells.

Some successes

Only very few people have been treated using stem cells. There have been some promising, but very early, results. Scientists from the University of Freiberg in Germany treated 60 patients who had heart disease. The patients were divided into two groups:

Group	Treatment	Improvement in the left ventricle after 6 months
1	injection of stem cells from the patient's own bone marrow into the heart muscle	7.0%
2	the best conventional treatment	0.7%

The scientists think that the stem cells

- turned into new blood vessel or heart muscle cells
- made the heart tissue secrete chemicals that encouraged growth of the patient's own heart cells

Questions

1 It may be possible to use a nucleus from a patient's own body cell to clone stem cells.

a Describe how scientists hope to do this.

b What is this type of cloning called?

c What is the main problem with this technology?

2 What are the two main problems with using adult stem cells for therapeutic cloning?

3 Why is bone marrow rich in stem cells?

Key words

therapeutic cloning

Find out about:
- how DNA controls which protein a cell makes

I Making proteins

The great number of jobs carried out by proteins means that they are very different from one another. The exact shape of a protein can be very important to how it works. Cells make proteins from about 20 different **amino acids**. They join them in chains of 50 to many thousands of amino acids. In each protein, the amino acids are joined in a particular order, but there are thousands of possibilities. The order of the amino acids fixes the way the chains of amino acids fold to form the three-dimensional shape of the protein.

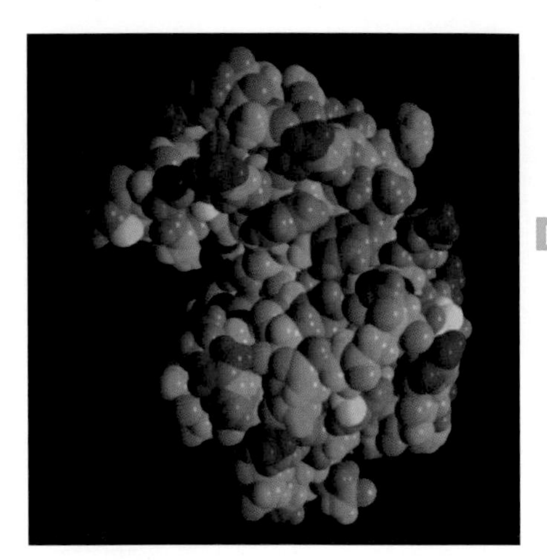

Enzymes work because of the shape of their active sites.

H The genetic code

When Crick and Watson discovered the structure of DNA, they were left with one major problem of the genetic code. They knew that DNA was the code for making proteins. But how could just four bases code for 20 amino acids?

If one base coded for one amino acid, DNA could code for only four amino acids. Two bases could code for 16.

In 1961, Crick worked out that a three-base code for each amino acid would work. This is called a **triplet code**. Different combinations of the four bases (A, T, G, and C) produce 64 triplet codes. So there is more than one code for each amino acid. There are also codes for start and stop. They mark the beginning and end of a gene.

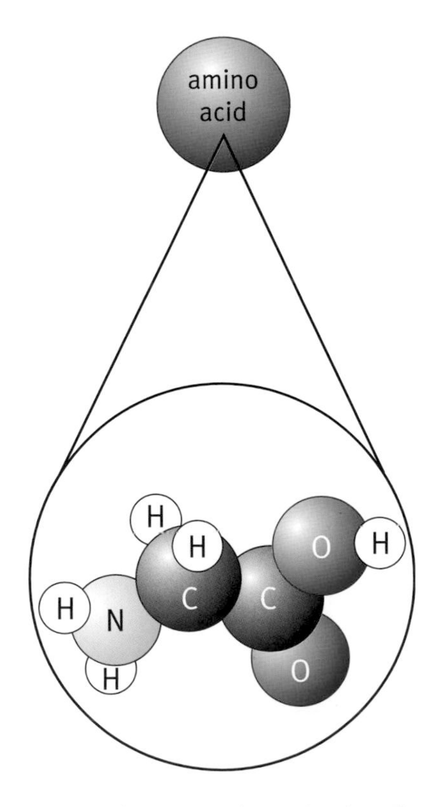

Amino acids are complex molecules. The diagrams on this page and the next represent the molecule simply.

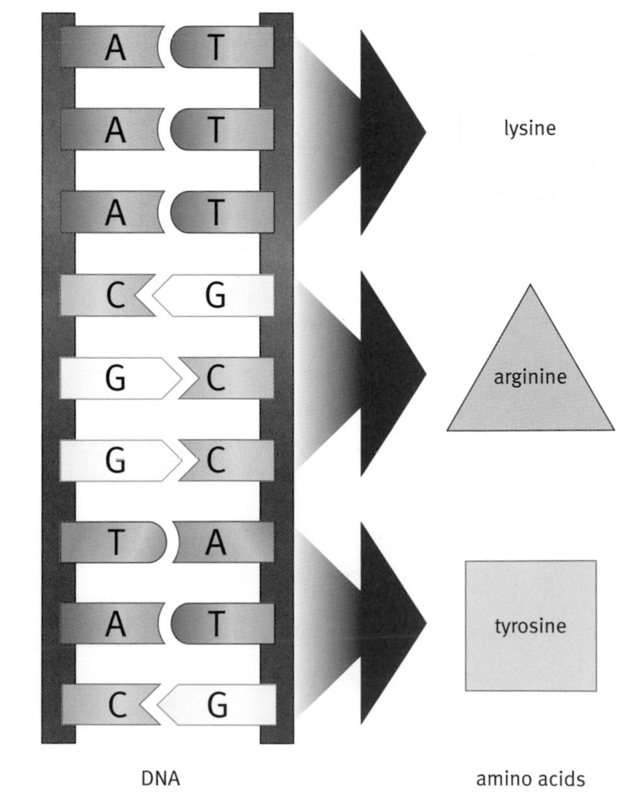

Three bases on the DNA code for each amino acid. For example, TTT codes for lysine.

Which part of a cell makes proteins?

DNA in the nucleus contains the genetic code for making the proteins. But tiny organelles in the cytoplasm of a cell, called ribosomes, actually make them. Genes cannot leave the nucleus. So how do ribosomes get the instruction for making a protein? A molecule small enough to get through the pores of the nuclear membrane transfers the genetic code to the ribosomes. This smaller molecule is called messenger RNA (**mRNA**).

H The differences between DNA and mRNA are that mRNA has

- only one strand
- the base U in place of the T in DNA

The diagram below shows how a protein is made.

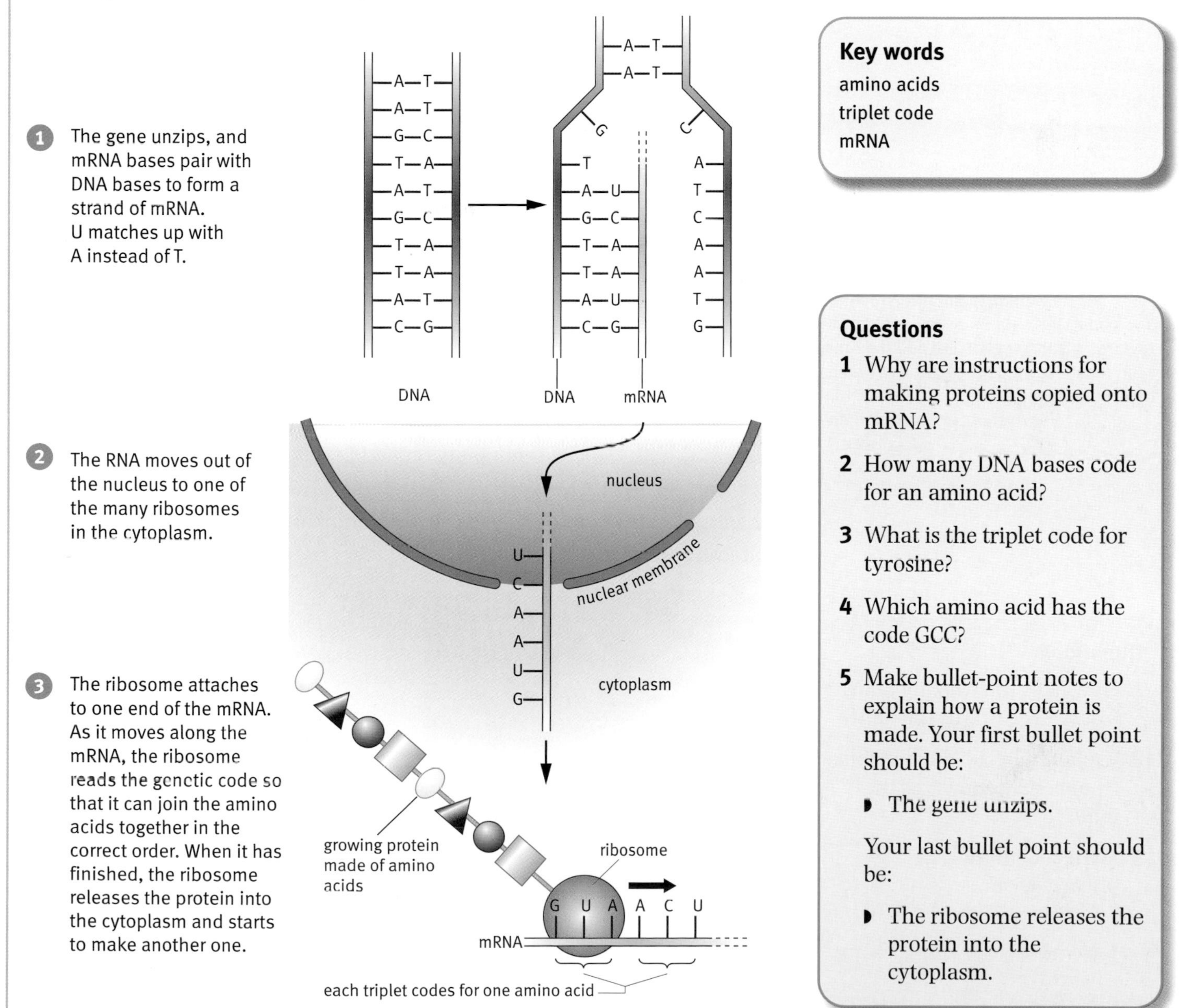

DNA makes protein with a ribosome's help (schematic).

Key words

amino acids
triplet code
mRNA

Questions

1 Why are instructions for making proteins copied onto mRNA?

2 How many DNA bases code for an amino acid?

3 What is the triplet code for tyrosine?

4 Which amino acid has the code GCC?

5 Make bullet-point notes to explain how a protein is made. Your first bullet point should be:

- The gene unzips.

Your last bullet point should be:

- The ribosome releases the protein into the cytoplasm.

Find out about:

- why plants grow towards light

J Phototropism

Plants rooted in soil cannot move from place to place – not even the 'walking palm' tree in the picture below.

The walking palm tree, *Socratea durissima*, in Costa Rica. New roots grow towards a sunny patch and pull the stem and leaves towards the light. Hormones control the direction of root growth. Older roots in the shade die.

You may have noticed that plants on windowsills seem to be bending towards the light. They are not moving, but growing. When the direction the light comes from affects the direction of plant growth, it is called **phototropism**.

This houseplant has grown towards the window to increase the amount of light falling on its leaves.

Questions

1. How does the walking palm tree grow towards the light?
2. Explain why a plant benefits from bending towards the light.
3. Write a definition for phototropism.

Darwin's phototropism experiments

Charles Darwin experimented with phototropism. He showed that the young shoots of grasses

- normally grew towards light
- remained straight when he covered their tips

Foil covers different parts of the barley shoots.

In the experiment, shown in the picture above, covering the lower parts of the shoot did not stop bending towards the light. This shows that only the tip is sensitive to light. But the shoot bends below the tip – where cells are no longer dividing but are increasing in length.

Darwin did not know how bending towards light happened. Now scientists have found out that higher concentrations of auxins cause shoot cells to expand. The diagrams on the right explain how this causes phototropism.

Questions

4 Look at the diagram showing phototropism experiments in barley. Suggest where:

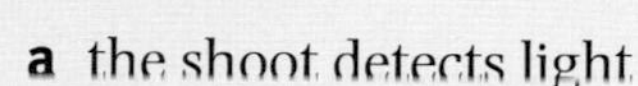

a the shoot detects light

b the cells are growing very quickly to cause bending

5 a Which way will shoots A and B grow?

b Will A or B grow taller? Explain your answer.

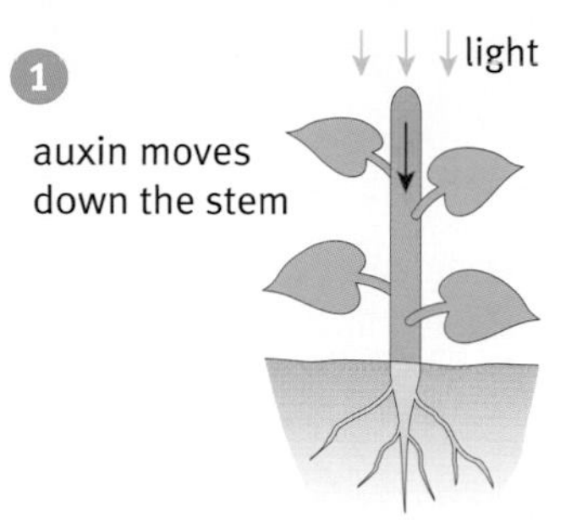

When a growing shoot gets light from above, the auxins spread out evenly and the shoot grows straight up.

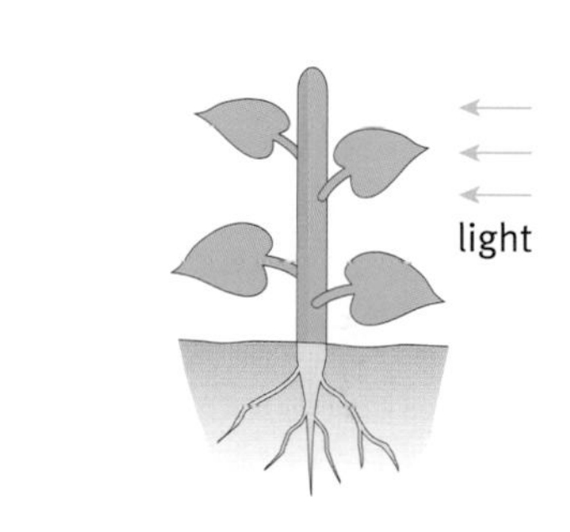

When light comes from one side, the auxins move over to the shaded side.

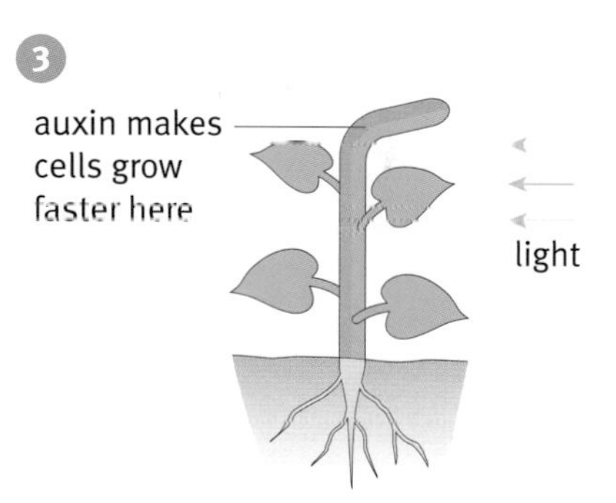

The shoot grows faster on the shaded side, making the shoot bend towards the light.

How auxins explain phototropism

Key words

phototropism

B5 Growth and development

Summary

In this module you have met ideas about cell division and growth, and explanations for how a single fertilized egg cell can produce all the different cells needed for a new animal.

The cell cycle

- Cells go through cycles of growth and division:
  - During cell growth the number of organelles increases, and chromosomes are copied.
  - During cell division copies of the chromosomes separate, and then the cell divides.
- Cell division by mitosis produces two new cells identical to the parent cell.
- Cell division by meiosis leads to four gamete cells with half the chromosome number of the parent cell.

Making proteins

- Genes in the nucleus of a cell are instructions for making proteins.
- Copies of genes are carried from the nucleus to the cytoplasm, where proteins are produced.
- Chromosomes are made of the chemical DNA, which has a double helix structure that allows it to be reproduced accurately.
- DNA molecules are made up of four bases, which always pair up in the same way.
- The order of bases in a gene is the code for joining up amino acids in the correct order to make a particular protein. H

A new human

- When two gametes join at fertilization, a zygote (fertilized egg cell) is produced that contains a set of chromosomes from each parent.
- Zygotes divide by mitosis to make an embryo.
- Up to the eight-cell stage of a human embryo, all the cells are identical embryonic stem cells.
- Embryonic stem cells can produce any sort of cell required by the organism.
- Cells become specialized because many genes are not active, so the cell only produces the particular proteins that it needs.
- It is possible to reactivate some inactive genes in the nucleus of body cells. H
- Adult and embryonic stem cells have the potential to produce cells needed to replace damaged tissues. H

Plant development

- Plant cells specialize into plant tissues (e.g. xylem, phloem) and organs (e.g. roots, leaves, flowers).
- Some plant cells (meristem cells) stay unspecialized throughout the plant's life, so they can develop into any type of plant cell.
- Mersitem cells divide to make new cells that increase the plant's height, width, and length of roots.
- Meristem cells can re-grow whole parts of the plant, which means that gardeners can make new plants from cuttings.
- Plant hormones (e.g. auxins) affect how plant cells develop.
- Plants grow towards light; this response is called phototropism.
- Auxin builds up in cells on a shoot's shady side, which makes the cells grow longer, so the shoot bends towards the light. H

Questions

1 This question is about the cell cycle of a human body cell.

a Write down what happens during:

i growth phase

ii cell division

b Copy and complete the table below, to describe mitosis in a human body cell.

Why mitosis happens	growth
Where mitosis happens	
Number of cells after division	
Number of chromosomes in new cells	

c Draw a similar table to describe meiosis.

2 **a** Describe the shape of a DNA molecule.

b Draw a flow chart to explain how a gene controls the production of a protein. The first and last statements are:

- Genes are the code for producing a protein. They are kept in the nucleus.
- The new protein is made.

3 **a** Name the four bases that make up a DNA molecule.

b Describe how bases pair up in a DNA molecule.

c Draw a flow chart to explain how a DNA molecule is copied. The first and last statements are:

- A DNA molecule is made of two strands.
- Two new DNA molecules have been made.

4 **a** Name two tissues from:

i an animal

ii a plant

b Name the unspecialized cells in:

i an animal

ii a plant

c Explain why many plants can re-grow whole organs if they are damaged, but most animals cannot.

5 Genes are instructions for making proteins. All human body cells contain the same genes. But cells only make the proteins that they need to do their specialized job.

Explain how cells with the same genes can become specialized.

6 **a** Name the plant hormone found in rooting powders.

b Explain why young plants grown on a window sill grow towards the light.

B6

Why study the brain and mind?

The human brain allows our species to survive on Earth. It gives us advantages of intelligence and sophisticated behaviour. Without our complex brains our species may easily have died out thousands of years ago.

The science

Animals respond to stimuli in order to survive. The central nervous system, the brain and spinal cord, coordinates millions of electrical impulses every second. These impulses determine how we think, feel, and react – our behaviour. Some drugs can affect our behaviour by interfering with the way nerve cells carry impulses.

The structure of our brains allows human beings to learn from experience and recall large amounts of information from our memory stores. Scientists have come up with some models for memory, but so far none can really fully explain how it works.

Biology in action

Research into the brain is helping to develop new treatments for some diseases. Knowing more about how we learn could help many people, for example children with learning difficulties, and adults recovering from brain damage.

Brain and mind

Find out about:

- how organisms respond to stimuli
- how nerve impulses are passed around your body
- how your brain coordinates your senses
- how you learn new skills
- how scientists are finding out about memory

Find out about:

- what behaviour is
- how simple behaviour helps animals survive

A What is behaviour?

Imagine you are sitting outside. The temperature drops, and you get cold. You start to shiver.

Shivering is a **response** to the change in temperature. A change in your environment, like a drop in temperature, is called a **stimulus.** Eating is a response to the stimulus of hunger. Scratching is a response to an itch. Shivering, eating, and scratching are all examples of **behaviour.**

You can think of behaviour as anything an animal does. The way an animal responds to changes in its surroundings is important for its survival.

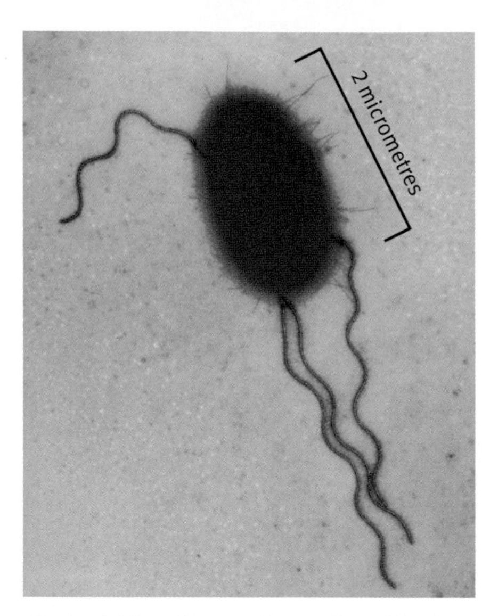

Escherichia coli bacteria are found in the lower gut of warm-blooded animals. They detect the highest concentration of food and move towards it.

Simple behaviour

Simple animals always respond to a stimulus in the same way. For example, woodlice always move away from light. This is an example of a **simple reflex** response. Reflexes are always **involuntary** – they are automatic. Reflexes are important because they increase the animal's chance of survival. The photographs in this section all show reflexes.

Why are simple reflexes important?

Simple reflex behaviour helps an animal to:

- find food, shelter, or a mate
- escape from predators
- avoid harmful environments, for example extreme temperatures

Woodlice move away from light, so you are most likely to find them in dark places.

Single-celled *Amoeba* move away from high concentrations of salt, strong acids, and alkalis.

Simple reflexes usually help animals to survive. But animals that only behave with simple reflexes cannot easily change their behaviour, or learn from experience. This is a problem if conditions around them change. Their simple reflexes may no longer be helpful for survival.

When a giant octopus sees a predator, it rapidly contracts its body muscles. This squirts out a jet of water to push the octopus away from danger. The octopus may also release a dark chemical (often called 'ink') which hides its escape.

When the tail of the sea hare *Aplysia* is pinched, the muscle contracts quickly and strongly. This reflex helps the animal escape from the spiny lobster that preys on it.

Earthworms have some of the fastest reflexes in the animal kingdom. A sharp tap from a beak on its head end is detected in the body wall. Rapid contraction of the worm's muscles pulls it back into its burrow. But this time the bird was too quick.

Have you ever tried to swat a fly? It has a very fast response to any movement that its sensitive eyes detect.

Complex behaviour – a better chance of survival

Complex animals, like mammals, birds, and fish, have simple reflex responses. But a lot of their behaviour is far more complicated. It includes reflex responses that have been altered by experience. Also, much of their behaviour is not involuntary – they make conscious decisions. For example, if it gets very cold, you do not just rely on your reflexes to keep you warm; you decide to put on extra clothes.

Because complex animals can change their behaviour when environmental conditions change, they are more likely to survive.

Key words

response
stimulus
behaviour
simple reflex
involuntary

Questions

1 Write a sentence to explain each key word on this page.

2 Describe an action you did today that:
- you did not have to think about and you have never had to learn to do
- you have learned to do but you can now do without thinking
- you had to think about while you were doing it

3 Which of the actions you described is an example of conscious behaviour, and which is most likely to be a simple reflex response?

Find out about:

- reflexes in newborn babies
- simple reflexes that help you to survive

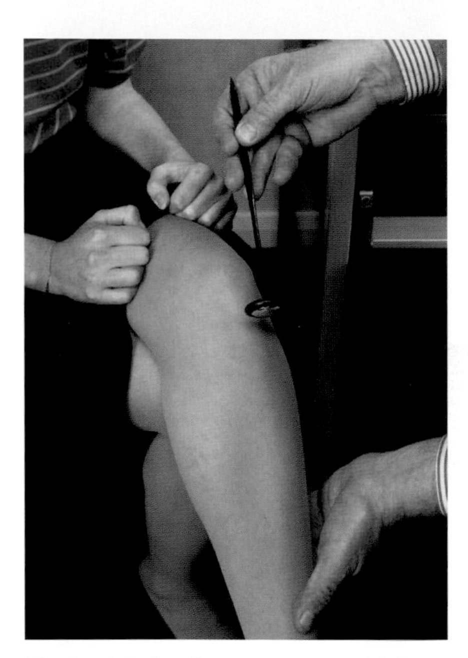

The knee jerk reflex causes your thigh muscle to contract, so your lower leg moves upwards. Doctors may test this and other reflexes when you have a health check. Try standing still with your eyes closed. You will notice this reflex helping you to balance.

B Simple reflexes in humans

Behaviour in humans and other mammals is usually very complex. But simple reflexes are still important for survival. For example:

- When an object touches the back of your throat, you gag to avoid swallowing it. This is the gag reflex.
- When a bright light shines in your eye, your pupil becomes smaller. This **pupil reflex** stops bright light from damaging the sensitive cells at the back of your eye.

Newborn reflexes

When a baby is born, the nurse checks for a set of **newborn reflexes**. Many of these reflexes are only present for a short time after birth. They are gradually replaced by behaviours learned from experience. In a few cases these reflexes are missing at birth, or they are still present when they should have disappeared. This may mean that the baby's nervous system is not developing properly.

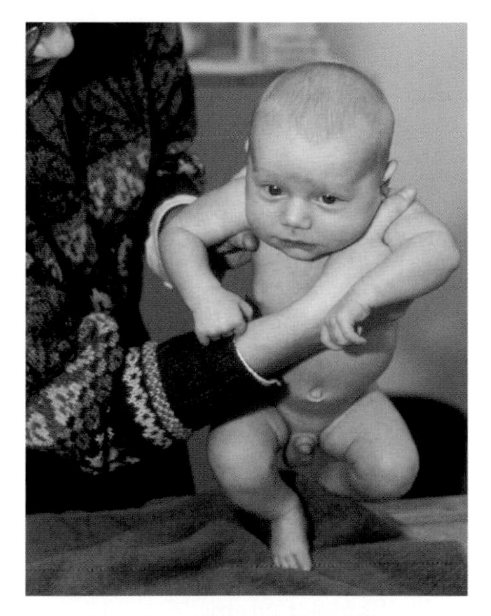

Stepping. If you hold a baby under his arms, support his head, and allow his feet to touch a flat surface, he will appear to take steps and walk. This reflex usually disappears by 2–3 months after birth. It then reappears as he learns to walk at around 10–15 months.

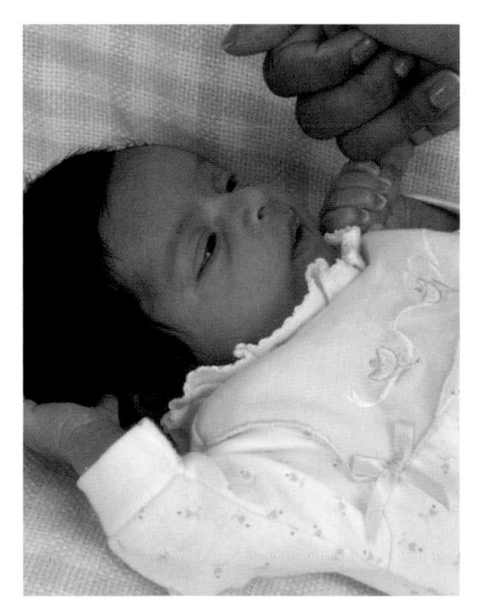

Grasping. When you put your finger in a baby's open palm, the baby grips the finger. When you pull away, the grip gets stronger. This reflex usually disappears by 5–6 months. If you stroke the underneath of a baby's foot, its toes and foot will curl. This reflex usually disappears by 9–12 months.

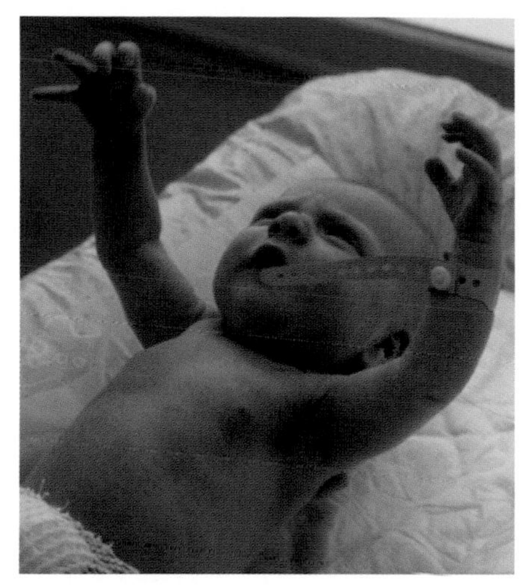

Startle. This is also called the Moro reflex, named in 1687 after the Italian scientist Artur Moro. It usually happens when a baby hears a loud noise or is moved quickly. The response includes spreading the arms and legs out and extending the neck. The baby then quickly brings her arms back together and cries. This reflex usually goes by 3–6 months.

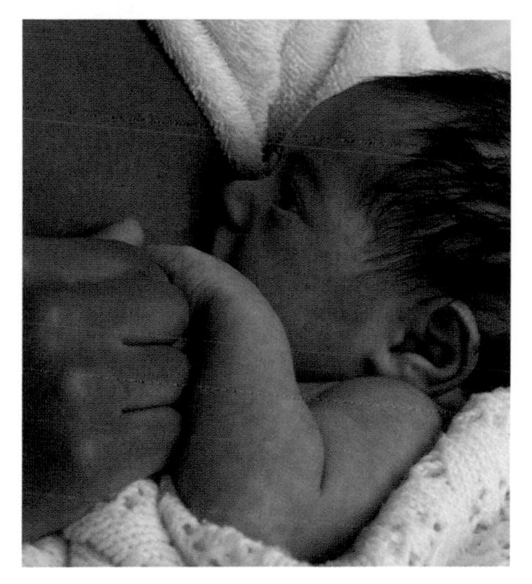

Sucking. Placing a nipple (or a finger) in a baby's mouth causes the sucking reflex. It is slowly replaced by voluntary sucking at around 2 months.
Rooting. Stroking a baby's cheek makes her turn towards you, looking for food. This reflex helps the baby find the nipple when she is breast feeding. The rooting reflex is gone by about 4 months.

Swimming. If you put a baby under six months of age in water, he moves his arms and legs while holding his breath.

Sudden infant death syndrome (SIDS)

Sudden Infant Death Syndrome, or cot death, is tragic and unsolved. In the UK about 7 babies a week die from SIDS. This is 0.7 deaths for every 1000 live births.

It is likely that there are many different causes of cot death. Some people think that it could be because a baby's simple reflexes have not matured properly. This is how doctors think this may happen:

- When a fetus detects that oxygen in its blood is low, its reflex response makes it move around less. This makes sense because the less it moves around, the less oxygen it will use up in cell respiration.
- This response changes as the baby matures. When an older baby or child's airways are covered, for example, by a duvet, the baby moves more. He turns his head from side to side. He also pushes the obstruction away. So now the response to low oxygen is more activity, not less.
- If the newborn baby has not grown out of the fetal reflex, he may lie still if his bedcovers cover his airways. He is more likely to suffocate.

Doctors now advise mothers to put babies onto their backs to sleep, and not to use soft bedding like duvets. This way their faces are less likely to become covered.

Key words

pupil reflex
newborn reflexes

Questions

1. Describe two reflexes in:
   - **a** adult humans
   - **b** newborn babies
2. How do you think the startle reflex helps a baby to survive?
3. Why are premature babies more at risk from SIDS than babies born at the correct time?

Find out about:

- different parts of your nervous system
- how reflexes are controlled

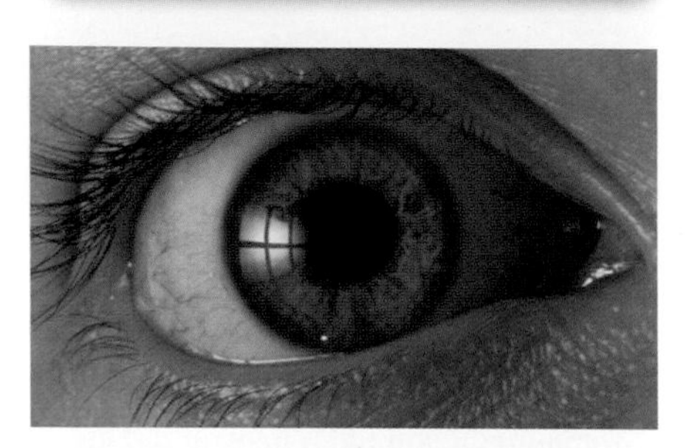

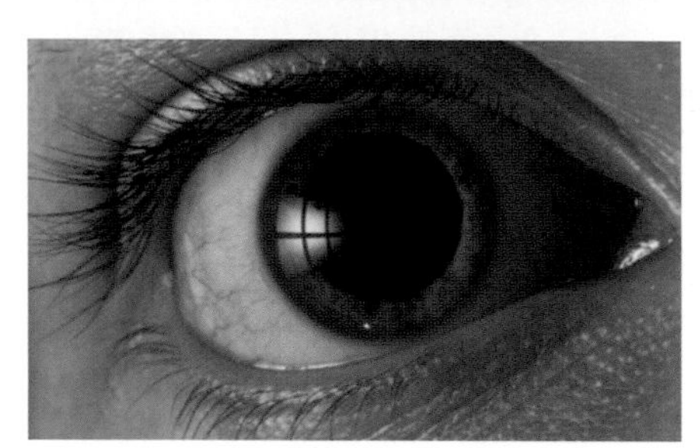

Muscles in the iris cause the pupil to change size depending on the brightness of light entering the eye. (The pupil size controls the amount of light intensity that is the stimulus.)

Key words

nervous system
nerve impulses
reflex arc
receptor
sensory neuron
central nervous system
coordinates
motor neuron
effector
peripheral nervous system
neurons
axon
fatty sheath

C Your nervous system

Walk out from a dark cinema on a bright afternoon and your pupils will become smaller. This pupil reflex prevents bright light damaging your eye. Like all reflexes, this behaviour is coordinated by your **nervous system.**

Cells in your nervous system carry **nerve impulses.** These nerve impulses allow the different parts of the nervous system to communicate with each other.

The reflex arc

In a simple reflex, impulses are passed from one part of the nervous system to the next in a pathway called a **reflex arc.** The diagram below shows this pathway for the pupil reflex:

- The stimulus is detected by a **receptor** cell.
- Nerve impulses are carried along a **sensory neuron** to your **central nervous system.**
- The central nervous system is made up of your brain and spinal cord. It **coordinates** your body's responses.
- Nerve impulses are carried along a **motor neuron** to an **effector.**
- The effector carries out the response to the stimulus.

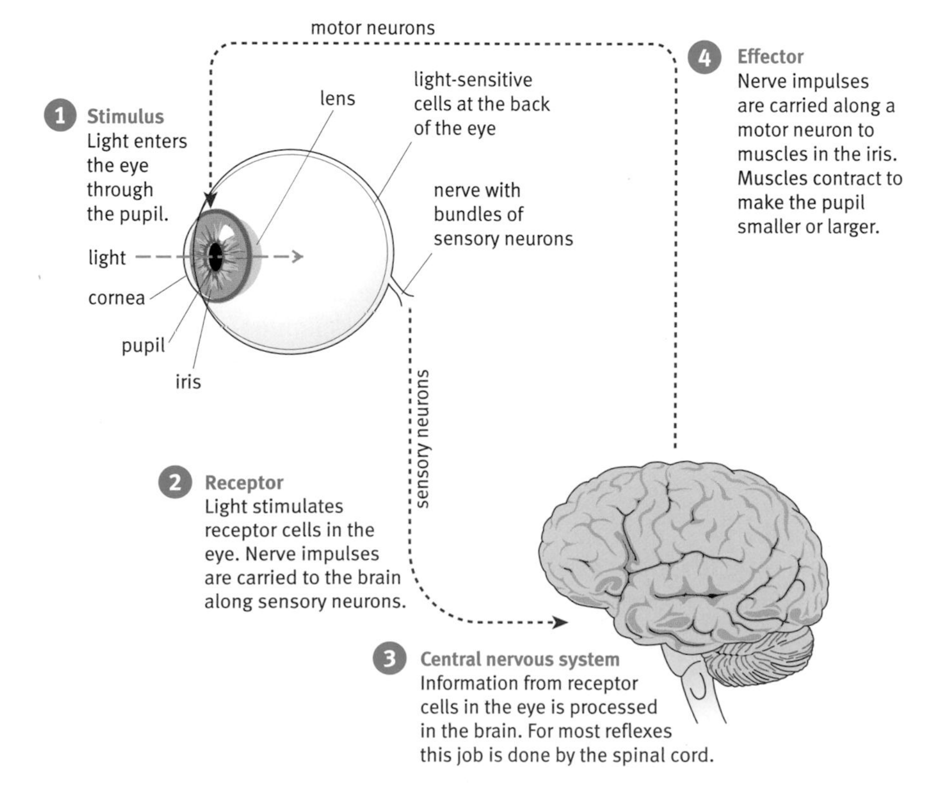

An example of a reflex arc

Peripheral nervous system

Many nerves link your brain and spinal cord to every other part of your body. These nerves make up the **peripheral nervous system.**

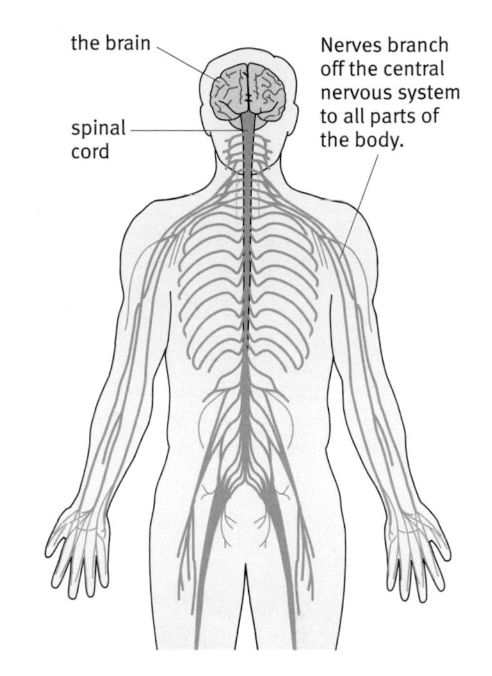

The peripheral nervous system links the brain and spinal cord with the rest of the body.

Nerves and neurons

Nerves are bundles of specialized cells called neurons. Like most body cells, **neurons** have a nucleus, a cell membrane, and cytoplasm. They are different from other cells because the cytoplasm is shaped into a very long thin extension. This is called the **axon**, and it is how neurons connect different parts of the body.

Axons carry electrical nerve impulses. Like wiring in an electrical circuit, the axons must be insulated from each other. The insulation for an axon is a **fatty sheath** wrapped around the outside of the cell. The fatty sheath increases the speed that impulses move along the axon.

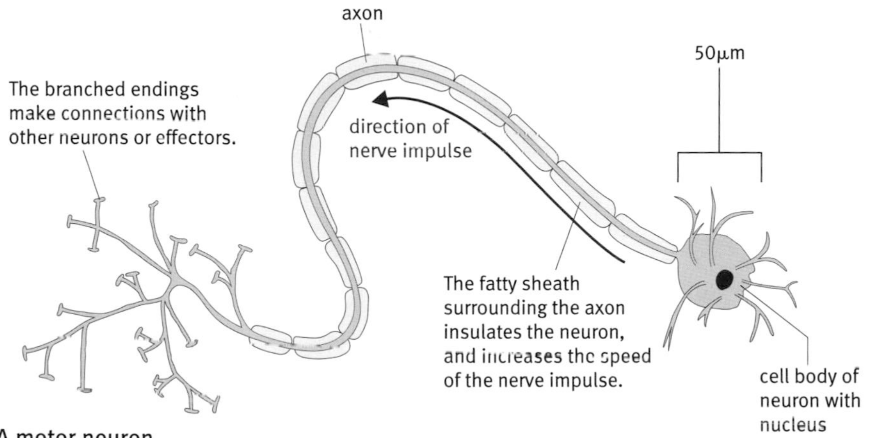

A motor neuron

Central nervous system

Your central nervous system (CNS) coordinates all the information it receives from your receptors. Information about a stimulus goes to either your brain or your spinal cord. In a reflex arc the CNS directly links the incoming information from receptors with the effectors that will carry out the necessary response.

Questions

1 Describe the difference between:

- the job of a sensory neuron and a motor neuron
- an axon and a neuron
- the central nervous system and the peripheral nervous system

2 Draw a labelled diagram to show a reflex arc for the newborn grasp reflex shown on page 152. This reflex is coordinated by the spinal cord.

Receptors

You can only respond to a change if you can detect it. Receptors inside and outside your body detect **stimuli**, or changes in the environment.

Some animals have receptors that are triggered by stimuli which humans cannot detect. For example, falcons can detect an object 10 cm across when they are 1.5 km away. The rattlesnake's heat-sensitive receptors can detect a warm mouse 40 cm away. The receptors in your eyes and skin could not detect these stimuli.

Some sharks can detect when the concentration of food in the water is only 0.1 parts per billion. This concentration certainly would not trigger your taste or smell receptors. Sharks also have receptors that are stimulated by the tiny electrical currents given off by their prey.

A shark's sensitive receptors help it to locate its prey.

The falcon's eyes can spot the slightest movement on the ground.

Body heat gave the position of this rattlesnake's latest meal.

Detecting stimuli inside and out

You can detect many different stimuli, for example sound, texture, smell, temperature, and light. Different types of receptors each detect a different type of stimulus. Receptors on the outside of your body monitor the external environment. Others monitor changes inside your body, for example core temperature and blood sugar levels.

Sense organs

Some receptors are made up of single cells, for example pain receptors in your skin. Other receptor cells are grouped together as part of a complex sense organ, for example your eye. Vision is very important in humans and most other mammals. Light entering our eyes helps us humans produce a three-dimensional picture of our surroundings. This gives us information about objects such as their shape, movement, and colour.

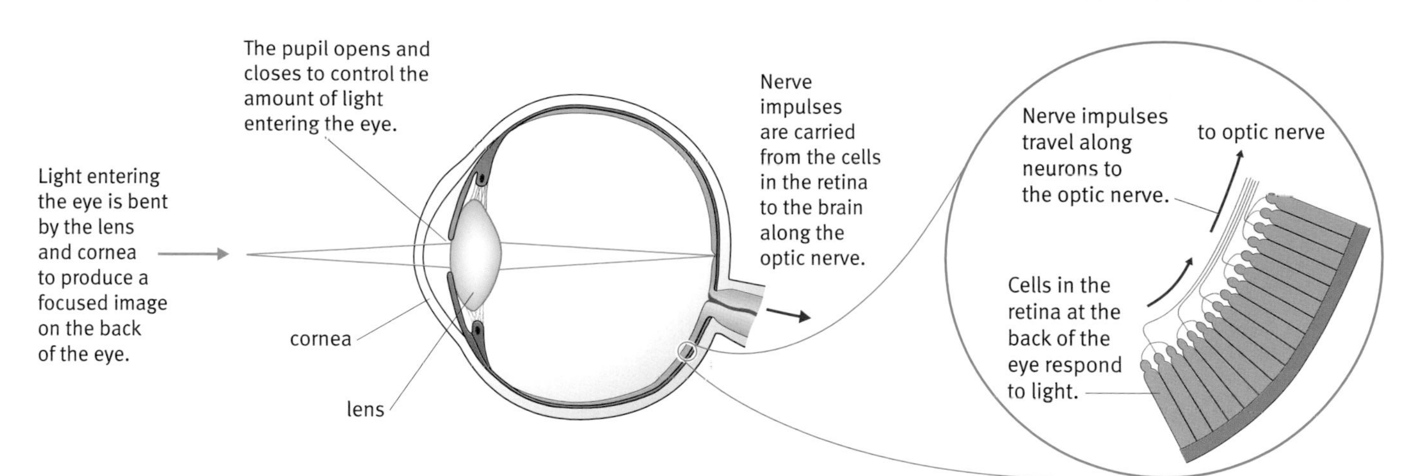

Light is focused by the cornea and lens onto light-sensitive cells at the back of the eye. These cells are receptors. They trigger nerve impulses to the brain.

Effectors

The body's responses to stimuli are carried out by effector organs. Effectors are either **glands** or **muscles**. When impulses arrive at these effectors they cause:

- glands to release chemicals, for example hormones, enzymes, or sweat
- muscles to contract and move part of the body

Key words

stimuli
glands
muscles

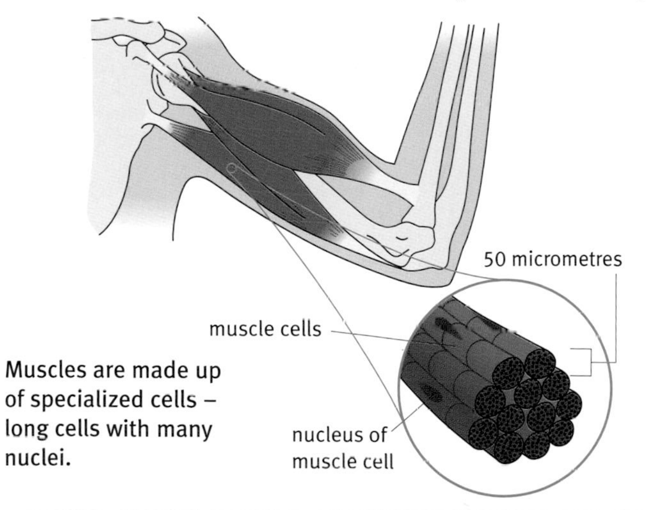

Muscles are made up of specialized cells – long cells with many nuclei.

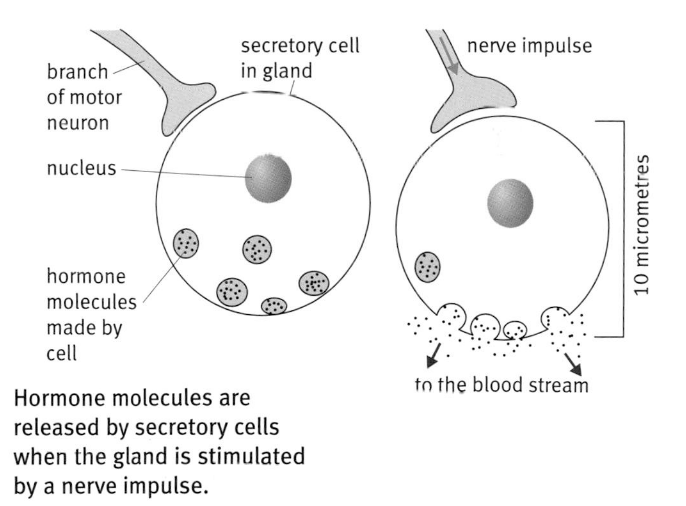

Hormone molecules are released by secretory cells when the gland is stimulated by a nerve impulse.

Questions

3 What kind of stimulus can a shark detect that we cannot?

4 Name the two different types of effectors and say what they do

5 Which receptors would you use in order to thread your trainers with new laces?

6 Which effectors are you using when you:

- text a friend?
- cry?
- run a race?

7 Some people suffer from a disease where tiny clusters of light receptor cells in different parts of the eye become damaged. How would this affect what the person sees?

Conscious control of reflexes

Reflexes are involuntary actions. Most of them are coordinated by your spinal cord. Even those that are coordinated by your brain, for example your pupil reflex, are involuntary. You do not think about these responses – your brain does not have to make a decision. They happen automatically, because they are designed to help you survive.

But sometimes a reflex may not be what you want to happen. Some reflexes can be modified by conscious control. Imagine picking up a hot plate. Your pain reflex makes you drop it. But if your dinner were on the plate, you can overcome this reflex and hold onto the plate until you were able to put it down safely. The conscious control of your brain overcomes the reflex response. The diagram below explains how this happens.

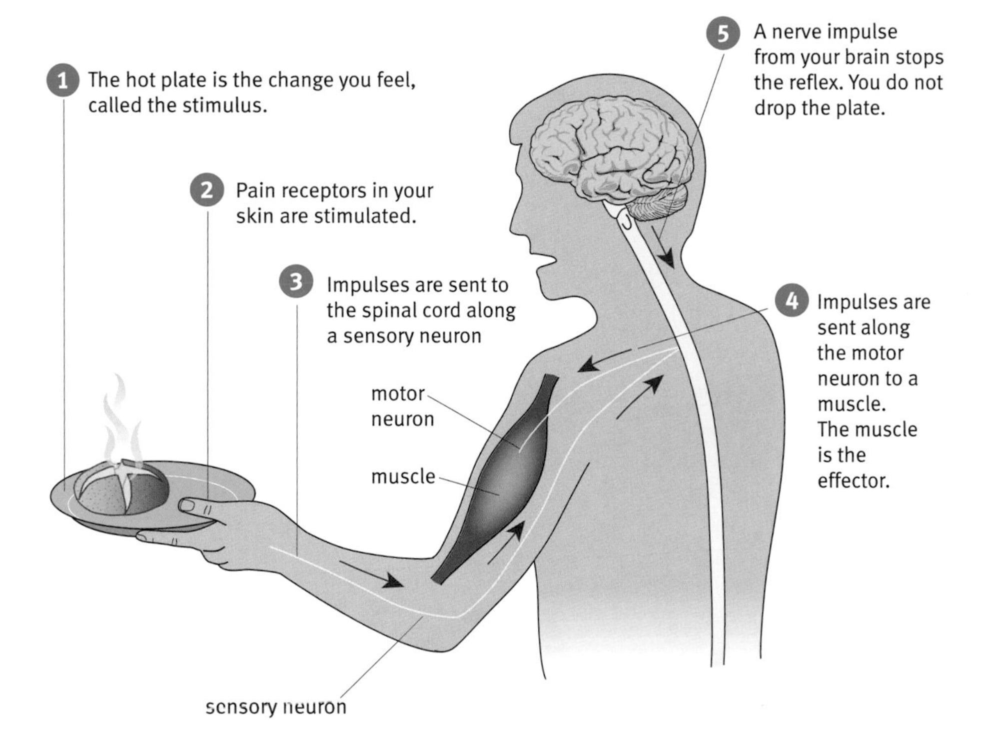

Overcoming a reflex by conscious control. Some reflexes can be modified by conscious control. If you pick up a hot plate, your reflex response is to drop it. But a second nerve impulse travels up your spine to your brain, then back down to cause a muscle movement stopping this reflex. You can put the plate down quickly without dropping it because the conscious control from your brain overcomes the reflex.

Questions

8 List two reflexes you can overcome, and two that you can not.

9 A tiny baby urinates whenever its bladder is full. Draw a labelled diagram to show how nerve impulses from the brain overcome this reflex when he is older.

It's all in the mind – more complex behaviour

Most human reflex arcs are coordinated by the spinal cord. Only reflexes with receptors on the head are coordinated by the brain. A reflex arc only has simple connections between a sensory neuron and a motor neuron.

Connections in the brain usually involve hundreds of other neurons with different connections. Using these complex pathways, your brain can process highly complicated information, such as music, smells, and moving pictures. Different parts of the brain also store information (memory) and use it to make decisions for more complicated behaviour.

You can find out more about the brain, and complex behaviour like learning and memory, later in this module.

Hundreds of neurons interact to coordinate the responses you make when you are receiving this many stimuli.

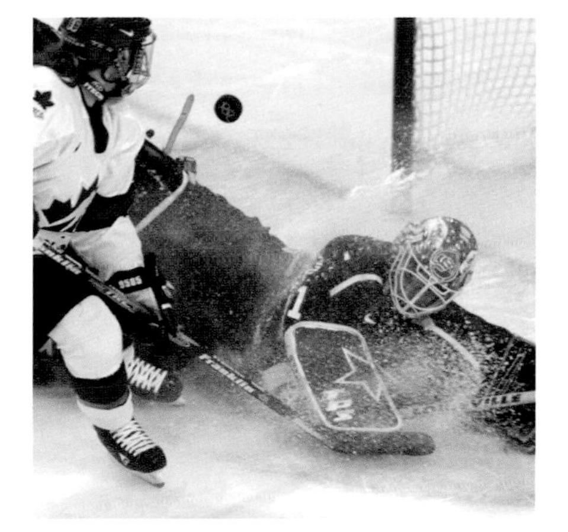

Hundreds of complex pathways in the brain have to be used to succeed in this fast-moving sport.

This complicated behaviour involves highly complex pathways in the brains of both these animals.

A day in the life of a neuroscientist

Steven Rose is a neuroscientist – a scientist who studies how the brain works. Steven is interested in how complex behaviour is controlled.

He studies the behaviour of chicks and looks for changes in chemicals in their brains as they learn new skills.

From his work with chicks, Professor Rose found that a particular chemical, APP, is needed for learning and memory. If this chemical is changed or absent, animals cannot remember something new.

A small amount of APP can rescue the memory of animals with symptoms similar to Alzheimer's disease (loss of memory). This research is leading to a new approach to drug development for Alzheimer's.

Find out about:

- how nerve impulse pass from one neuron to the next

Quick responses to stimuli are also essential in real life. They help you survive by avoiding danger.

Curare is a very powerful toxin. It is used on the tips of blowpipe darts.

D Synapses

Think about playing a fast sport or computer game. You need very quick reactions to win. Nerve impulses give you fast reactions because they travel along axons at 400 metres per second.

Mind the gap

Neurons do not touch each other. So when nerve impulses pass from one neuron to the next, they have to cross tiny gaps. These gaps are called **synapses**. Some drugs and poisons (toxins) interfere with nerve impulses crossing a synapse. This is how they afffect the human body.

How do nerve impulses cross a synapse?

Nerve impulses cannot jump across a synapse. Instead, chemicals are used to pass an impulse from one neuron to the next. The diagram below explains how this works.

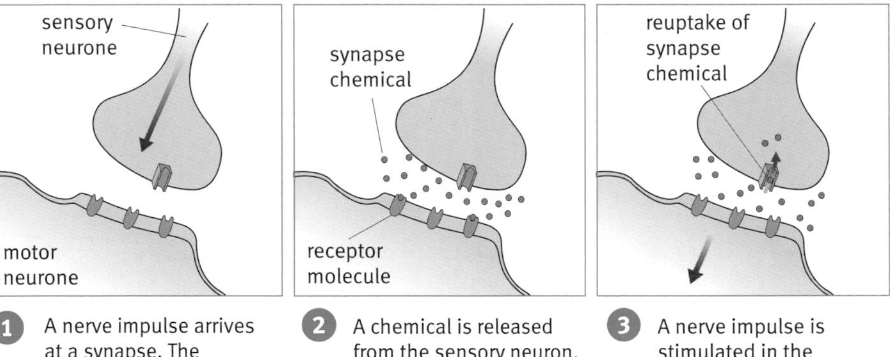

1. A nerve impulse arrives at a synapse. The direction of the impulse is shown by the arrow.
2. A chemical is released from the sensory neuron. It diffuses across the synapse. The molecule is the correct shape to fit into **receptor molecules** on the membrane of the motor neuron.
3. A nerve impulse is stimulated in the motor neuron. The chemical is absorbed back into the sensory neuron to be used again.

How a synapse works

Do synapses slow down nerve impulses?

The gap at a synapse is only about 20 **nanometres** (nm) wide. The synapse chemical travels across this gap in a very short time. Synapses do slow down nerve impulses to about 15 metres per second. A nerve impulse still travels from one part of your body to another at an incredible speed.

Being human – just chemicals in your brain?

The way we think, feel, and behave does involve a series of chemicals moving across synapses between neurons, but there is more to behaving like a human than chemicals in your brain. These processes are very complicated. Scientists researching how your nervous system works are only just beginning to understand the brain.

Serotonin

Serotonin is a chemical released at one type of synapse in the brain. When serotonin is released, you get feelings of pleasure. Pleasure is an important response for survival. For example, eating nice-tasting food gives you a feeling of pleasure. So you are more likely to repeat eating, which is essential for survival.

Lack of serotonin in the brain is linked to depression. Depression is a very serious illness. At least one person in five will suffer from a depressive illness at some point in their life. They feel very unhappy for many days on end and often find it difficult to manage normal everyday things like working, studying, or looking after their family.

How do some drugs affect the brain?

Prozac is the name of an antidepressant drug. These drugs can be helpful for treating people with depression. Prozac causes serotonin concentration to build up in synapses in the brain. So a person may feel less unhappy. The diagram explains how Prozac works. Like all drugs, Prozac can have unwanted effects. A doctor will consider all the factors very carefully before prescribing treatment.

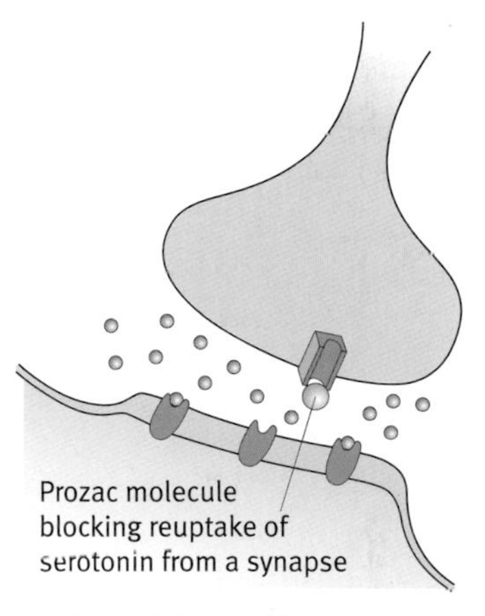

Feelings of depression can be caused by too little serotonin in the brain. Prozac works by blocking reuptake of serotonin.

Ecstasy

Ecstasy is the common name for the drug MDMA. Ecstasy works in a similar way to Prozac. People who have taken Ecstasy say that it can give them feelings of happiness and being very close to other people. Studies on monkeys suggest that long-term use of Ecstasy may destroy the synapses in the pleasure pathways of the brain. Permanent anxiety and depression might result, along with poor attention span and memory. For some people the harmful effects of Ecstasy are more immediate. Ecstasy interferes with the body's temperature control systems. It also slows down production of the hormone ADH in the brain. These effects can be fatal. You can read more about ADH in Module B4 *Homeostasis*.

Questions

1. Write down a sentence to describe a synapse.
2. Draw a flow diagram to describe what happens when a nerve impulse arrives at a synapse.
3. Explain how the release of serotonin in the brain helps us survive.
4. Some drugs (like curare) block the receptors on the motor neuron at a synapse. Explain how this would affect muscles that the motor neuron is linked to.
5. Explain how Prozac may help a person who has depression.

Key words

synapses
receptor molecules
nanometres
serotonin
Prozac
Ecstasy

Find out about:

- the structure of your brain
- how scientists learn about the brain

E The brain

Think of some things you did today. Getting up, deciding what to have for breakfast, travelling to school, talking and listening to friends. All of this complex behaviour has been controlled by your brain. But exactly how it happens is still being researched. Scientists who study the brain are called **neuroscientists**. Neuroscience is a fairly 'new' science. This means that scientists have only recently started to investigate how the brain works.

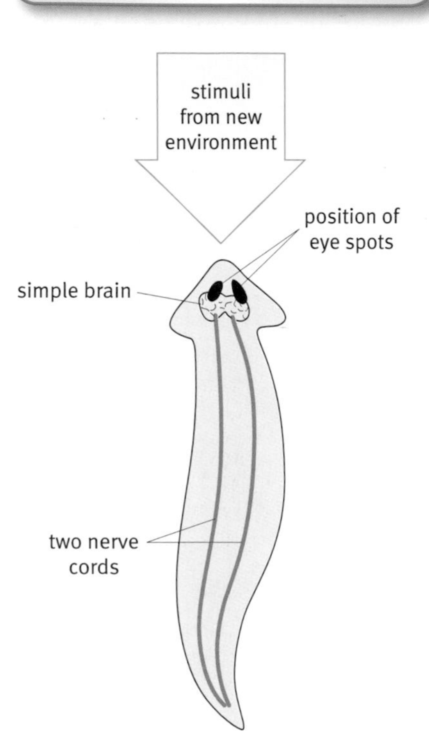

Sense organs on the flatworm head detect light and chemical stimuli. A simple brain processes the response.

Simple animals

Neurons carry electrical impulses around your body. Quite simple animals have a larger mass of neurons at one end of their body – the head end. This end reaches new places first, as the animal moves. These neurons act as a simple brain. It processes information coming from the receptors on their head end.

Looking at how the brains of simple animals work can help scientists begin to understand more complicated brains.

Complex animals

More complex behaviour like yours needs a much larger brain. So your brain is made of billions of neurons. It also has many areas, each carrying out one or more specific functions all in the same organ. Your complex brain allows you to learn from experience, for example how to behave with other people.

Average human brain statistics
width: 140 mm
length: 167 mm
height: 93 mm
weight: 1.4 kg

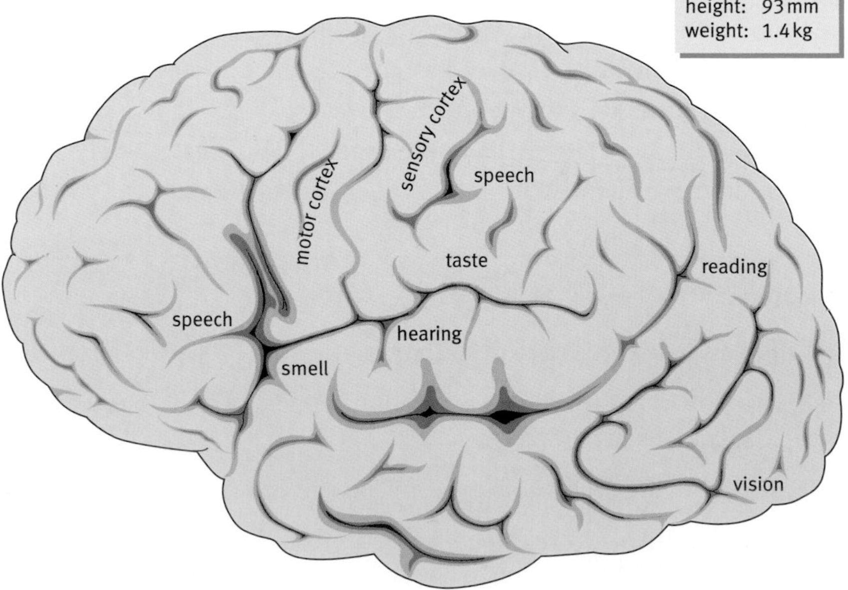

The human cerebral cortex is a highly folded region. Although it is only 5 mm thick, its total area is about $0.5\,m^2$. This map of the cerebral cortex shows the regions responsible for some of its functions.

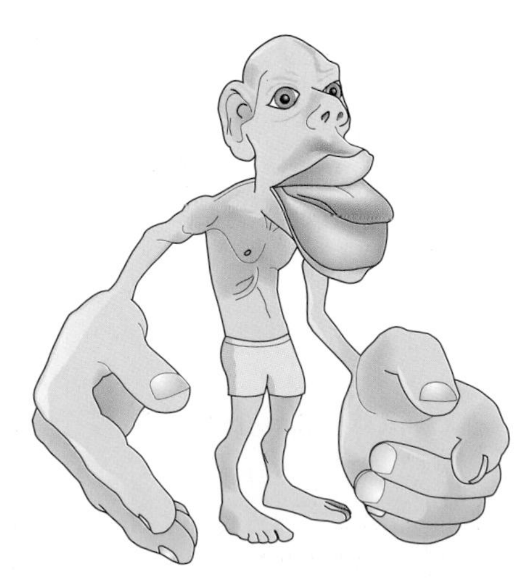

This is a diagram called the 'sensory homunculus' . Each body part is drawn so that its size represents the surface area of the sensory cortex that receives nerve impulses from it.

The conscious mind

When you are awake, you are aware of yourself and your surroundings. This is called **consciousness**. The part of your brain where this happens is the **cerebral cortex**. This part is also responsible for intelligence, language, and memory. Brain processes to do with thoughts and feelings happen in the cerebral cortex and are what is called your 'mind'.

The cerebral cortex is very large in humans compared with other mammals. Studying what goes on in this part of the brain helps us to understand what it means to be human.

Key words

neuroscientists
consciousness
cerebral cortex

Finding out about the brain

In the 1940s a Canadian brain surgeon, Wilder Penfield, was working with epileptic patients. Penfield carried out operations on his patients. He applied electricity to the surface of their brains in order to find problem areas. The patients were awake during the operations. There are no pain receptors in the brain so they did not feel pain.

Penfield mapped the motor cortex by stimulating the exposed brain during open brain surgery. Regions of this brain have been identified and labelled in a similar way.

Penfield watched for any movement the patient made as he stimulated different brain regions. From this information he was able to identify which muscles were controlled by specific regions of the motor cortex.

Injured brains

Scientists study patients whose brains are partly destroyed by injuries or diseases like strokes. These studies provide useful information about brain function.

Brain imaging

Modern imaging techniques such as magnetic resonance imaging (MRI) scans provide detailed information about brain structure and function without having to open up the skull. MRI can be used to show up which parts of the brain are most active when a patient does different tasks. These scans are called functional MRI (fMRI) scans.The active parts of the brain have a greater flow of blood.

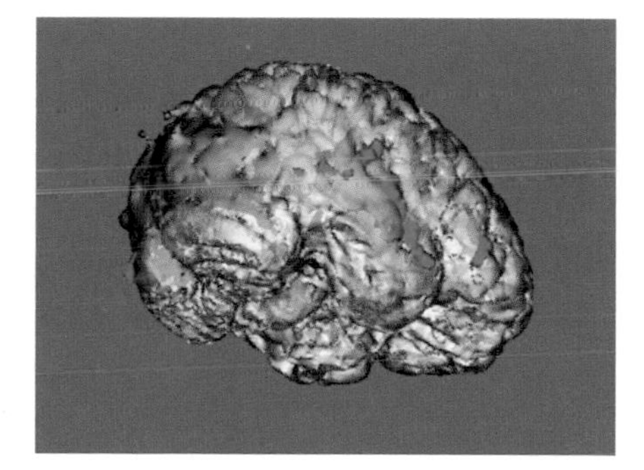

This functional MRI scan shows up areas of the brain that are active as a patient carries out a specific task. This patient was reading out loud.

Questions

1 What is your brain made up of?

2 Why is a complex brain so important for survival?

3 What four functions of the brain happen in the cerebral cortex?

4 Describe three methods that scientists have used to map the cortex.

5 Explain how the structure of your brain gives it such a large surface area.

6 Explain why it is necessary for blood flow to increase to parts of the brain that are very active.

7 Compare the diagram of the brain with the functional MRI scan. How can you tell that the person was reading aloud?

Find out about

- how conditioned reflexes can help you survive

F Learned behaviour

The lion cub below is just a few weeks old. She was born with reflexes that are helping her to stay alive. But much of her behaviour for example, how to hunt, or how to get on with other lions in the pride, she will learn from her mother. This is learned behaviour.

Learned behaviour is just as important for this lion cub's survival as reflexes.

Being able to learn new behaviour by experience is very important for survival. It means that animals can change their behaviour if their environment changes.

Conditioned reflexes

In 1904 the Russian scientist Ivan Pavlov won a Nobel Prize for his study on how the digestive system works. In his research Pavlov trained a dog to expect food whenever it heard a bell ring. The diagrams on the left explain what happened.

Pavlov's experiment

Adding a stimulus that produces the same response as a reflex action is a type of learning called **conditioning.**

Pavlov's dog salivated when presented with food.

The food is the stimulus and salivation is the response.

Pavlov rang a bell while his dog was eating its food.

After a while the dog salivated when it heard the bell, even if no food was around.

The stimulus of hearing the bell became linked with food.

Learning to link a new stimulus with a reflex action allows animals to change their behaviour. This is called a **conditioned reflex.**

Conditioning aids survival

Conditioned reflexes can help animals survive. For example, bitter-tasting caterpillars are usually brightly coloured. A bird that tries to eat one learns that these bright colours mean that caterpillars will have a nasty taste. After a first experience the bird responds to the colours by leaving them alone. So this helps the caterpillars survive.

'Warning' colours protect this caterpillar from predators.

If the brightly coloured insect is also poisonous, this reflex will help the bird survive as well. If other very tasty insects have similar colours and patterns, the bird does not eat them because of this conditioned response. You might have been caught out by this too – harmless yellow-and-black striped hover flies sometimes alarm people who have been stung by a wasp.

Conditioning your pet

Open a can of soup in your kitchen. If you have a dog or a cat, this sound may get them very interested. But they are not hoping for soup! The animal's reflex response to food has been conditioned. It has learned through experience that the sound of a tin being opened may be followed by food being put into its dish.

If a cat only uses its basket when you are taking it to the vet for an injection, it may become conditioned to link the basket with a frightening experience. The cat will then always be frightened by the stimulus of the basket. It will fight to keep out of the basket, even if you are only trying to take it to a new home.

Goldfish become conditioned to expect food when they see you in the room. They swim to the front of the bowl when you appear. The goldfish are linking the stimulus of seeing you with the original stimulus – food in the water.

Questions

1 Draw a flow diagram to explain how a cat can become conditioned to expect food when it hears a bathroom shower being run.

Use the key words from this section in your answer

2 Adverts often have glamorous, funny, or exciting images and catchy tunes. Write down a list of photos and tunes from adverts that remind you of things you could buy. How is conditioning involved in making us more likely to buy these products?

Key words

conditioning
conditioned reflex

Find out about

- how human beings learn new things
- explanations that scientists have for how your memory works

G Human learning

When humans and other mammals experience something new, they can develop new ways of responding. Experience changes human behaviour, and this is called learning.

Intelligence, memory, consciousness, and language are complex functions carried out by the outer layer of the brain, which is called the cerebral cortex. These functions are all involved in learning.

How does learning happen?

Neurons in your brain are connected together to form complicated **pathways**. How do these pathways develop? The first time a nerve impulse travels along a particular pathway, from one neuron to another, new connections are made between the neurons. New experiences set up new neuron pathways in your brain.

If the experience is repeated, or the stimulus is particularly strong, more nerve impulses follow the same nerve pathway. Each time this happens, the connections between these neurons are strengthened. Strengthened connections make it easier for nerve impulses to travel along a pathway. As a result, the response you produce becomes easier to make.

The brains of human babies develop new nerve pathways very quickly. Your brain can develop new pathways all your life. This means you can still learn as you get older, though more slowly

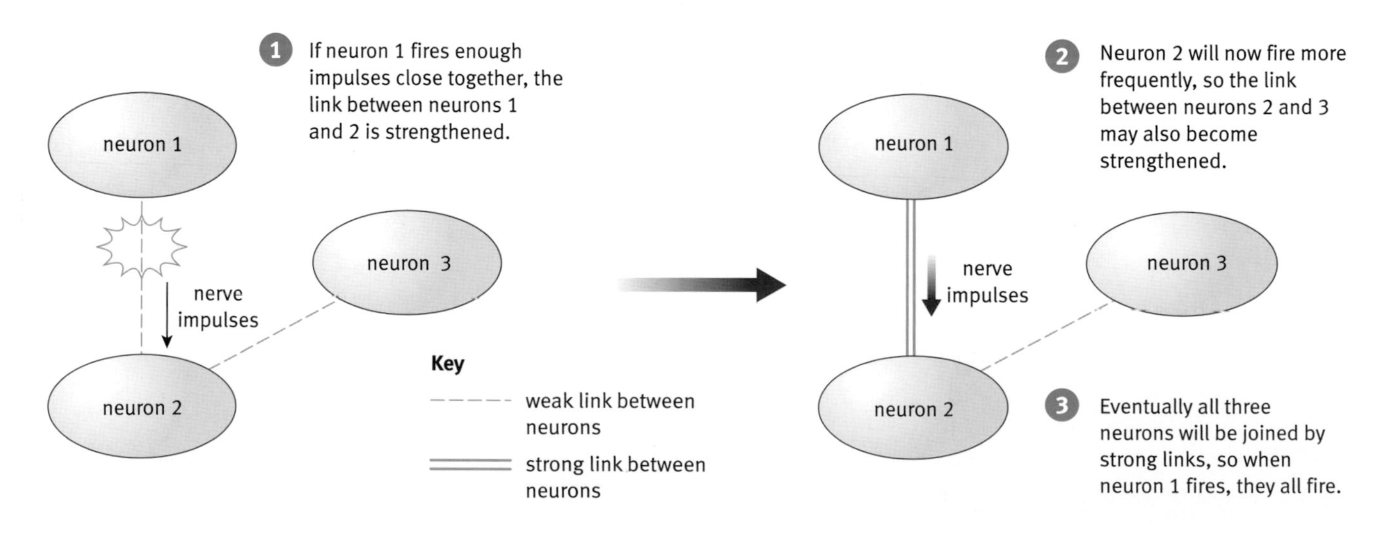

Nerve pathways form in a baby's brain as a result of a stimulus from its environment. Repeating the stimulus strengthens the pathway. The baby then responds in the same way each time it receives the stimulus. Some neurons in the brain do not take part in any pathway. Many of these unused neurons are destroyed.

Repetition

Repetition helps you learn because it strengthens the pathways the brain uses to carry out a particular skill. Perhaps you have learned to ride a bicycle, play a musical instrument, perform a new dance sequence or touch type. To do these things you created new nerve pathways then strengthened them through repetition.
This made it easy for you to respond in the way that you practised.

For example, Marie is a gymnast. When she has to learn new movements she stands still and imagines going through the motions – the position of her body and muscles being used at each stage. Visualization works because thinking about using a muscle triggers nerve impulses to that muscle. This strengthens the pathways the impulse takes. After a period of visualization, the actual movement is a lot easier to perform.

Marie visualizes new movements to help her learn them.

Age and learning

You learn to speak through repetition because you are surrounded by people talking. Children learn language extremely easily up to the age of about eight years. Their brains easily make new neuron pathways in the language processing region. As we get older it becomes harder for this part of the brain to make new pathways.

Feral children

In 1799, in southern France, a remarkable creature crept out of the forest. He acted like an animal but looked human. He could not talk. The food he liked and the scars on his body showed he had lived wild for most of his life. He was a wild, or **feral**, child. The local people guessed that he was about twelve years old and named him Victor.

Victor was taken to Paris. He lived with a doctor who tried to tame him and teach him language. At first people thought Victor had something wrong with his tongue or voice box. He could only hiss when people tried to teach him the names of objects. He communicated in howls and grunts.

Victor never learned to say more than a couple of words. By the time he was found, the time in his development when it was easy to learn language had passed.

Key words

neurons
pathways
repetition
feral

Questions

1 Write a few sentences to explain how you learn by experience Use the key word on this page in your answer

2 Explain why repeating a skill helps you learn it.

3 Write a list of skills you could practise by visualization.

Find out about:

- short-term and long-term memory
- the multi-store model of memory
- the working-memory model

H What is memory?

Psychologists are scientists who study the human mind. They describe **memory** as your ability to store and retrieve information.

Short-term and long-term memory

Read this sentence:

- As you read this sentence you are using your **short-term memory**.

  Short-term memory lasts for about 30 seconds in most people. If you have no short-term memory you will not be able to make sense of this sentence. By the time you get to the end of the sentence, you will have forgotten the beginning.

Think about a song you know the words to:

- To remember the words you use your **long-term memory**.

  Verbal memory is *any information* you store about words and language. It can be divided into short and long-term memory. Long-term memory is a lasting store of information. There seems to be no limit to how much can be stored in long-term memory. And the stored information can last a lifetime.

Different memory stores work separately

People with advanced Alzheimer's disease suffer short-term memory loss. They cannot remember what day it is, or follow simple instructions. But they may still remember their childhood clearly.

Some people lose long-term memory because of brain damage or disease. Their short-term memory is normal. This evidence is important because it shows that long-term and short-term memory must work separately in the brain.

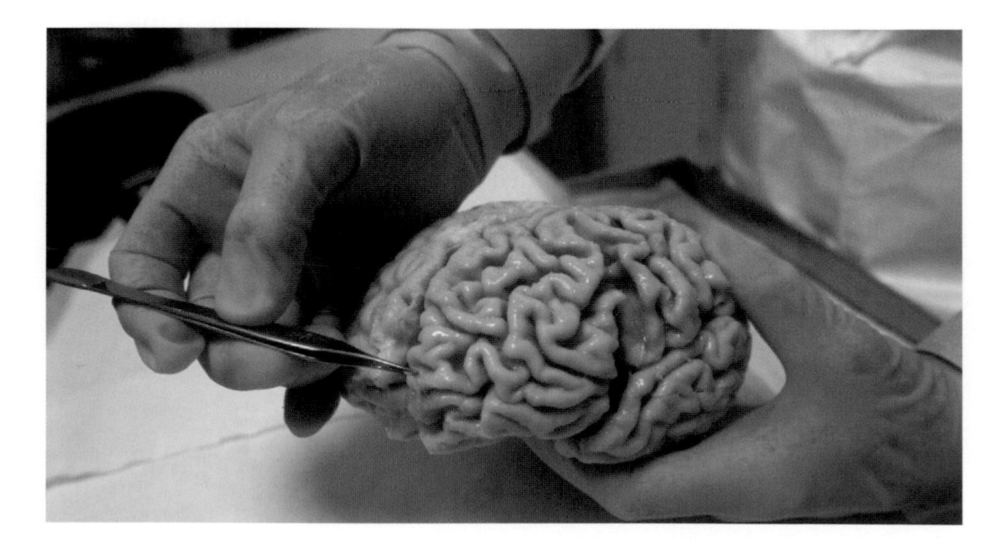

The 'Nun Study' at the University of Kentucky, USA, has had the participation of 678 School Sisters of Notre Dame. They ranged in age from 75 to 106 years. The sisters have allowed scientists to assess their mental and physical function every year and to examine their brains at death. The study has led to significant advances and discoveries in the area of Alzheimer's disease and other brain disorders.

Sensory memory store

You can also use a sensory memory store to store sound and visual information for a short time. When you wave a sparkler on bonfire night it leaves a trail of light. You can even write shapes in the air that other people can see. You see the trail because you store each image of the sparkler separately for a short time. The ability to store images for a short time makes the separate pictures in a film seem continuous. You can store sound temporarily in the same way.

Your sensory memory stores each image of the sparkler separately for a short time. This makes the whole shape seem continuous.

Questions

1. Write down one sentence to describe memory.
2. What is the difference between short-term and long-term memory?
3. Explain why a person with advanced Alzheimer's disease is unable to do simple things like go shopping or cook for themselves.
4. Give one piece of evidence that short-term and long-term memory are separate.

Key words

memory
short-term memory
long-term memory

How much can you store in your short-term memory?

Cover up the list of letters below with a piece of paper. Move the paper down so you can see just the top row. Read through the row once, then cover it up and try to write down the letter sequence. Then go down to the next row and do the same. Find out how many letters you can remember in the correct order.

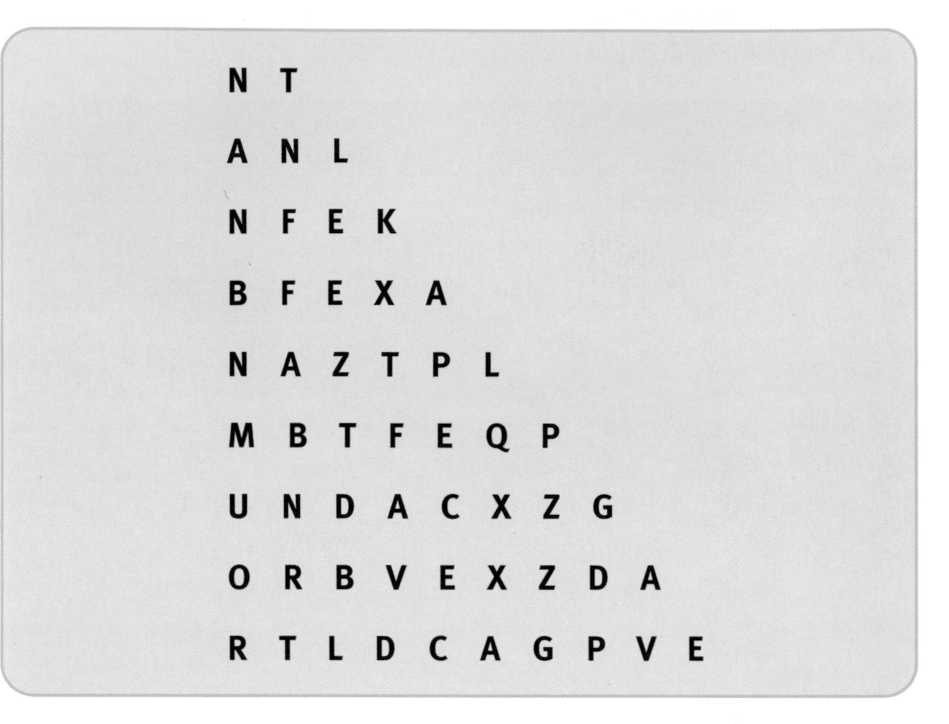

If you remembered more than seven letters in a row correctly, then you have excellent short-term memory. Short-term memory can only store about seven items. When you are remembering letters in a list, each letter is an 'item'. To remember more letters, chunk them into groups.

For example, the row O R B V E X Z D A has nine letters. Chunk these into groups of three: 'ORB' 'VEX' 'ZDA'.

The nine letters are easy to remember because now they are only three items. Three items doesn't overload your short-term memory

Models of memory

Key words

models of memory
multi-store model

Trying to remember word lists is a way of testing your memory. Memory tests can tell us what memory can and cannot do. But they do not explain how the neurons in the brain work to give you memory. Explanations for how memory happens are called **models of memory**.

The multi-store model: memory stores work together

Read through the list of words below once. Then cover the page and try to write down as many of the words as you can remember. They can be in any order.

dog, window, film, menu, archer, slave, lamp, coat, bottle, paper, kettle, stage, fairy, hobby, package

How many did you remember? If this type of test is carried out on large numbers of people, a pattern is seen in the words they recall. People often remember the last few words on the list and get more of them right. They also recall the first few words on the list quite well.

When you look at a list of words:

- Nerve impulses travel from your eyes to your sensory memory.
- Some sensory information is passed on to your short-term memory. Only the information you pay attention to is passed on. You will not be able to remember words you have not noticed.
- If more information arrives than the short-term memory can hold then some is lost (forgotten). You will not remember these words either.
- Some information is passed to your long-term memory. These are words you will remember – usually the first few words on the list.
- The last information your short-term memory receives will still be there when you start to write down the list. So these are also words you will remember usually the last few words on the list.

This use of sensory, short-term and long-term memory is known as the **multi-store model of memory**.

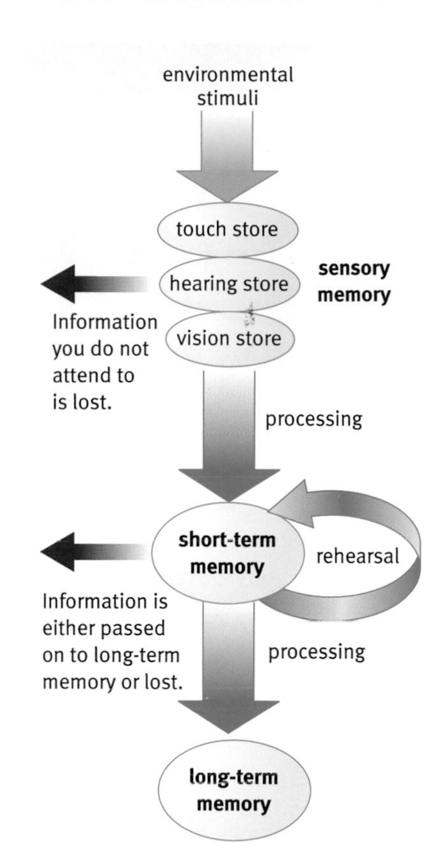

The multi-store model of memory can be used to explain how some information is passed to the long-term memory store and some information is lost.

Questions

5 You read the menu on a board inside a café. When you try to tell your friend sitting outside all the choices, you forget some. Why can you not remember everything on the list?

6 'So far none of the models of memory can explain completely how your memory works.' What must an explanation do to be accepted by scientists?

Rehearsal is one technique actors use to learn their lines

Rehearsal and long-term memory

Look at this row of letters:

R T I D A C G P E V

There are too many letters in this row for you to store them separately in your short-term memory. Given time you would probably repeat the letters over and over until you remembered them. Rehearsal is a well-known way of memorizing things. An actor can memorize a sonnet (a fourteen-line poem) in around 45 minutes. Psychologists think that rehearsal moves information from your short-term memory to your long-term memory store.

The working-memory model

In 1972, two psychologists, Fergus Craik and Robert Lockhart concluded that the multi-store memory model was too simple. They suggested that rehearsal is only one way to transfer information from short-term to long-term memory.

Rehearsed information is processed and stored rather than lost from short-term memory. Craik and Lockhart argued that you are more likely to remember information if you process it more deeply. They suggested that if you understand the information, or it means something to you, you will process it more deeply.

For example, if you can see a pattern in the information, you process it more deeply. So:

AAT, BAT, CAT, DAT, EAT

is much easier to remember than:

DAT, AAT, EAT, CAT, BAT

You also process information more deeply if there is a strong stimulus linked to the information, for example, colour, smell, or sound.

An active working memory

Short-term memory is now seen as an active '**working memory**'. Here you can hold and process information that you are consciously thinking about. Communication between long-term and working memory is in both directions. This way you can retrieve information you need, and also store information you may need later.

Putting it into practice

You can apply what the psychologists have discovered to your own school work.

- *Repetition:* If you are struggling to remember a piece of information you have read, read it several times.
- *Rehearsal:* Read sections of what you have to learn that are short enough to keep in your short-term memory. Make notes from memory to help move the information to your long-term memory.
- *Active memory:* Use highlighter pens to pick out key facts. Add to your old notes as you learn new topics.

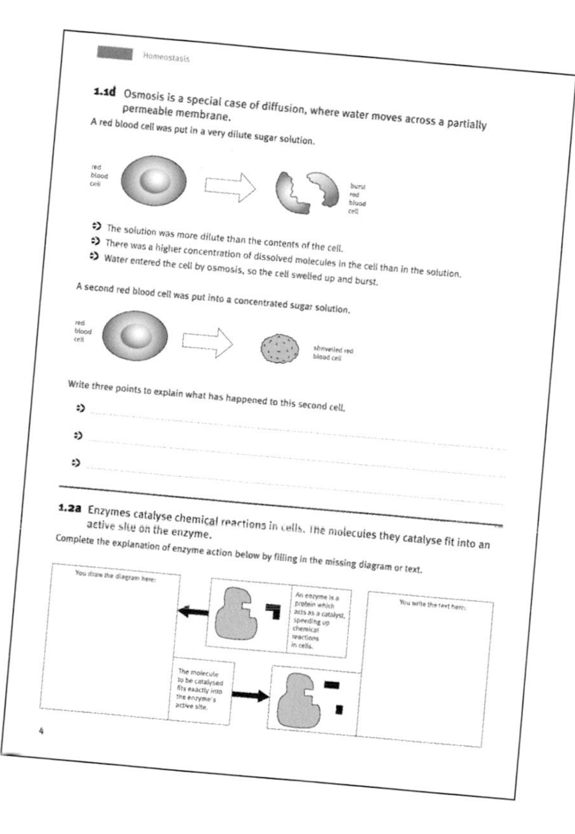

Homeostasis

1.1d Osmosis is a special case of diffusion, where water moves across a partially permeable membrane.

A red blood cell was put in a very dilute sugar solution.

red blood cell → burst red blood cell

- The solution was more dilute than the contents of the cell.
- There was a higher concentration of dissolved molecules in the cell than in the solution.
- Water entered the cell by osmosis, so the cell swelled up and burst.

A second red blood cell was put into a concentrated sugar solution.

red blood cell → shrivelled red blood cell

Write three points to explain what has happened to this second cell.

- ....
- ....
- ....

1.2a Enzymes catalyse chemical reactions in cells. The molecules they catalyse fit into an active site on the enzyme.

Complete the explanation of enzyme action below by filling in the missing diagram or text.

You draw the diagram here:

An enzyme is a protein which acts as a catalyst, speeding up chemical reactions in cells.

You write the text here:

The molecule to be catalysed fits exactly into the enzyme's active site.

4

Ben is doing homework. He uses both his long-term and working memory.

Key words

working memory

Questions

7 Make two lists of 10 different things to buy from a supermarket. Try to remember one list. Put the second list into 'families' e.g. tins, bakery, cleaning, and memorize it. Which list is easier to remember and why?

8 Give an example of something you can remember because a strong stimulus is linked to the information:

a a colour **b** a smell **c** a sound

9 Write down an example of something you have memorized through rehearsal, for example the directions to the cinema, or a complicated set of moves in a computer game. How has rehearsal helped you to remember this?

10 Explain why using a highlighter pen to pick out key facts makes your revision more successful.

B6 Brain and mind

Summary

In this module you have explored how your nervous system controls your responses to changes in the environment. These may be simple responses or involve more complex behaviour.

A mammal's nervous system

- The nervous system is made up of the central nervous system (CNS) and peripheral nervous system.
- The CNS coordinates an animal's responses via sensory and motor neurons.
- Receptors and effectors can be single cells or parts of complex organs.

Nerve impulses

- The structure of a neuron means that it can carry electrical impulses.
- Neurons are separated by tiny gaps called synapses.
- Some drugs and toxins affect the transmission of impulses across synapses.
- Impulses pass across synapses via chemical transmitters, which bind to specific receptor molecules. H

Reflex actions

- Living things respond to stimuli to help them survive.
- Simple reflexes travel through reflex arcs.
- Animals that only have simple behaviour are at a disadvantage because they cannot respond to new situations.
- Conditioned reflexes increase an animal's chance of survival. H
- The brain can modify some reflex responses. H

The brain

- The human brain has billions of neurons.
- New neuron pathways form during development.
- During learning some pathways become more likely to transmit impulses than others, so some skills may be learnt through repetition.
- The variety of potential pathways in the brain means that mammals can adapt to new situations.
- Some learning can only happen at particular times in development. H

Memory

- Scientists have used different methods to map the brain's cerebral cortex.
- Memory is the storage and retrieval of information.
- Memory can be divided into short-term and long-term memory.
- Scientists have suggested several models to explain memory, but so far none of them can account for all the observations.
- Humans are more likely to remember information if they can see a pattern in it. H

Questions

1 A car is waiting at the traffic lights. As soon as the light turns green, the driver's leg muscle moves his foot. He presses on the accelerator and the car moves forward.

a Name the stimulus, the receptor, and the effector in this response.

b As the driver accelerates, some dust blows into the car, making him sneeze. Explain why sneezing is an example of a reflex action.

2 All living things use reflexes. Simple animals use reflexes to control all their behaviour. But more complex animals also make conscious decisions.

a Name two reflexes seen in human babies.

b Name two reflexes also seen in human adults.

c Explain why reflexes are important for survival. Use these key words in your answer: *stimulus*, *involuntary*, *response*

d Explain the disadvantage of having only reflex behaviour.

3 A person stands on a drawing pin. Pain receptors in the skin detected the stimulus. They quickly move their foot upwards.

a How does this reflex help the person to survive?

b Draw a diagram to describe this reflex arc. Label the diagram with notes to explain what is happening at each stage. (The central nervous system in this reflex is the spinal cord.)

4 **a** Draw a labelled diagram of a motor neuron.

b Describe two jobs done by the fatty sheath.

c Name the gap between two neurons.

d Explain how nerve impulses pass from one neuron to the next. H

5 Many different kinds of nasty-tasting caterpillars have black-and-yellow markings. H

a Explain how birds learn to avoid eating prey with black-and-yellow markings. H

b Suggest what might happen if a nice-tasting caterpillar had black-and-yellow markings. H

6 Your brain has billions of neurons. It allows you to learn from experience.

a Explain how you learnt a new skill, e.g. riding a bike. Include these key words in your answer: *neuron pathway*, *impulses*, *repetition*

b Suggest why it is easier to learn some skills at a certain age. H

7 **a** Name four functions of the cortex in the human brain.

b What methods have scientists used to find out this information?

8 **a** What memory are you using to:

i read this sentence?

ii remember the words of a song?

b Scientists have come up with several models to describe how memory works.

i Describe one of these models.

ii Explain why none of the models scientists have come up with so far is a full explanation of memory.

Why study biology?

Biology is the study of living things. Biologists explore all aspects of life – from the chemical reactions of individual cells to the interactions within an ecosystem. Astrobiologists are involved in the search for life beyond Earth. Knowledge of biology informs any science which affects living things, such as the medical and environmental sciences. Engineers sometimes copy design solutions from the natural world, and modify natural organisms to make products we can use.

The science

No organism lives in isolation. Understanding life processes such as photosynthesis and respiration helps to explain how living things interact with each other. Some biologists study human body systems, and develop new technologies with the potential to improve our quality of life.

Biology in action

Biotechnologists modify microorganisms to produce medicines, and engineer plants to give higher yields. Geneticists analyse genes to find family links, and look for genes which affect our risk of certain diseases, or which may affect our response to certain medicines. Sports scientists help people maintain good health and improve their fitness. And if you suffer a sporting injury, physiotherapists treat damage to the skeleton and muscles.

Biology across the ecosystem

Find out about:

- how living things depend on other species for survival
- photosynthesis and heterotrophic nutrition
- modern genetic technologies
- respiration – releasing energy from food
- transport around the human body
- the structure of the human skeleton and how it moves

Topic 1 Interdependence

Wherever you go on Earth, you are never alone. Other species are everywhere – in the rock and soil beneath your feet, the air around you, even inside and on the surface of your body. You depend on some of these organisms for your survival. Some of them depend on you. This is the **interdependence** of life.

Life exists in even the harshest of environments.

Ultimate dependence – the Sun

All life on Earth depends on the Sun. It provides energy

- to keep the Earth's atmosphere warm
- to drive the production of food chemicals

All living things need a continual supply of energy. Plants capture energy from the Sun during photosynthesis. They use photosynthesis to produce food chemicals. This energy flows through the rest of the ecosystem as the plants are eaten or decompose.

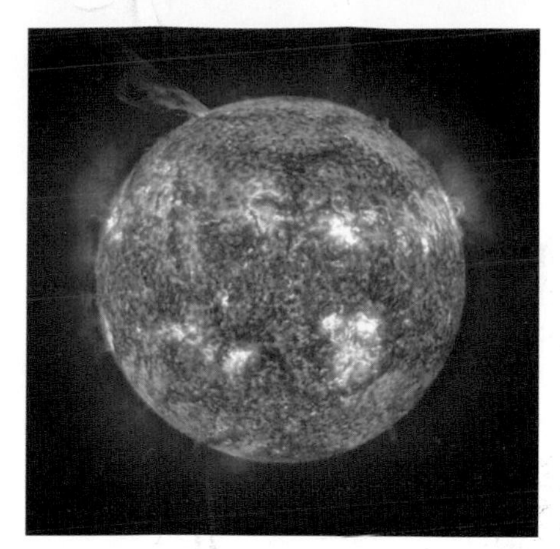

The Sun gives out 386 billion billion megawatts of energy every second. It would take 2500 million of the largest power stations on Earth a year to generate this amount.

Webs of life

The need for energy from food is behind many interactions between living things. Food webs show the story of food in an ecosystem – what eats what. This helps us understand more about how organisms affect each other.

Energy passes from the rose bush to the aphids and then to the ladybird.

Nature's recycling

Living things are made up of a rich diversity of chemicals. Many of these are **organic** compounds, such as glucose. Organic compounds contain two or more carbon atoms. Almost all organic compounds are made by living things. Biochemists study biological molecules, and develop processes to make them artificially.

When living things die, the carbon in their bodies must be recycled to produce the bodies of new organisms. Microorganisms play a crucial role in recycling carbon and other elements on Earth.

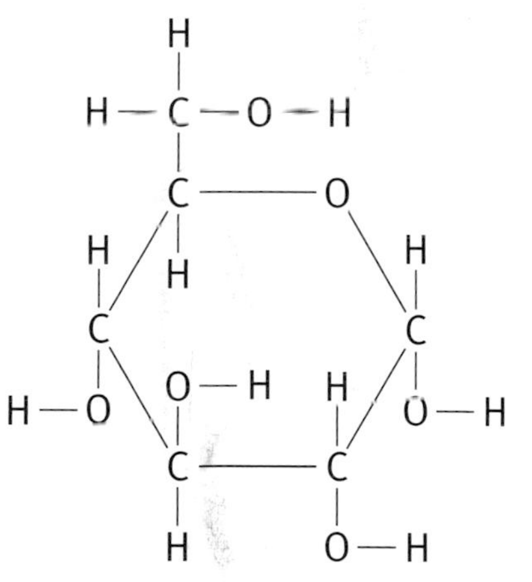

The formula for a glucose molecule

Key words

interdependence
organic

Find out about:

- how photosynthesis captures energy for life on Earth

1A Harvesting the Sun

How is the Sun's energy harnessed by living organisms?

Plants capture energy from sunlight. This may look as easy as sunbathing, but the process is one of nature's cleverest tricks. Many complex chemical reactions are involved.

This sunbather is enjoying warmth from the Sun, but he does not rely directly on the Sun's energy for his life processes. The lizard is using energy from the Sun to raise its body temperature. The plant is harnessing light energy to drive food production.

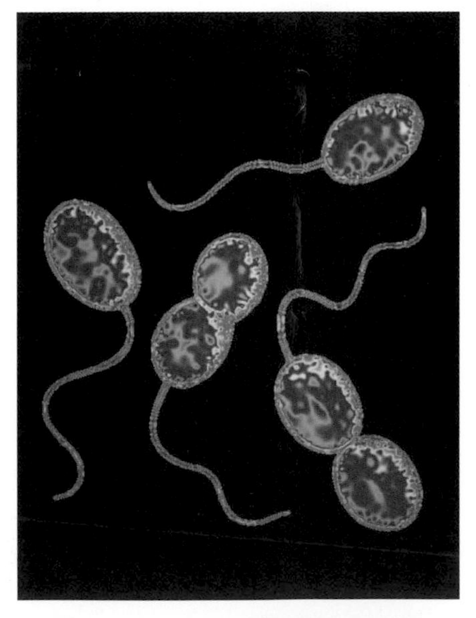

This bacterium uses sulfur from the environment for chemical reactions that release energy. (× 7500)

Plants use energy from sunlight to build up carbon dioxide and water into organic compounds such as glucose. This is the process of **photosynthesis**. Energy from sunlight is transferred to chemical energy in the new compounds. So these compounds act as an energy store. They can be used as food molecules which are broken down to release energy. They also provide raw materials to build new plant cells for growing.

Living things that can make their own food are called **autotrophs**. Most autotrophs are types of plant, but there are some bacteria which also make their own food chemicals. These bacteria do not use energy from light to drive this process. Instead they obtain energy from chemical reactions using raw materials from their environment.

Most living things cannot make their food

Animals have to eat. They cannot make their own food, so they need to take in organic molecules. Organisms that rely on other living things for food are called **heterotrophs**.

Just passing through

Some plant material passes to other organisms as they eat the plant. Energy stored in the chemicals making up the plant cells is passed on to other organisms along food chains. In this way energy moves through the ecosystem. This is similar to how elements like carbon and nitrogen move through an ecosystem. But there is a big difference.

Carbon and nitrogen are always being recycled in an ecosystem. When an animal or plant dies, its organic compounds are broken down by decay. The carbon and nitrogen atoms become part of a new organism in the cycle.

When energy is passed through an ecosystem, some is lost at every stage of a food chain. This is because all living things release energy from food by respiration. Land plants use half the organic compounds they make for respiration. Some of the energy released in respiration is transferred to the environment. This energy warms the air or water around the plants and animals, and eventually radiates back into space.

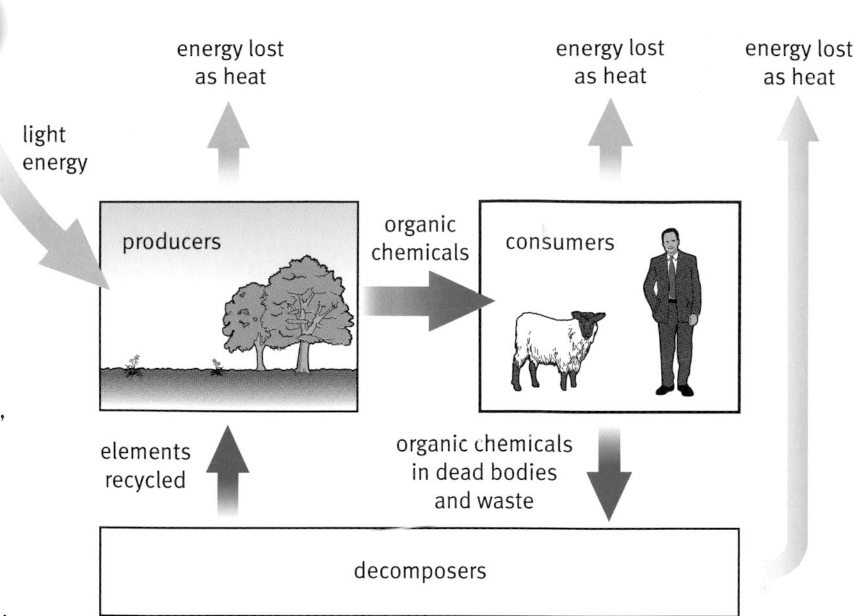

Transfer of energy and elements through an ecosystem

Questions

1 Give two reasons why sunlight is essential for life on Earth.

2 Write down definitions for the terms
  a autotroph
  b heterotroph

3 Identify two autotrophs and two heterotrophs on this double page.

4 Explain how energy is transferred from
  a the grass to sheep
  b the tree to decomposers

(5) Some animals rely on energy from the Sun to raise their body temperature. Others, including human beings, do not use the Sun's energy in this way. Why is this?

Key words

organic
photosynthesis
autotroph
heterotroph

Find out about:

- how photosynthesis captures energy for life on Earth

1B Trapping light energy

Less than one-billionth of the Sun's energy reaches the plants on Earth. Plants convert only about 1–3% of this light energy into new plant material. This might sound very small, but remember that the Sun's energy output is enormous. So this 1–3% is still enough energy to drive life on Earth.

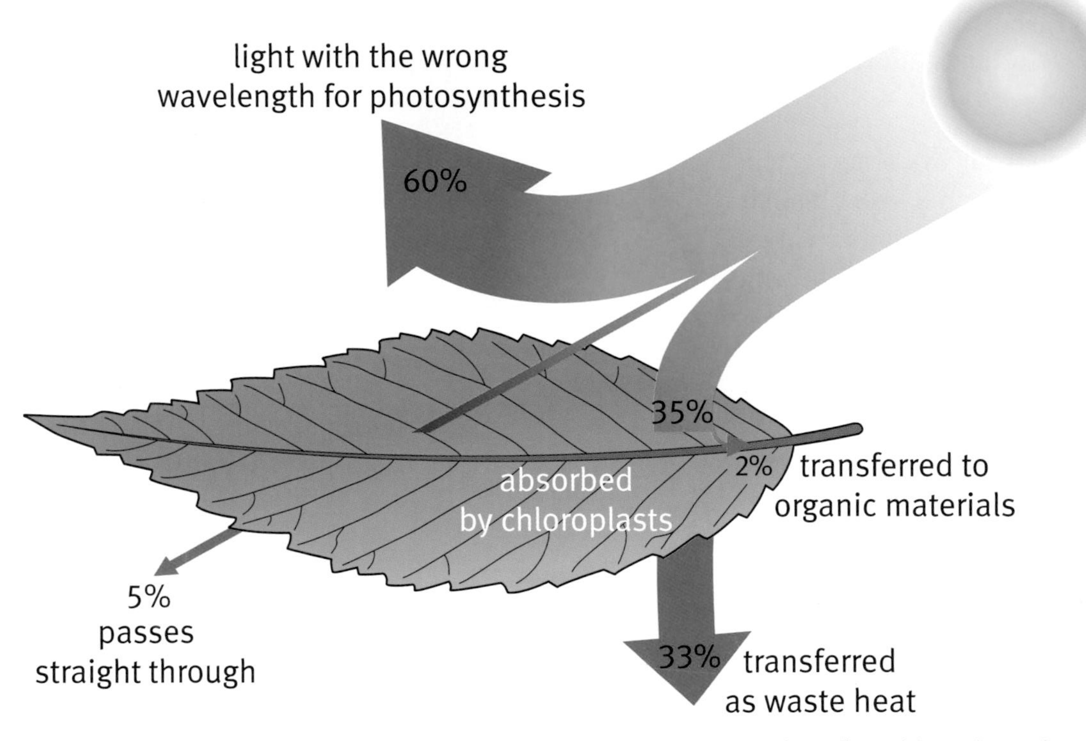

Most of the light energy reaching a leaf is reflected from the surface, is transferred as waste heat, or passes straight through the leaf.

What happens during photosynthesis?

The chemical equation for photosynthesis is:

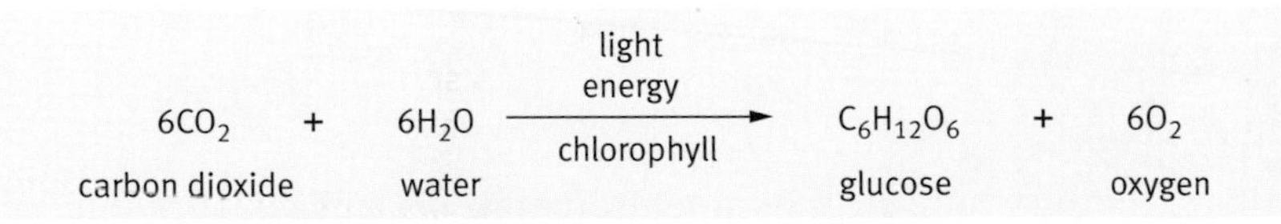

$$6CO_2 + 6H_2O \xrightarrow[\text{chlorophyll}]{\text{light energy}} C_6H_{12}O_6 + 6O_2$$

carbon dioxide + water → glucose + oxygen

Do not let the equation mislead you. The reaction does not happen in one go – it has lots of smaller steps. The equation is a convenient way of summing up the process.

A glucose molecule is made up of carbon, hydrogen, and oxygen atoms. So glucose is a **carbohydrate**.

Photosynthesis takes place in **chloroplasts**. They contain a green pigment called **chlorophyll**. Chlorophyll absorbs light and uses the energy to kick-start photosynthesis.

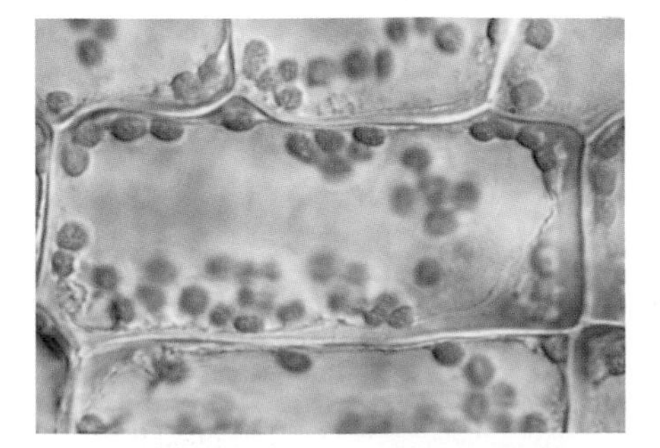

Chloroplasts contain the green pigment chlorophyll and the enzymes that are needed for photosynthesis. (× 2000)

The light energy splits water molecules into hydrogen and oxygen atoms. The hydrogen is combined with carbon dioxide from the air to make glucose. The oxygen is released as a waste product. It passes out of the plant into the air. Given enough raw materials, light, and the right temperature, a large tree can make 2000 kg of glucose in a day.

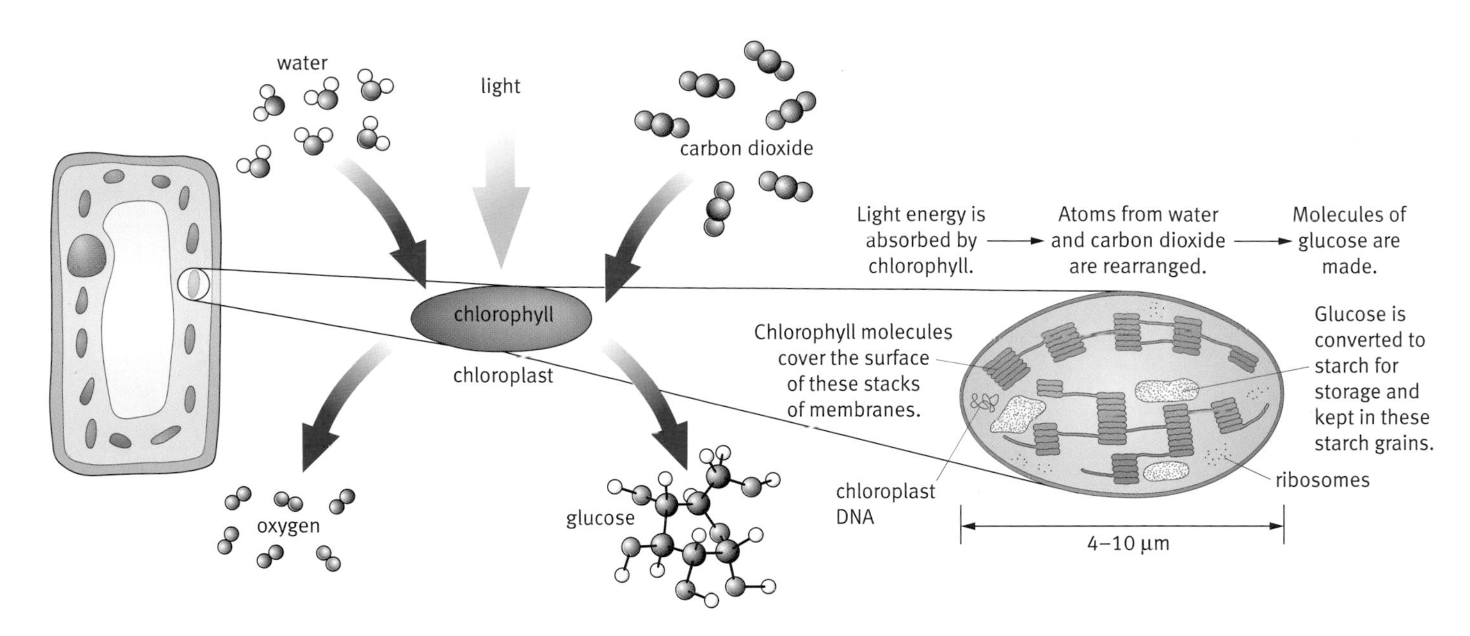

Glucose is made by photosynthesis.

Why are plants green?

Most leaves are green because of the chlorophyll in plant cells.

Chlorophyll absorbs energy from visible light. Visible light is part of the electromagnetic spectrum.

White light is split into its different wavelengths in a prism.

Not all of visible light can be used for photosynthesis. Chlorophyll absorbs red light and blue light. So these are the most useful parts of the light spectrum for photosynthesis. Green light is not absorbed. It is reflected by the chlorophyll. This is why most leaves are green.

Key words

carbohydrate
chloroplast
chlorophyll

Questions

1 Draw a diagram to show the flow of chemicals in and out of leaves during photosynthesis.

2 Describe the main stages in making glucose by photosynthesis.

3 Write down the word equation that sums up photosynthesis.

4 What property of glucose makes it a carbohydrate?

5 Explain why only a small percentage of light energy from sunlight is used in photosynthesis.

6 Explain why chlorophyll is green in colour.

Find out about:

- what hapens to the glucose made by photosynthesis

1C Using glucose from photosynthesis

Glucose made during photosynthesis is used by plant cells in three ways.

(1) Making other chemicals needed for cell growth

Like all larger living things, plants are made up of different types of cell. But every cell is made up of the same basic chemicals. The diagram below shows the chemicals in a typical plant cell.

Glucose has to be converted into the other chemicals needed for cell growth. Cells need other carbohydrates, as well as **fats** and **proteins**. Two important carbohydrates in plants are **cellulose** and **starch**. Both of these are large molecules made up of thousands of glucose molecules linked together. Large molecules made up of many smaller molecules of the same type linked together are called **polymers**. So cellulose and starch are both polymers of glucose.

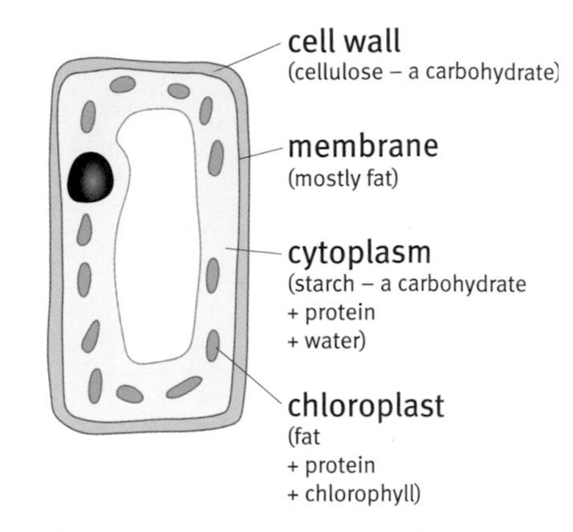

A plant cell contains many different chemicals.

Cellulose molecules are long straight chains of glucose molecules. These chains are linked together to make a strong material for cell walls.

(2) Storing energy in starch molecules

Sometimes photosynthesis produces glucose faster than the plant needs it at the time. This extra glucose is converted into starch. Starch is a storage molecule. When photosynthesis cannot keep up with the demand for glucose, the starch can be converted back. Starch may be stored in leaf cells, but some plants have special organs such as the tubers of a potato which have cells that are filled with starch.

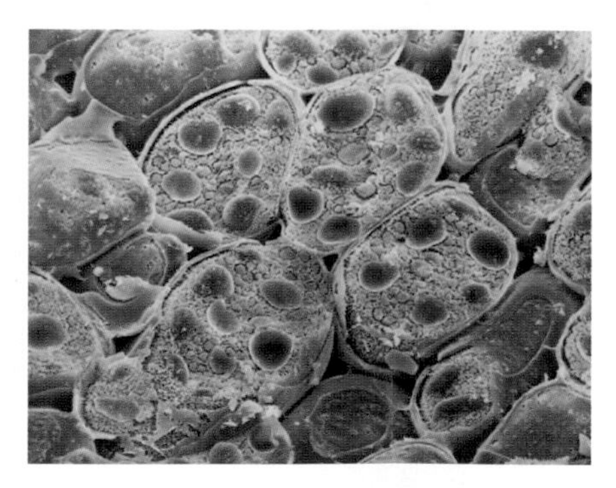

Starch grains in a plant cell store glucose as starch (× 200).

(3) Releasing energy in respiration

Plant cells use glucose in **respiration**. The glucose molecules are broken down, releasing the energy stored in the molecules. The equation for respiration is:

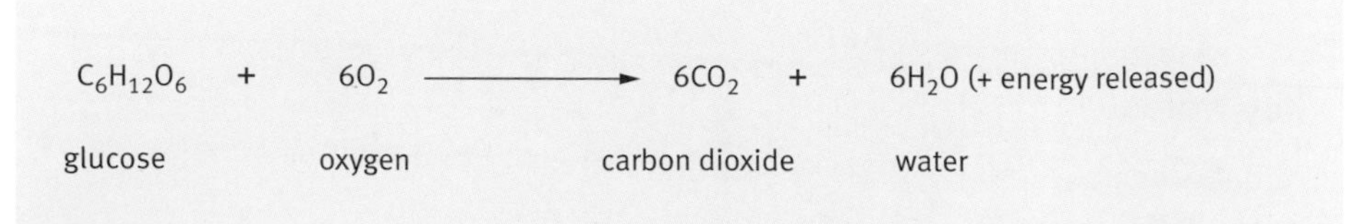

This energy is used to power chemical reactions in the cells, such as converting glucose to cellulose, starch, or proteins.

Making proteins needs nitrogen

On pages 136 and 142–3 you learnt about the structure of proteins. Remember that proteins are long chains of amino acids. Every amino acid contains atoms of carbon, hydrogen, oxygen, and at least one nitrogen atom. To make amino acids, nitrogen must be combined with carbon, hydrogen, and oxygen atoms from glucose.

Most of the Earth's nitrogen is in the air, but plants mainly take in nitrogen from the soil as **nitrate ions**. The nitrate ions are absorbed by the plant roots.

Nitrates are not the only minerals that plants need. For example, they need magnesium to make chlorophyll, and phosphates to make DNA. But they need nitrogen in the largest quantities, because proteins are used to build cells and to make enzymes.

Nitrates contain this group of atoms:

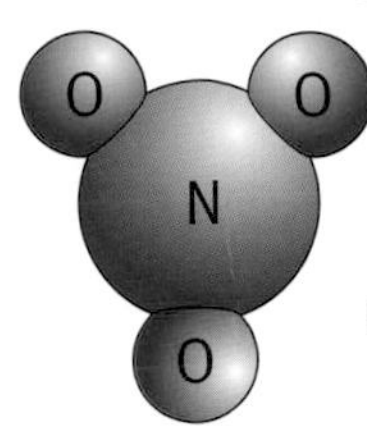

(This group of atoms has a negative electrical charge on it.)

Nitrate ions are found dissolved in soil water, and in rivers and seas.

H Nitrate ions are absorbed by active transport

Nitrate ions are at higher concentration inside root cells than outside. This means the ions will move by diffusion out of the cell, not into it. To move nitrate ions into the root cells, active transport is needed. This requires energy. You can read more about how this is done on page 110.

Why is glucose stored as starch?

Glucose is a soluble carbohydrate. Cell cytoplasm is mainly water. So glucose made by photosynthesis dissolves in the cytoplasm.

This causes a problem for cells. If the concentration of glucose in the cell cytoplasm becomes too high, then too much water moves into the cell. This happens by osmosis. So the glucose cannot be stored in cells because it upsets the cell's **osmotic balance**, as shown in this diagram:

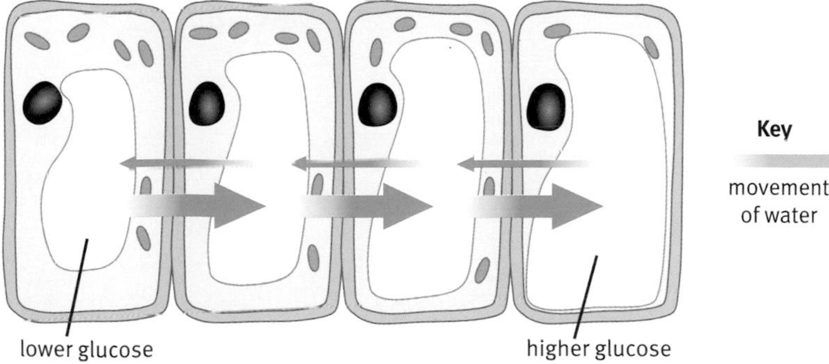

Too much dissolved glucose in a plant cell would cause overall water movement from neighbouring cells by osmosis.

Large carbohydrates like starch are insoluble. They have very little effect on the osmotic balance of a plant cell. This makes them ideal for storing glucose. The starch is kept in small membrane-surrounded bags in the cell called starch grains. You can read more about osmosis on page 109.

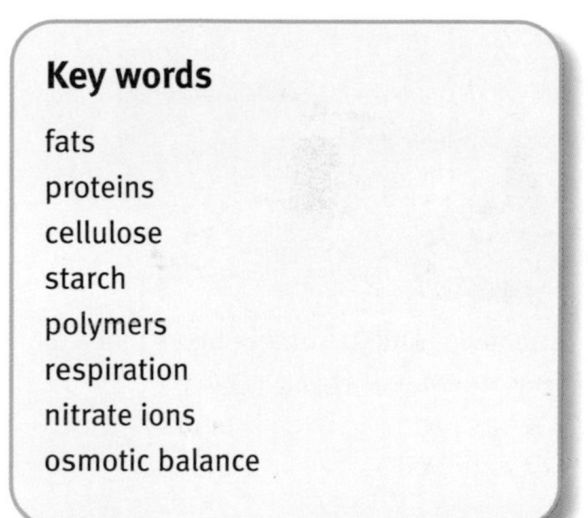

Key words

fats
proteins
cellulose
starch
polymers
respiration
nitrate ions
osmotic balance

Questions

1. Glucose from photosynthesis has three roles in the plant cell. Explain what these are.
2. Why do plant cells need a source of nitrate ions?
3. Explain why starch is needed to store glucose in plant cells.
4. (circled) Water moves into root cells by osmosis. What does this tell you about the water in the soil?
5. (circled) Leguminous plants, such as clover, are able to survive in soil low in nitrate ions. Explain why.

Find out about:

- what limits the rate of photosynthesis

1D The rate of photosynthesis

The conditions inside the greenhouse in the photograph on the left are kept under very careful control. The tomato plants growing here have the optimum conditions for photosynthesis. They are making glucose at their highest rate, so they are growing quickly. All this is planned by the farmer so that the **yield** from the tomato plants will be as high as possible. Yield is the amount of product the farmer has to sell.

Intensive tomato farming takes place all year round in this greenhouse.

All reactions speed up when the temperature rises, and photosynthesis is no exception. The greenhouse is kept warm at 26 °C. This is the optimum temperature for photosynthesis to take place in these plants.

Faster photosynthesis – light

Other factors have an effect on the rate of photosynthesis. Light energy drives photosynthesis, so increasing the amount of light a plant receives increases the rate of photosynthesis.

The diagram shows an experiment to investigate how changing light intensity affects the rate of photosynthesis in a piece of pondweed. The results from the experiment are shown in the graph. The graph shows that

- At low light intensities, increasing the amount of light increases the rate of photosynthesis.
- At a certain point increasing the amount of light stops having an effect on rate of photosynthesis.

barrel of syringe
plastic tube
board
plunger of syringe
lamp
thermometer
test tube
beaker of water (heat shield)
Oxygen from photosynthesis collects in the end of the capillary tube.
clamp
stand
The distance between the lamp and the pondweed can be changed and the light intensity measured with a light probe.
pond weed
After a set amount of time, the syringe is used to draw the gas into the capillary tube so the length of the bubble can be measured.

This experiment investigates the effect of light intensity on the rate of photosynthesis.

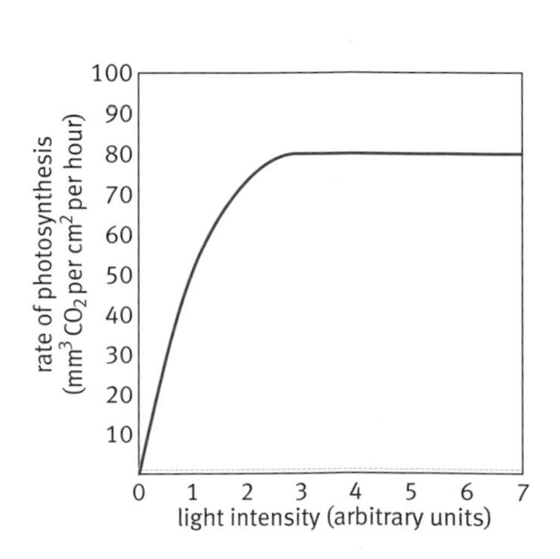

Why does the rate not keep on rising?

Photosynthesis needs more than light energy. Extra light makes no difference to the rate of photosynthesis if the plant does not have the carbon dioxide, water, or chlorophyll to use the energy to the full. The temperature must also be high enough for photosynthesis reactions to speed up. Increasing the light intensity stops having an effect on the rate of photosynthesis because one of these other factors is in short supply. This factor is called the **limiting factor**.

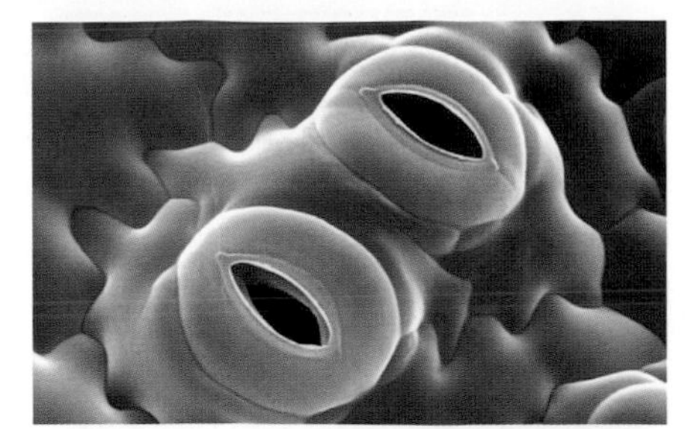

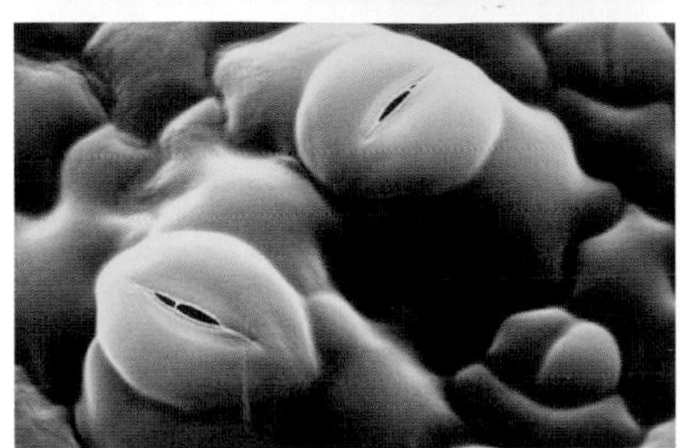

Top: The stomata on the underside of leaves open to allow gases to move in and out of the leaf. *Bottom:* They close to conserve water. (× 400)

H Limiting factors

In a British summer the limiting factor for photosynthesis is often water. Stomata close to prevent water diffusing out of the leaves, and this also has the unavoidable consequence of reducing carbon dioxide diffusion into the leaf.

The graph below shows the effect of increasing light intensity on the rate of photosynthesis at two different carbon dioxide concentrations. At 0.04% CO_2, more light increases the rate of photosynthesis up to a point, until light is no longer the limiting factor. Increasing the CO_2 level to 0.4% makes the rate of photosynthesis higher – carbon dioxide must have been the limiting factor. But even this graph levels off as another factor becomes in short supply.

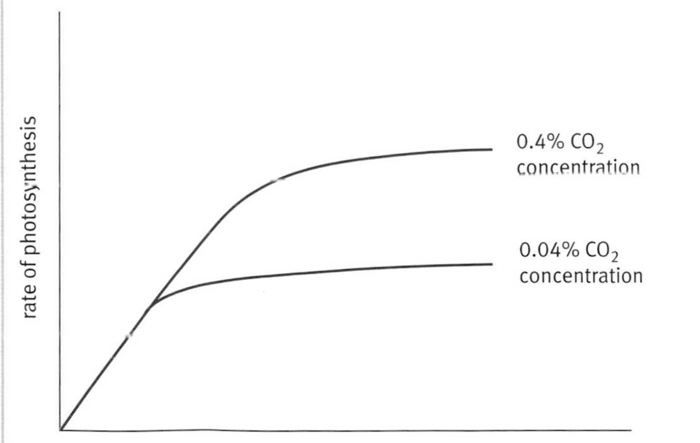

At the higher carbon dioxide concentration, photosynthesis takes place faster. But the rate still levels off. Another factor must be limiting photosynthesis.

Carbon dioxide levels in the greenhouse

Carbon dioxide forms 0.04% of normal air. Levels over about 1% are toxic to plants and animals. The levels in the tomato greenhouse are kept at 0.1%. Raising the concentration higher than this has no effect on the rate of photosynthesis. So it would not be cost effective for the farmer to add more carbon dioxide than this to the greenhouse.

Key words

yield
limiting factor

Questions

1 Write down four factors that can affect the rate of photosynthesis.

2 Explain what is meant by a limiting factor.

3 Suggest a factor that could be limiting bluebells growing on a woodland floor in spring.

4 a Sketch a graph showing how increasing light intensity affects the rate of photosynthesis.

b Label the graph line 0.04% CO_2.

c Draw a second line on the graph to show the rate of photosynthesis for the same plant at 0.1% CO_2.

Find out about:

- how respiration and photosynthesis occur together in a plant
- carbon dioxide levels in the atmosphere

1E Balancing respiration and photosynthesis

There is only a certain amount of carbon on Earth. Much of the carbon is in molecules that make up the bodies of living things. A lot is also in the atmosphere and oceans as carbon dioxide, and in molecules of fossil fuels.

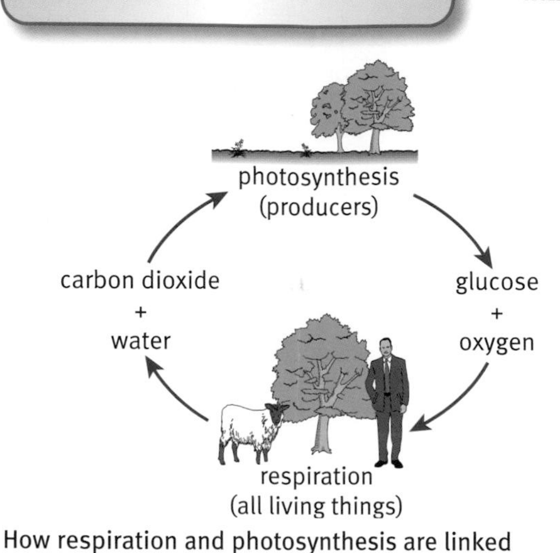

How respiration and photosynthesis are linked

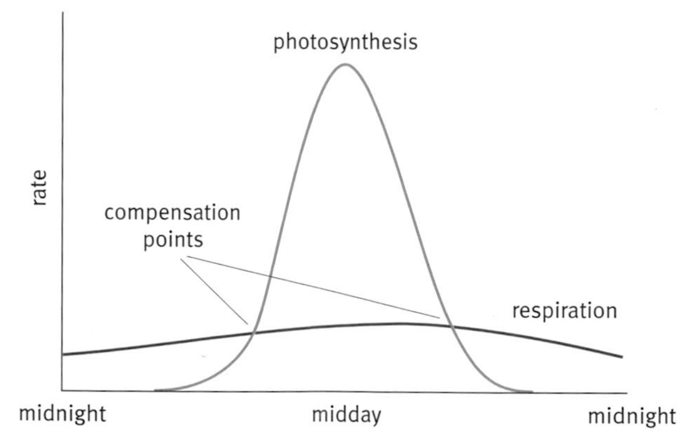

At the compensation points, there is no net movement of carbon dioxide into or out of the plant.

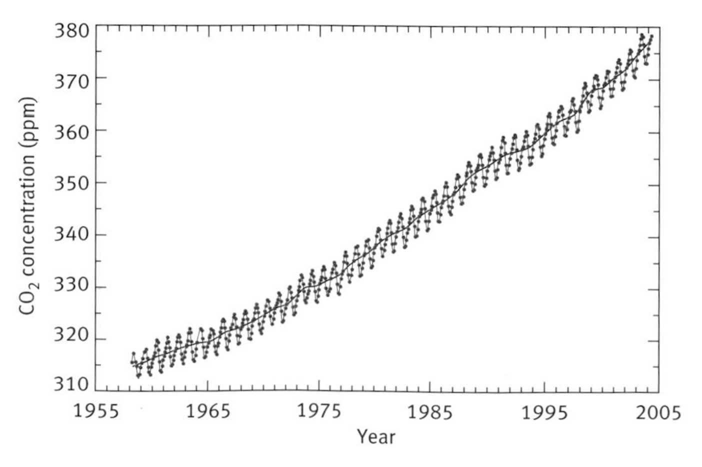

Carbon dioxide concentrations have been recorded at Mauna Loa in Hawaii since 1958. They rise and fall each year, but the overall trend has been an increase of about 1.5 ppm per year since 1980.

Carbon dioxide is taken out of the atmosphere by photosynthesis. The carbon is used to produce glucose molecules. The glucose is broken down during respiration. This releases carbon dioxide back into the atmosphere.

Compensation point

Plants only carry out photosynthesis when they are in light. But respiration happens 24 hours a day. So for part of each 24-hour period, plants are actually net producers of carbon dioxide:

- Plants take in carbon dioxide for photosynthesis during the day.
- They make carbon dioxide in respiration 24 hours a day.
- During the day they take in more carbon dioxide for photosynthesis than they make in respiration.

When photosynthesis and respiration are taking place at the same rate, glucose production and use in the plant are balanced. This is called a plant's **compensation point**.

Rising carbon dioxide levels

Carbon dioxide is also added to the atmosphere by the burning of wood and fossil fuels, such as oil, gas, coal, and petrol. The stages in recycling carbon on Earth are explained in the diagram of the **carbon cycle** on the opposite page.

If the amount of carbon dioxide released into the air does not balance the amount taken up by photosynthesis, atmospheric carbon dioxide levels will change.

Most scientists agree that the average carbon dioxide level in the atmosphere is rising. The official figure in 2005 was 0.04% carbon dioxide. It is expected to be 0.05% by the end of the century.

Why is the level of carbon dioxide rising?

Increasing carbon dioxide levels in the air are considered by most scientists to be due to human activity. The main sources of this additional carbon dioxide are the combustion of fossil fuels, cement manufacture, and loss of rainforests. Trees store carbon in their wood, and this is released as carbon dioxide when forests are cut down and burned. The loss of rainforest also reduces the plants available to take carbon dioxide out of the atmosphere for photosynthesis.

Why is the extra carbon dioxide not used up?

Some of the extra carbon dioxide from human activity is stored in the oceans, but almost half stays in the atmosphere. But this extra CO_2 is not used up by photosynthesis. If the Earth's vegetation behaved like plants in a greenhouse, the extra carbon dioxide would increase the rate of photosynthesis and so be used up. Tests have been carried out on maize, rice, soya bean, and wheat – the four main food crops. The results have shown a smaller increase in the rate of photosynthesis than expected with higher CO_2 levels. There were other factors limiting the rate of photosynthesis, so the plants could not use the extra carbon dioxide.

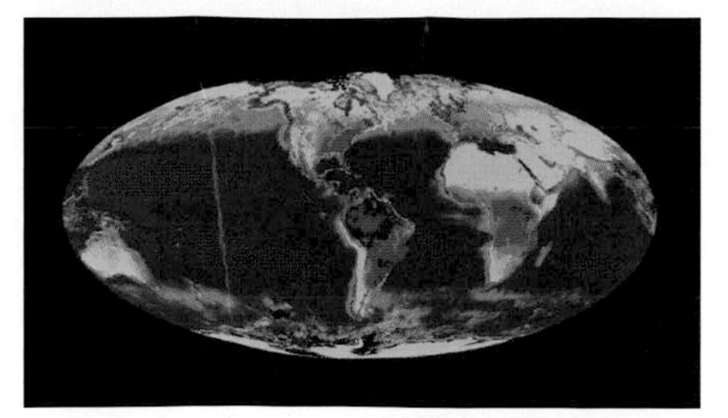

A false-colour satellite image showing the distribution of vegetation on land and phytoplankton in the oceans. The colours represent chlorophyll densities: from red (most dense) through yellow and blue to pink (least dense) in the oceans, and from dark green to pale yellow on land.

Key words

compensation point
carbon cycle

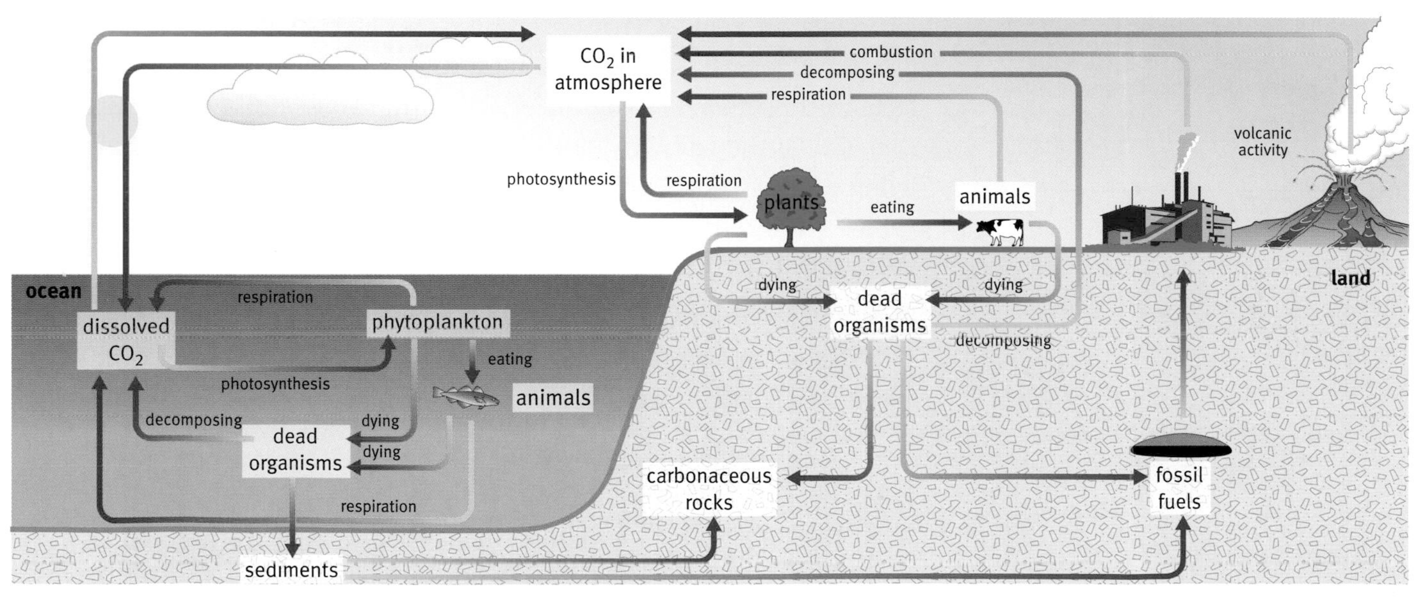

The carbon cycle

Questions

1 List ways in which carbon dioxide is

a added to the atmosphere

b taken out of the atmosphere

2 How is human activity causing the amount of CO_2 in the atmosphere to rise?

3 At what times of the day does a plant reach its compensation point?

4 Explain what is happening inside the plant at this point.

Find out about:

- how energy and nutrients move through food webs
- what pyramid diagrams show

1F Feeding relationships

Food chains

Food chains like the one on page 87 all follow the same pattern. They start with autotrophs. These are the **producers** which use carbon dioxide to make organic chemicals. These chemicals then pass to the heterotrophs. First herbivores eat the producers, and then the herbivores are eaten by carnivores.

One very important group of heterotrophs is often overlooked in a food chain. **Decomposers** may be smaller and less attractive than other organisms, but they do an essential job. A lot of nutrients and energy in an ecosystem pass through them.

Any ecosystem has many food chains interlinked in a food web.

Pyramid of numbers

One way of summing up how living things in an ecosystem are interlinked is to use a pyramid diagram.

The simplest diagram is a **pyramid of numbers**. The different living things in a given area are counted. They are grouped into different **trophic levels**: producers, primary **consumers** (herbivores), secondary consumers, and tertiary consumers. A diagram showing the number of living things at each level is then drawn. For example, a pyramid of numbers for all the living things in an open field is shown opposite.

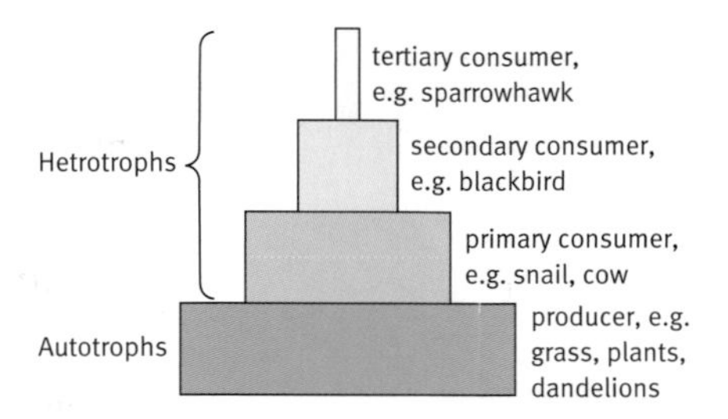

A pyramid of numbers.

Why a pyramid shape?

A pyramid of numbers usually shows more producers than herbivores. This makes sense because not all the nutrients in the producers end up in herbivores. So herbivores do not get all the energy stored in producers.

- Most herbivores don't eat the whole plant. For example, they may leave the roots.
- Not all of the plant material will be digested by the herbivores – some will pass out of their bodies in faeces.
- Plants use at least half of the glucose they make for respiration. Some of the energy released during respiration is used to make new plant cells. Energy stored in the chemicals of these cells is passed along the food chain to the herbivores. But some of the energy released in respiration escapes into the environment as heat energy, so this is not passed on to herbivores.

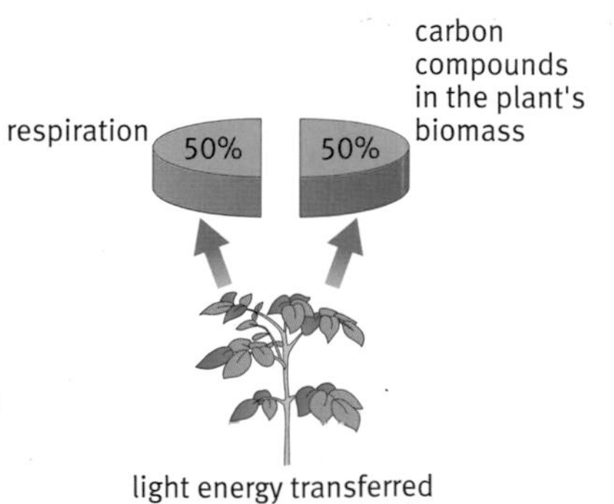

Plants use at least half the carbohydrate they make for respiration.

Energy use in consumers

Herbivores and carnivores break down food molecules in respiration as well. Some of this energy is used for growth, where food molecules become part of the structure of new cells.

Animals also use energy released by respiration for other life processes, for example, movement and keeping warm. So the number of organisms usually gets less at each level of an ecosystem. This is because on average only about 10% of the energy at each stage of a food chain gets passed on to the next level. The rest

- is used for life processes in the organism, e.g. movement, keeping warm
- escapes into the environment as heat energy
- is excreted as waste – passes to decomposers
- is uneaten – passes to decomposers

The diagram shows the efficiency of energy flow through an ecosystem.

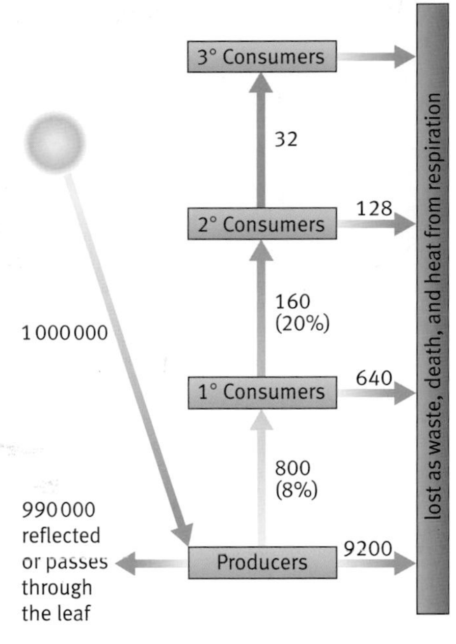

Energy flows through an ecosystem.

When pyramids of numbers are a different shape

Pyramids of numbers may not always show the shape you expect. For example, look at the pyramid of numbers for an oak woodland.

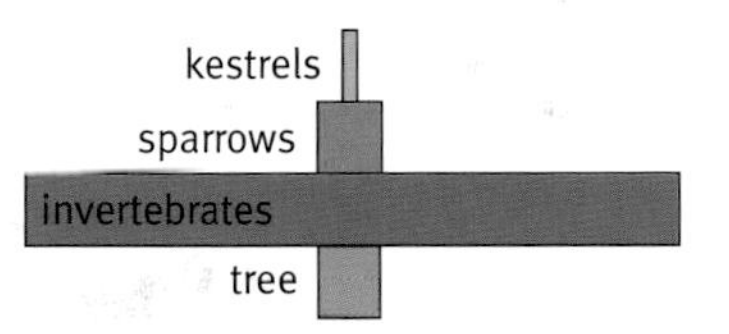

Oak woodland pyramid of numbers

Because oak trees are so large, each one has many small herbivores feeding on it. There may be thousands of caterpillars on every oak tree in this ecosystem. Using a **pyramid of biomass** instead shows the situation more clearly.

This diagram shows the biomass of all the organisms in an ecosystem. Biomass is the total mass of the living organisms. Obviously each oak tree has a much larger biomass than each herbivore (e.g. a caterpillar!), so the pyramid of biomass for this ecosystem has a normal pyramid shape.

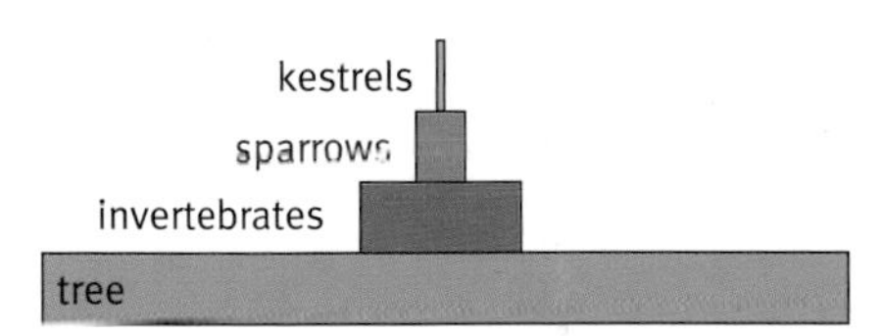

Oak woodland pyramid of biomass

Questions

1 Ecosystems contain different trophic levels. Look at the food web on page 87. Identify two

a producers **b** primary consumers (herbivores)

c secondary consumers **d** tertiary consumers

2 Explain why the energy in a producer will not all be transferred to a carnivore in the same food chain.

3 Calculate the percentage efficiency of energy transfer between the secondary and tertiary consumers in the ecosystem above (shown in the diagram at the top of page).

4 Explain why a pyramid of biomass may give a more accurate picture of an ecosystem than a pyramid of numbers.

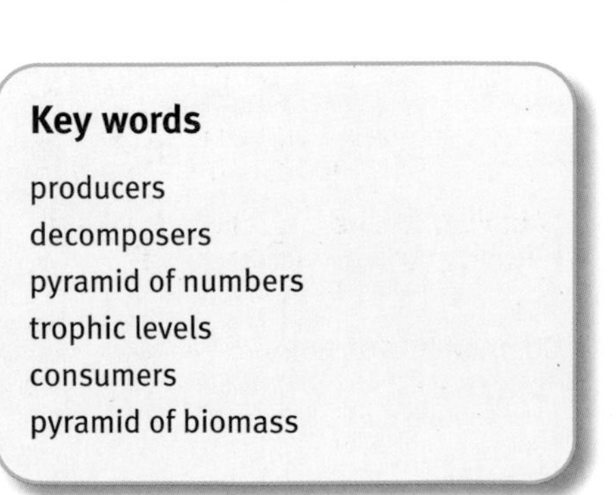

Key words

producers
decomposers
pyramid of numbers
trophic levels
consumers
pyramid of biomass

Canada geese migrate to the UK in winter and leave in the spring.

When pyramids of biomass are a different shape

Untangling a food web is a complicated business. Not all pyramids of biomass will be the same shape. For example, some organisms are only in the ecosystem for part of the year. They may be missing when the measurements are taken.

Other organisms in an ecosystem may be eaten almost as fast as they are reproducing. Any pyramid records only a 'snapshot' of the ecosystem at the time when the measurements are taken. If the measurements were taken over a longer period of time, the results could be very different. For example, look at the pyramid for an ocean ecosystem.

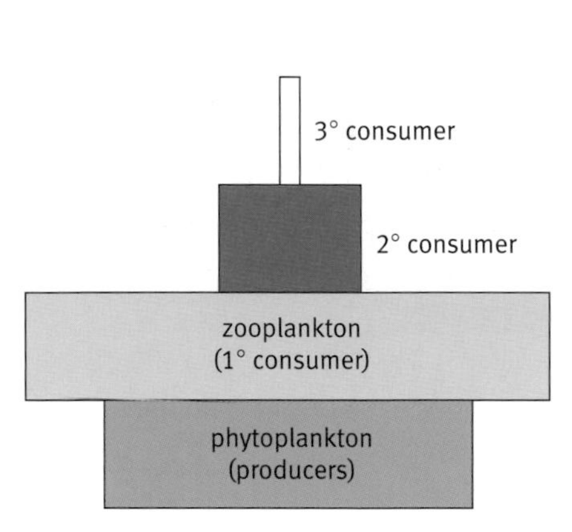

Ocean pyramid of biomass

This pyramid shows that at the time these measurements were made there was less phytoplankton producer biomass than the zooplankton feeding on them. Common sense tells you that this could not work, but the pyramid was correct when the measurements were taken. Phytoplankton have a much faster reproduction rate than zooplankton. Over a year the total biomass of phytoplankton in this ecosystem will be greater than the biomass of the zooplankton feeding on them.

Pyramids of biomass also have other limitations. For example, living things contain different amounts of water. Biomass can only be compared fairly if the material is dried first.

Our impact on ecosystems

At the moment humans are using about 10–20% of the biomass produced on land. Some is used for food and the rest as a source of animal feed, fuels, and raw materials for building. Tropical swamps and rainforests produce more biomass than any other natural ecosystems. Their removal limits the biomass left for other animals. You can read more about the impact of deforestation and habitat change on page 86.

Underground in an ecosystem

The Earth's total biomass is difficult to measure. A lot of it is under water or in the soil. Soil can take thousands of years to form, which makes it virtually non-renewable, so soil conservation is crucial. If vegetation is cleared from an area, the soil is no longer held in place by the plants' roots. Rain and winds can wash and blow the soil away.

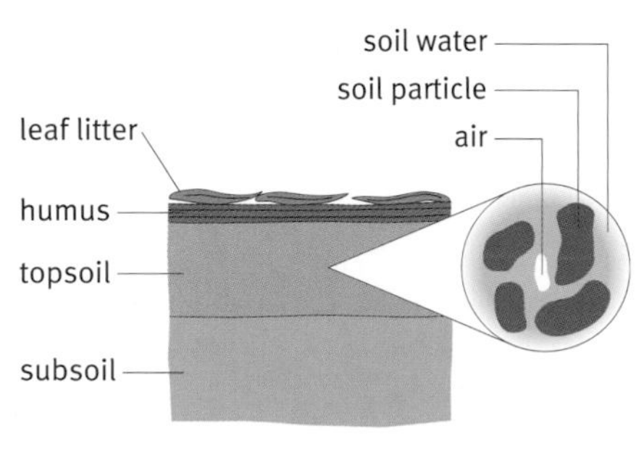

Components of soil

Soil has four main components:

- inorganic particles of sand, silt, and clay
- air
- water (with dissolved mineral ions)
- biomass (living organisms and decaying material)

The air and water in soil are essential for life underground. Air provides oxygen for aerobic microorganisms. Water is held within soil by the small inorganic particles of sand, silt, and clay. Organic material in the soil also has a very important role in holding water in the soil. Some of the minerals forming inorganic particles in soils are soluble. These gradually dissolve in the soil water, providing mineral ions which are taken up by plant roots.

Most of the soil's biomass is not as obvious as this earthworm. A fertile soil contains plenty of decaying plant materials and a huge variety of microorganisms.

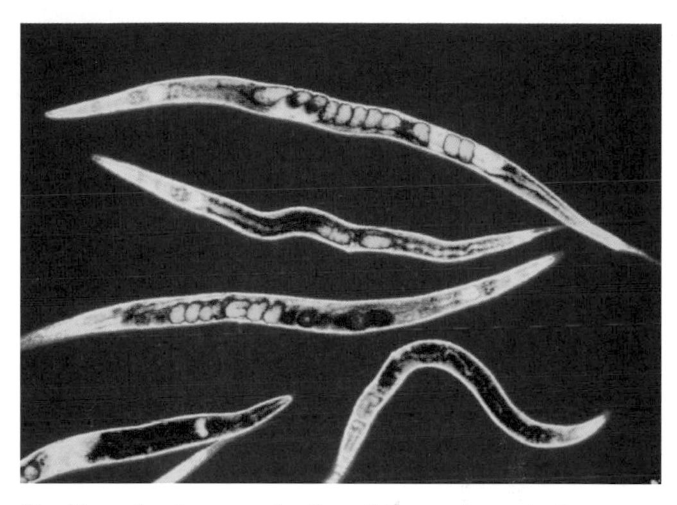

The films of water around soil particles are home to these microscopic worms and the bacteria they feed on. (× 120)

Underground biomass

Soil has two sorts of biomass – living organisms and decaying material. A fertile soil is packed with living roots, invertebrates, fungi, and bacteria. There can be a billion bacteria in one gram of soil, forming the bulk of its biomass.

The biomass in decaying material is attacked by a huge range of bacteria and is unable to resist this onslaught. Between them the different species of bacteria have enzymes for a wide range of possible chemical reactions. These convert the decaying biomass into inorganic raw materials, such as carbon dioxide and nitrate ions, which plants can recycle.

Soil microbes	Approximate number per gram of soil
bacteria	1 000 000 000
fungi	1 000 000
nematodes	500 000
flatworms	100 000
insects	5000

Questions

5 Explain why the pyramid of biomass for the ocean ecosystem is not a typical pyramid shape.

6 Describe the different components of soil.

(7) a Explain why it is more sensible to compare dry mass than wet mass of biomass in an ecosystem.

b Why do you think it is not always possible to use dry mass when measuring biomass?

Life in the soil

A single gram of soil contains millions of microorganisms. These bacteria and fungi play a vital role in plant growth. Lucy Gilliam is using genetic techniques to investigate the biodiversity of soil. She also investigates the importance of microorganisms in soil to the rest of the ecosystem.

Lucy works as a microbial ecologist at Rothamsted Research. Rothamsted is the largest agricultural research centre in the UK. She is part of a team looking at the different types of bacteria and fungi in soil. 'People think it's just dirt,' says Lucy, 'but we've got a great diversity of life under our feet and it's very important to understand how the soil ecosystem works.'

Soil bacteria improve the soil

Microbes improve the condition of soil. They provide essential nutrients to plants, and even help break down pollutants like oil or petrol. Some bacteria and fungi in the soil cause plant diseases but others help plants to resist disease. Soil bacteria are also very important for recycling of chemicals in the ecosystem. For example, soil bacteria have an essential role in the carbon and nitrogen cycles.

We need to learn more about the different microorganisms in soil. Scientists believe that a wide diversity of microorganisms is important to maintaining the health of the soil. They think that soils with a wide diversity of bacteria are more productive and have greater resistance to stresses such as pollution or flooding. Human activities, such as intensive agriculture, tend to reduce the biodiversity of the soil.

Lucy Gilliam at work in the lab

Even the smallest speck of soil contains tens of thousands of different species of bacteria. Most of them are only visible under a powerful electron microscope. Lucy studies the area of soil surrounding the roots of the plants. This is called the rhizosphere.

Around the roots

A recent experiment involved analysing the diversity of bacteria surrounding the roots of potato plants. Lucy took samples of the soil from the rhizosphere and extracted the genetic material. She analysed this DNA to identify different species of microorganisms. These data showed that growing different varieties of potato plant affects the amounts and types of bacteria in the soil.

'We are discovering new species of microbes in soils all the time and we really don't understand what they're doing there,' explains Lucy. 'We've only just got the tools to tap into this diversity of bacteria to figure out how it works.'

Lucy enjoys the fact that there is more to learn. 'It's really exciting work. I love trying to discover something that no one has ever known before.'

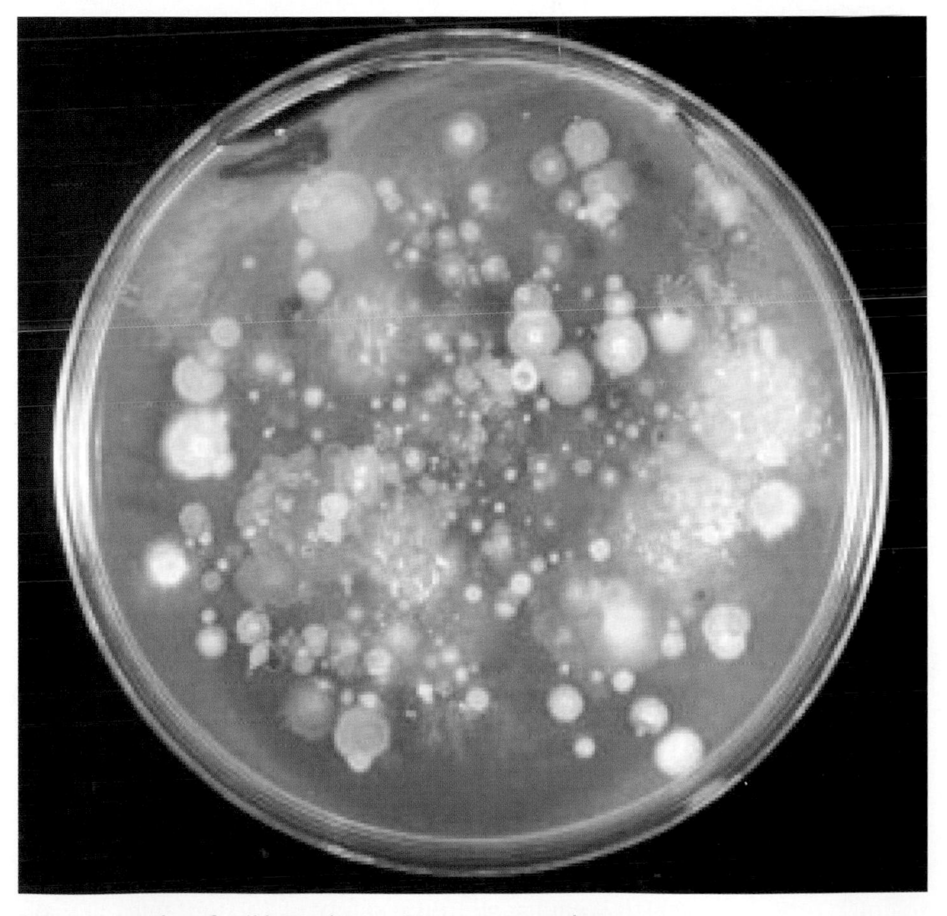

Different species of soil bacteria growing on an agar plate

Topic 2 Further up the food chain

Most species of living thing cannot produce their own food. They must eat other organisms for nutrients and energy. These organisms are called **heterotrophs**, and they have evolved a variety of feeding relationships.

Clown fish and giant anemone.

Living in harmony

The clown fish and anemone have a close relationship, based on food and survival. The clown fish sleeps, eats, and lays its eggs between the tentacles of the giant anemone. The fish is covered in a slime which protects it from the anemone's poisonous tentacles.

Living in the tentacles protects the fish from predators. It darts out quickly to catch other fish, which it eats back in the safety of the tentacles. Bits of this food fall into the tentacles, providing a meal for the anemone. When both organisms benefit from a relationship it is called **mutualism**.

Parasites

Not all relationships are as pleasant as mutualism. Some organisms feed off others without offering anything in return. They are **parasites**, for example, insects that feed on plant sap or blood, or worms that grow inside plant roots or inside our guts. These parasites may cause disease, or threaten people's lives by destroying the crops and animals they rely on for food. A lot of scientific ingenuity is devoted to fighting organisms that cause disease.

A parasite cannot survive without its host. Over millions of years the host has gradually evolved. The parasite's own evolution must have been closely linked to that of its host, so it could survive these changes in its environment.

Key words

- heterotrophs
- mutualism
- parasites

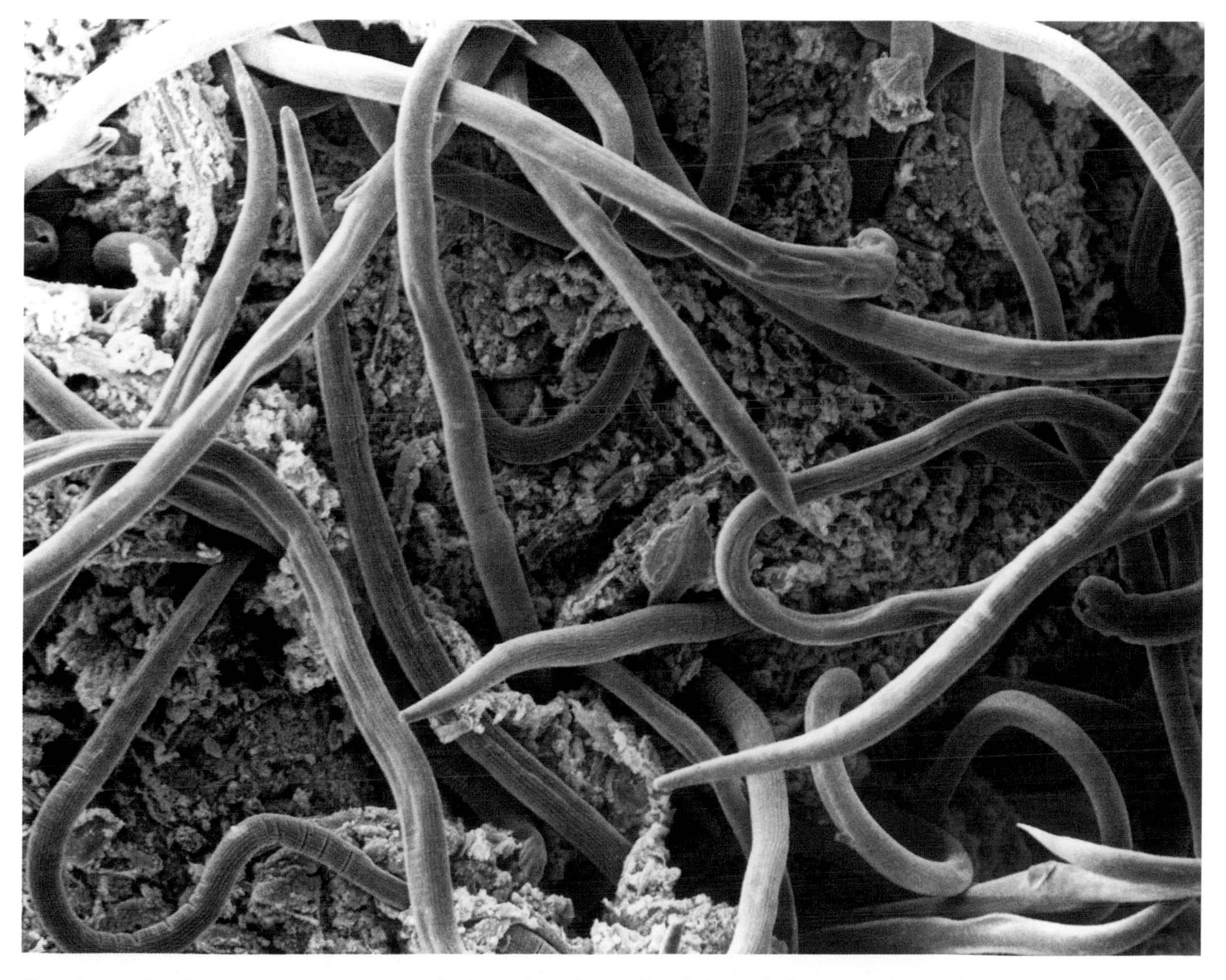

Threadworms are the most common worm parasite in northern Europe. The photograph shows threadworms in a human gut. (× 110)

Find out about:

- examples of symbiosis and commensalism

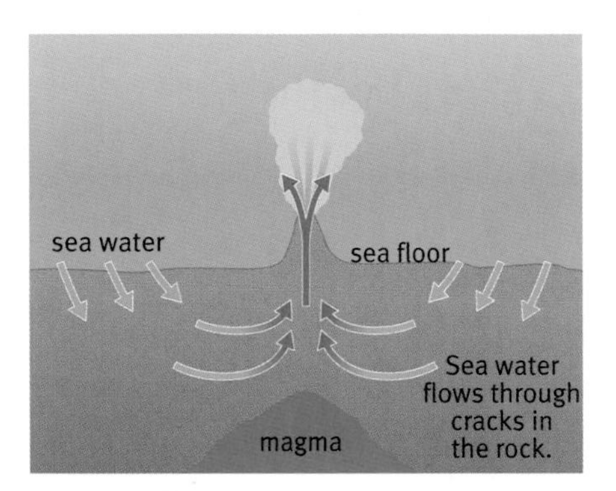

A hydrothermal vent. The average water temperature is 12 °C, but superheated water can gush out at 400 °C. Away from the vent the temperature drops to 2 °C.

2A Living together

Heterotrophs all rely on getting ready-made organic compounds to provide their food. But they have several different ways of doing this.

Dotted along the mid-ocean ridges are hydrothermal vents. Superheated water from the vents flows out into the ocean. Clustered around each vent are colonies of tube worms. They do not have mouths or guts – but closer investigation explains why these aren't needed.

Inside the worms are species of autotrophic bacteria. The bacteria convert carbon dioxide to glucose, using chemical energy instead of light. The worms have specialized blood which transports sulfide as well as oxygen. The worm's blood system transports sulfide to the bacteria, which use it to release energy.

Some scientists believe that hydrothermal vents are where life first originated on Earth. You can read more about this on page 80.

Mutual benefits

The conditions around a vent are harsh and unforgiving. The vent water is rich in minerals which the bacteria need, but the surrounding ocean contains very few nutrients. The flow of water from the vents shifts from moment to moment, so the tube worms provide an ideal environment for the bacteria.

Tube worms and the bacteria within them live in mutualism. The bacteria get a supply of sulfide and warmth. The worms get organic food molecules produced by the bacteria. The relationship benefits both organisms.

Tube worms are about 2 m in length. The total biomass of worms and bacteria in this ecosystem is 10 kg per m^2.

Mutualism is a common arrangement, for example:

- Sponges, sea anemones, and clams all have a close association with algae. The algal cells provide extra food in return for protection from herbivores.

The clam looks green because the soft lining of the shell's edges is full of algae.

- Leguminous plants have nitrogen-fixing bacteria in their root cells. The bacteria supply the plant with nitrates in return for the low-oxygen environment which they require, and sugar from plant photosynthesis.

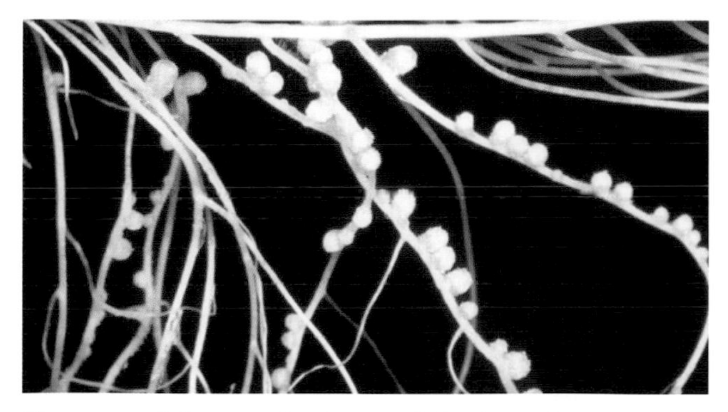

The root nodules contain nitrogen-fixing bacteria.

- Cows have cellulose-digesting bacteria in their gut. The bacteria digest cellulose in plant material eaten by the cows. This releases glucose which the cow can absorb from their gut. In return the bacteria have a constant supply of plant material and a warm environment.

Bacteria help these cows digest the cellulose in the grass.

Organisms do not have to live as one to be mutualistic. For example, some mutualistic species will swim into the mouth of a particular species of larger fish. The smaller animal clears the fish's mouth of parasites, which in return gives them a meal in return.

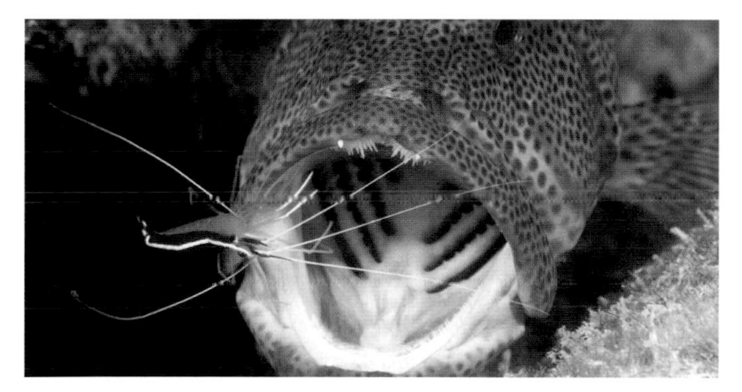

The red cod is a carnivore but it will not eat this shrimp.

One-sided relationships

Some species take advantage of others and give nothing in return. For example, sticky seed pods may cling to your clothes until you brush them off. This might be annoying, but there's no harm done, and the seeds catch a ride to a potential new home. This is an example of **commensalism**. Only one species benefits from the relationship, but the other is unharmed.

Key words

commensalism

Questions

1 Tube worms do not 'eat' in the way many other animal species do. Why are they classified as heterotrophs?

2 Explain the difference between mutualism and commensalism. Give one example of each relationship in your answer.

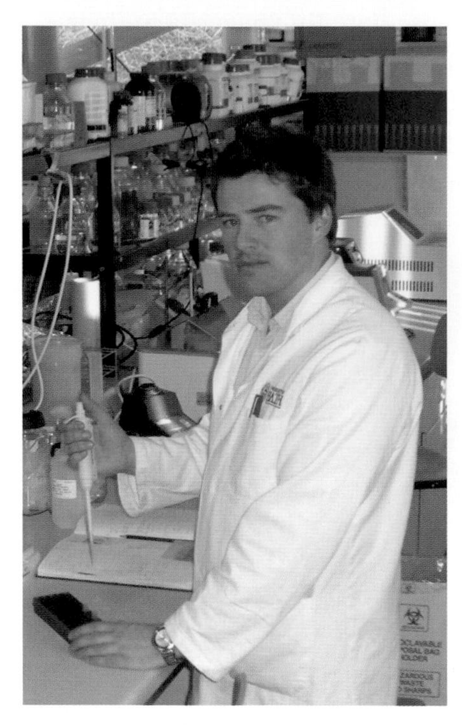

Dr Ben Heath of Bath University

Sex, reproduction, and parasites

'Sex and reproduction are usually thought of as being linked together, but to a biologist they are entirely separate and in one sense almost the reverse of each other,' explains Dr Ben Heath, a research scientist at Bath University.

'If you think about it, sexual reproduction is actually very inefficient. One female reproducing asexually can produce just as many offspring as a male and female reproducing sexually.'

So, why isn't the world filled with asexual organisms?

'Many ideas have been put forward by different scientists to explain why sexual reproduction happens in so many species. These ideas fall into two main categories:

- Sex helps to remove "bad genes" from a population.
- Sex helps to spread "good genes" through a population.'

Removing 'bad genes'

'In an asexual population all offspring are exact copies of their parents, so any "bad genes" or mistakes in the DNA of parents will be added to the mistakes of all previous generations and passed on. Eventually these mistakes can build up to a lethal level. This is a bit like making photocopies of photocopies and so on – the quality of the reproduction gets worse each time, and eventually you can't recognize the original document.'

Sexual reproduction mixes genes from two parents. This means that there is variation between the individuals in the population – they're not all the same, as they are in asexual reproduction. Individuals with harmful genes are less likely to survive, so these genes are less likely to be passed on to the next generation.

'Natural selection weeds out harmful genes so the population is better able to survive. In this way it is thought that over millions of years asexual populations have died out leaving the sexual ones we see all around us today.'

Spreading 'good genes'

'In addition to removing "bad genes", sex can also promote the spread of "good genes" and allow populations to adapt to their environment more rapidly. In a changing environment it can greatly increase the chance of having at least one offspring who is well adapted to the prevailing conditions. It's a bit like increasing your chances of winning a lottery by buying lots of tickets. If you're asexual you could still buy lots of tickets but they'd all have the same number, which wouldn't improve your chances of winning, now would it?'

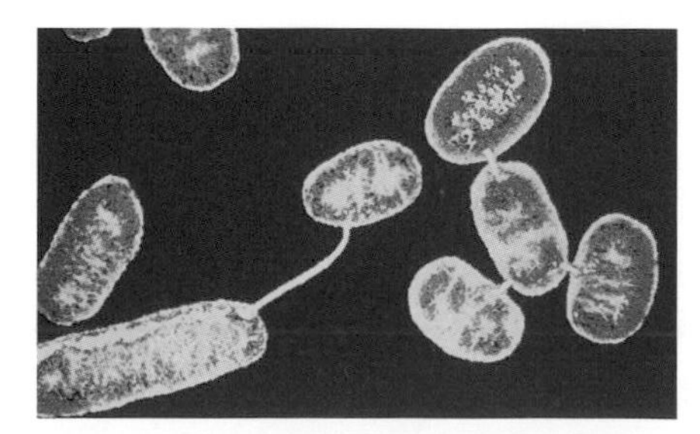

Plasmids normally move between bacteria in a primitive version of sex called conjugation. This is how antibiotic resistance genes can spread so quickly through a bacterial population.

Repelling parasites

Ben is interested in a third idea about the benefits of sexual reproduction. Some scientists think sex can help the fight against by parasites.

'The environment that an organism lives in and adapts to over many generations is also made up of other living things. These include many parasites that are constantly evolving new ways of overcoming their hosts' defences.

'These defences, which include the immune system in higher animals, must constantly change (like changing your password or the combination lock of a safe) in order to keep parasites out and minimize the harm they do.

'In this way every living organism is in a perpetual race against parasites. But it is a race they cannot win. The only way to survive is to keep "running" because their defences are constantly being overcome. Continually trying out new combinations of genes over the generations and throughout the centuries and millennia is the only way to stay one step ahead of the parasites.'

Ben's own research is on *Wolbachia*, which is a bacterial parasite that lives inside the cells of an estimated 76% of all insects, as well as some other important animal groups.

'*Wolbachia* is fascinating to evolutionary biologists, because it can hijack the reproductive system of its host in order to further its own transmission. In addition to shedding light on some fascinating areas of basic reproductive biology, there is currently great interest in using *Wolbachia* as a new way of combating important tropical diseases including malaria which affects 500 million people every year and kills a person every 15 seconds. Any approach that impacts on the transmission of malaria could potentially improve the lives of millions of the world's poorest people.'

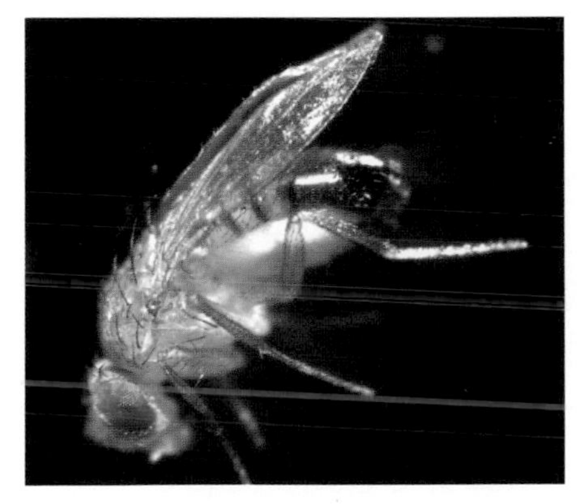

A male fruit fly (actual length 3.5 mm)

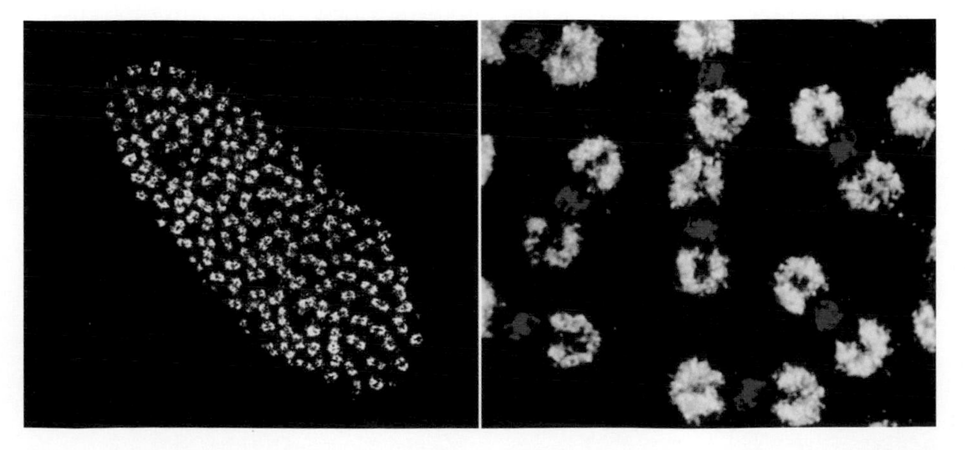

The micrographs show an early fruit fly embryo infected with *Wolbachia* bacteria. Nuclear DNA is stained red, and the *Wolbachia* are visible as yellow/green dots. The *Wolbachia* position themselves either side of the dividing DNA which ensures they get into both daughter cells during development. The entire embryo (left) is approximately 1 mm long. The right image shows an area 100 μm (0.1 mm) across.

Find out about:

- parasites and what makes them successful

2B Tapeworms and other parasites

Life on Earth exists on land, in water, and also in or on other living things. Every plant and animal on Earth is home to other organisms. Many of these organisms are parasites. Parasites are not good company. They benefit from the relationship, but cause harm to their host.

Mites living on human eyelashes. (× 170)

Human parasites

Parasites have several ways of getting in or onto the human body. They can be transferred

- by food or water
- through the nose, mouth, anus, and genital and urinary tracts
- by insect bites
- by burrowing under the skin

Most parasites are small and hidden, so it is hard to know they are there until symptoms start to appear. Some parasites can establish chronic infections which last for the host's entire lifetime.

Tapeworms

Perhaps the most gruesome human parasites are **tapeworms**. These live in the human gut and can be up to 9 metres long. In humans the adult tapeworm itself stays in the gut, which is the hollow tube running through the body from mouth to anus. The tapeworm competes with the **host** for digested food. The tapeworm does not enter the host's body tissues, so symptoms are usually just a mild stomach ache. However, in some cases the tapeworm larvae may move within the body to the brain and eyes, and cause very serious damage.

Parasites have specific features which enable them to survive in or on their host. Tapeworms have a number of adaptations:

- Their heads have suckers and stickers to grip the gut wall.
- They are protected from digestion by a thick, enzyme-resistant cuticle.
- They use anaerobic respiration so they can survive without oxygen.
- Each tapeworm has male and female sex organs so it can reproduce without a mate.
- They produce very large numbers of eggs (a cow tapeworm produces approximately 600 million eggs each year). This increases the chance that some will survive to find a new host.

Even a healthy human body houses an amazing number of other species – at least 200. Your own body cells are outnumbered by cells of other species by about 10:1.

Fortunately most of your residents are not parasites. The majority are bacterial cells which live in your large intestines and usually do not cause you any harm. If all the bacteria in your body could be gathered together, they would just squeeze into an empty drink can.

The most sophisticated adaptation of a tapeworm is its life cycle. This helps the tapeworm to infect new hosts. Human tapeworms use a second host (pigs, cows, or fish) to transfer them to new human hosts. When the eggs are eaten by this second host they hatch into larvae. The tapeworm larvae tunnel into the herbivore's blood and migrate to the muscle tissue. They form a cyst and wait. They will only become adult tapeworms if humans eat them in undercooked meat. Stomach acid then releases the larvae from the cyst.

Human tapeworms like this one may be 2 m to 7 m in length.

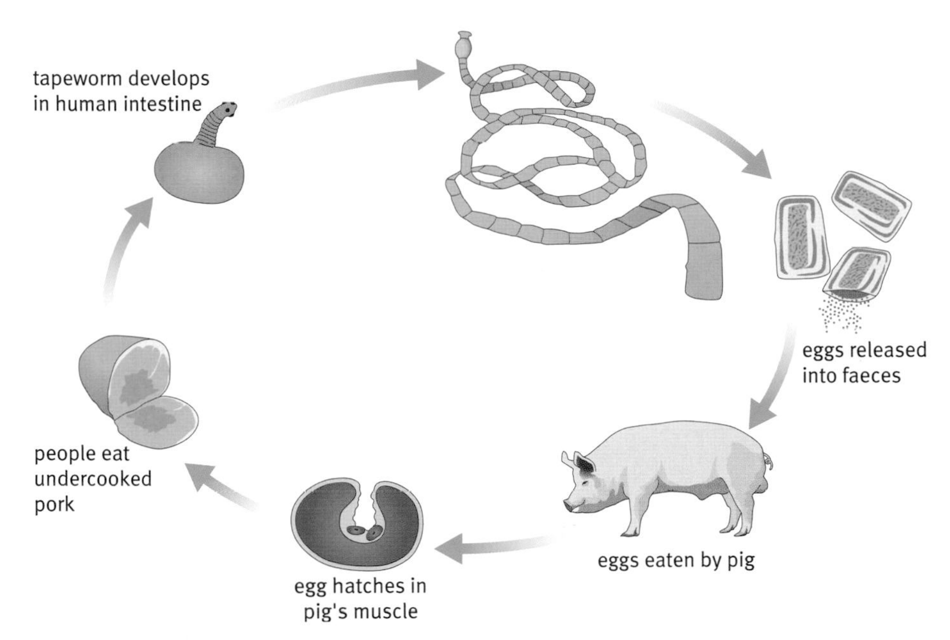

DNA evidence shows that tapeworms evolved alongside humans. As our ancestors changed their way of life the tapeworm adapted to use different herbivores to transmit their larvae.

This may seem a complicated lifecycle, but it is a very effective way for the tapeworm to spread from human to human. Humans are much more likely to eat infected meat than faeces. If human food is contaminated by infected faeces the lifecycle is short-circuited. The tapeworm larvae find their way to the person's muscles and brain. They form cysts which can be life threatening.

Parasitic worms

Parasites are an important cause of human disease. They also have an indirect effect on humans by damaging food production.

- A quarter of the world's human population carry some sort of parasitic worm. The worms feed on nutrients from a person's gut, causing malnutrition and other illnesses.
- Other worm species cause damage indirectly by limiting human food supplies. They cause illness in farm animals and also attack crop plants, blocking water and mineral uptake. This causes stunted growth, which reduces crop yields.

Key words

tapeworms
host

Questions

1 Write a short definition of a parasite relationship.

2 Describe how three features of a tapeworm enable it to survive as a parasite.

3 Explain how parasites cause damage to humans

a directly

b indirectly

(4) Scientists believe that the evolution of a parasite and its host are closely linked. Use the example of a tapeworm to suggest evidence for this view.

2C Parasites that cause disease

Find out about:

- malaria
- the relationship between malaria and sickle-cell anaemia

Malaria

The disease **malaria** is a major cause of ill health and death in countries where it thrives. Malaria is caused by a **protozoan**. This protozoan parasite is a single-celled animal. Blood-sucking mosquitoes carry this protozoan from host to host. As many as 300–500 million people worldwide are infected with this protozoan, and malaria kills more than two million people every year. The protozoan also relies on a complicated life cycle to survive.

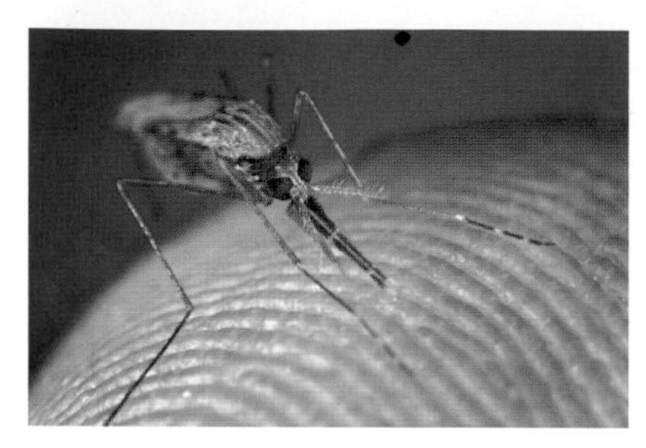

The parasite that causes malaria is transferred into human blood with a drop of saliva when a female *Anopheles* mosquito bites the skin.

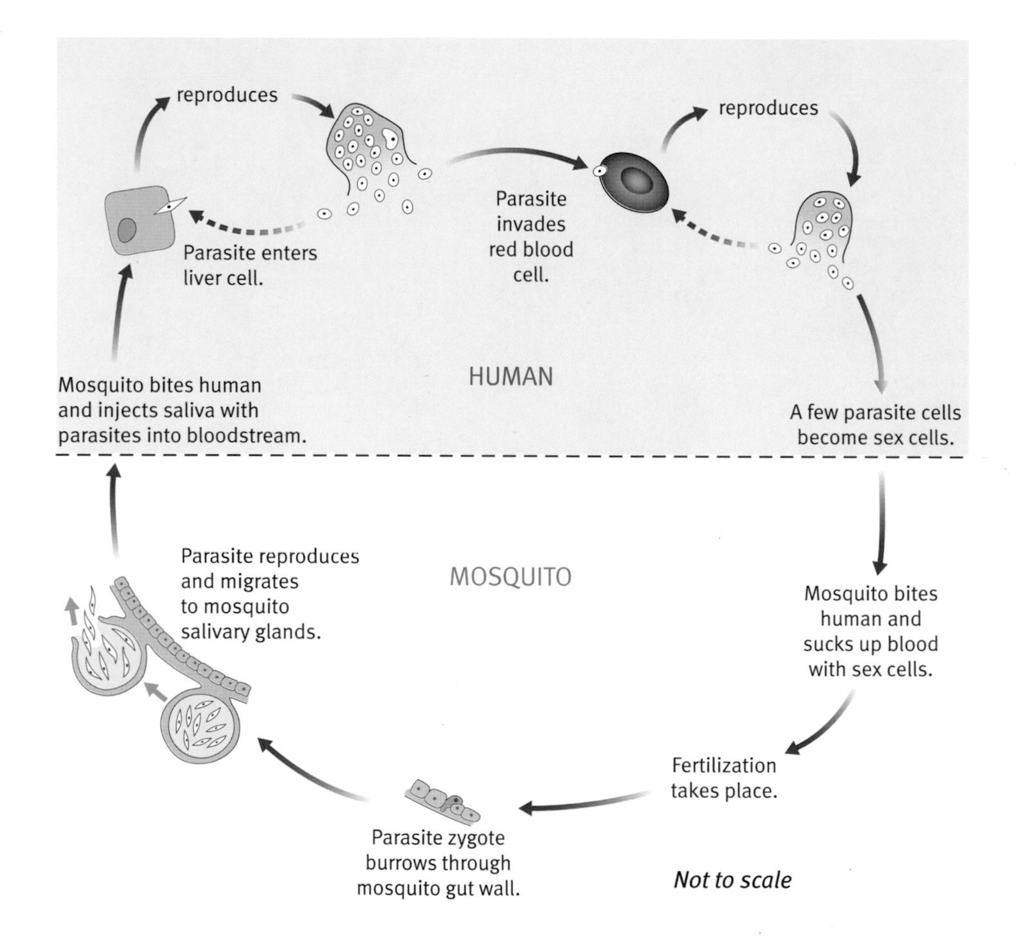

In the human body the parasites which cause malaria spend most time inside red blood cells, feeding on haemoglobin.

The malaria parasite's life cycle protects it from our immune system. The parasites have different markers on their cells at every stage of the life cycle. This makes it difficult for white blood cells to identify and attack them. The parasites also spend most of their time feeding inside our red blood cells. When one red blood cell is used up the parasites burst out and infect new cells. Toxins are released from the used-up cells, and these cause a very dangerous fever. You can read more about the human immune system in Module B2 *Keeping healthy*.

Key words

malaria
protozoan
sickle-cell anaemia
haemoglobin

Questions

1 Explain how the parasite that causes malaria is adapted to survive.

Sickle-cell anaemia

Sickle-cell anaemia is a genetic disorder. It is caused by a faulty allele of the gene which codes for the **haemoglobin**. Haemoglobin is the protein that carries oxygen molecules in red blood cells.

Sickle-cell haemoglobin proteins have a different shape from normal haemoglobin. When faulty haemoglobin gives up oxygen to body cells, the shape of the haemoglobin molecule changes. The haemoglobin molecules form long rods which stretch red blood cells into a rigid 'sickle' shape. These rigid red blood cells get stuck in small blood vessels, causing acute pain and extreme tiredness. This is called 'sickling'. Body cells don't get the oxygen they need, and over time this damages tissues and organs. 'Sickling' also damages red blood cells. The spleen removes damaged cells, but these cannot be replaced fast enough. Sickle-cell sufferers get severe **anaemia**.

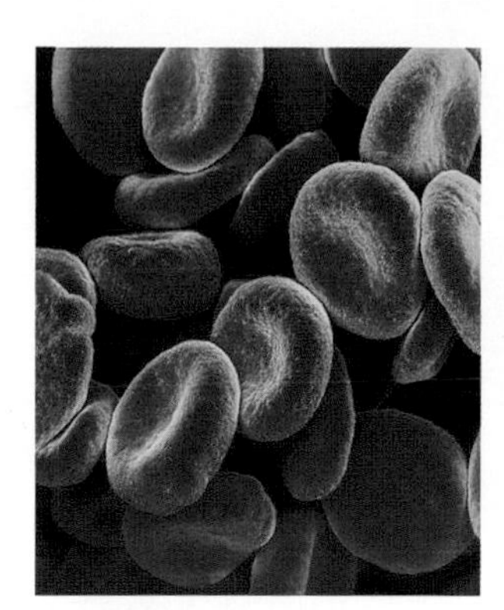

Red blood cells with normal haemoglobin

Red blood cells with sickle-cell haemoglobin

How is sickle-cell anaemia inherited?

The sickle-cell allele is recessive. A person must inherit the allele from both parents to have the disease.
A person with one sickle-cell allele and one normal allele is a carrier of sickle-cell anaemia. The diagram shows how two carriers of sickle-cell anaemia can pass the alleles on to their children.

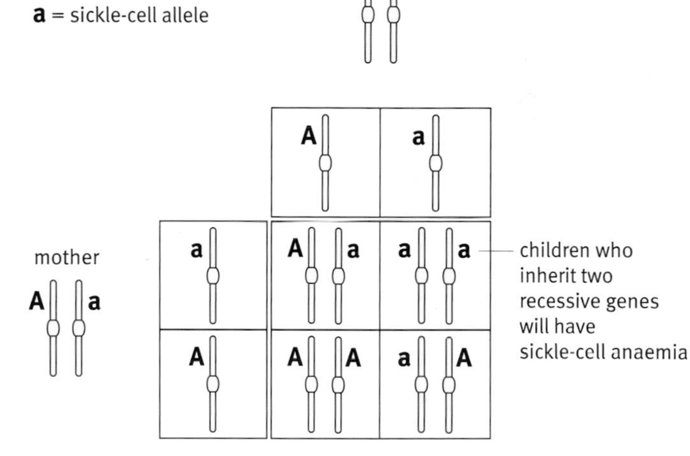

Children of two sickle-cell carriers each have a 25% chance of having the disease.

Sickle-cell anaemia and malaria

The sickle-cell allele occurs occasionally as a random mutation. Without proper management of the disease many sufferers would die young. So you might expect that the sickle-cell allele would not occur very often – it would have a low frequency in the population.

But people in parts of Africa have known for hundreds of years that sickle-cell anaemia is more common in areas where malaria is common – there is a correlation. However, the explanation for this observation is still uncertain. What is known is that carriers of the sickle-cell allele have a much higher resistance to malaria than people who have two normal haemoglobin alleles. Their condition is known as sickle-cell trait. In regions where malaria is common they have a greater chance of survival than non-carriers. They are more likely to survive and have children, passing on their genes, including the sickle-cell allele. Where malaria is common, natural selection favours people with sickle-cell trait. So the frequency of the sickle-cell allele is higher in these areas.

Questions

2 Describe the symptoms of sickle-cell anaemia.

3 Explain why the sickle-cell allele has a higher frequency in certain populations.

(4) Scientists are not certain how sickle-cell trait gives protection against malaria. Find out how this may work.

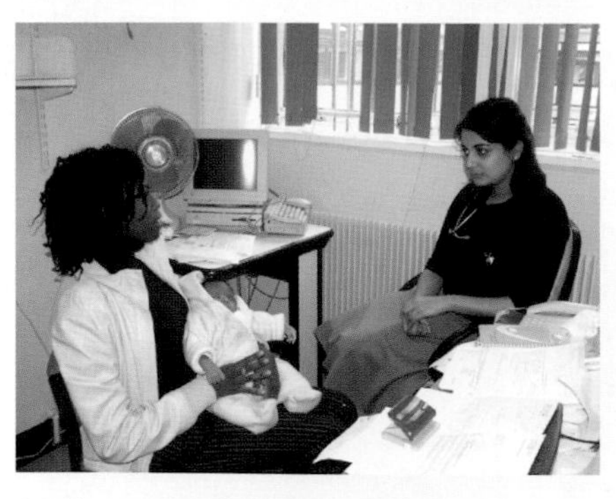

Emily and Dr Chaud

Sickle-cell anaemia

Dr Meera Chaud first met Emily in the accident and emergency (A & E) department at Whittington Hospital, London. Emily was in great pain with a **sickle-cell crisis**. She was also pregnant at the time, so Meera had to identify a drug that would help the pain, and be safe for the baby. Today Emily has come with her new baby, Akila, for a follow-up visit to the **haematology** clinic. This clinic helps patients with diseases of the blood system.

What are the symptoms?

Meera explains the symptoms of sickle-cell anaemia: 'Usually patients go to their doctor because they are in severe pain. Normal red blood cells are nice and round and flexible, so they can squeeze through capillaries quite easily. But sickle-shaped red blood cells don't bend. The cells stack together and block the capillaries. Blood cannot pass through the capillary, and the tissues behind the blockage are deprived of oxygen. The tissue has to use anaerobic respiration, which causes the pain. This is a sickle-cell crisis.' (You can read more about anaerobic respiration on pages 240–241.)

Emily with her partner Wayne and baby Akila. Wayne has no sickle-cell alleles; Emily has two sickle-cell alleles. Akila has inherited one sickle-cell allele and one normal allele. She has sickle-cell trait.

Sickle-cell anaemia may cause very serious problems. It can be particularly dangerous if a blockage happens in capillaries of the lungs. This stops blood from the body getting to the lungs to collect fresh oxygen. The whole body becomes deprived of oxygen, and the patient needs intensive care.

Unfortunately at the moment there is no cure for sickle-cell anaemia. Meera describes some of the work she does with patients. 'It's all about managing the symptoms so people can have the best quality of life. A lot of it is about making sure they know how to reduce the chance of a crisis. The important things are to avoid dehydration and strenuous exercise – but gentle exercise is fine. There's quite a static population of sickle-cell families in this part of London, so usually it's a lifelong relationship between the clinic and the patient. I think it's great, a joint achievement between the doctor and the patient. It's wonderful to see Emily so well today with her new baby.'

Emily's story

Emily is twenty-two years old. She was born in Uganda, and moved to London when she was eight.

'My mum found out that I had sickle-cell disease when I was six months old. I was diagnosed back home in Uganda. I wasn't able to crawl and my feet, legs, and arms swelled up and I didn't want to be cuddled. She went to the hospital and was told I had sickle cell.

'I was in and out of hospital a lot and missed a lot of school. My symptoms included stress, dehydration, exhaustion, and when my haemoglobin dropped really low I'd have to go into hospital. In Uganda I was treated with a saline drip and pain relief. Sometimes I was in hospital for months. Since arriving in the UK I've been in intensive care twice and had pneumonia.

'I've had to deal with so much pain that my threshold is quite high. This means that I can walk into A & E in severe pain, but not always show it. Once a doctor didn't believe I was in pain, as I was joking with my Mum to take my mind off the pain! Usually, though, I can discuss with the doctor how much pain relief I use.

'Little things spark it off. When I was younger I could never go swimming, and it turned out to be the chlorine in the water that was triggering my symptoms. I conquered that by going with my brother during the summer for me to get my body used to the smell of the chlorine. I try to find ways of managing my pain and ways of coping with the problems.'

'Both my parents have sickle-cell trait and so does one of my brothers, but the rest are fine. I've always known if I had a baby, they would inherit a sickle-cell allele from me. When I met Wayne, it's the first thing I told him, that I had sickle cell. It's important that he knows. It's hard and emotional as it doesn't just affect me, it affects the people close around me. Wayne was tested and we knew he was fine. Akila does not have sickle-cell disease, he has sickle-cell trait. When he grows up and knocks a girl off her feet, she'll have to be tested. But I enjoy life and I feel the future holds promise. I'm fine, I have baby Akila and couldn't ask for anything more.'

Becoming a doctor – Meera's story

'Well, I started by doing GCSE biology, but I didn't decide on medicine until I did my A levels. I did my first degree in Cambridge and then came to London for the clinical training you need to do as part of your course. After that I became a house officer – that's the junior doctor who runs around doing all the menial jobs. It's a steep learning curve, but it's also a very sociable and fun job.

'Now I've started medical rotation, where I work around the medical departments (the hospital departments that don't do operations or surgery). At the moment I'm in the haematology department, but to start with I worked in A & E, which is where I met Emily.

'If I pass what feels like a few thousand exams, I hope to be a registrar in about three years' time. By then I'll be twenty-eight. I would like to be a consultant in my thirties.'

Topic 3 New technologies

Living things make complex molecules far more efficiently than they can be made in a lab. For thousands of years people have harnessed microorganisms to make products such as drinks, bread, yogurt, and cheese. Microorganisms are added to the food and kept in the right conditions. As they grow, their by-products create the desired product. For example, yeast produces alcohol in beer-making. Some microorganisms also make non-food products, for example, natural antibiotics.

Fermenting beer

Making new products

The microorganisms used for these processes are carefully chosen from natural organisms. But it is now possible to alter the genes of a microorganism so that it produces an altered product, or even a completely new product. This is **genetic modification**. Genetically modified organisms include bacteria that make drugs and hormones for human use, and crops which have better resistance to disease.

Looking at genes

Genetic tests can give people information about the genes they carry. You read about some uses of genetic tests in Module B1 *You and your genes*. Genetic tests can also help to match a suspect to the scene of a crime, or show whether two people are related to each other. Many new applications of **DNA technology** are being developed.

The bacteria growing in this tank have been genetically modified to make a protein for use in drug manufacture.

Analysis of a person's DNA can provide useful information.

Key words

genetic modification
DNA technology

Find out about:

- how microorganisms can be grown to produce useful chemicals

3A Living factories

Bacteria and fungi make many useful organic chemicals very efficiently.

Microorganisms show amazing variety. The different species have a great variety of microbial enzymes at their disposal. Many of the proteins they make are difficult to produce in the laboratory. In biotechnology, microbial enzymes are harnessed on a huge scale. Scientists look for the correct species of microorganism and ensure that they have optimum conditions for growth.

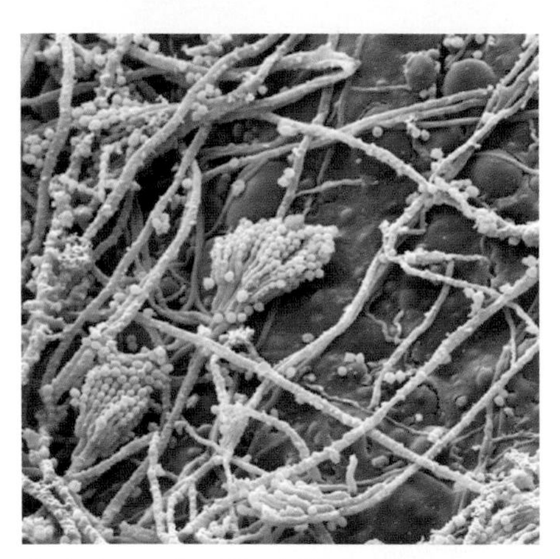

A photograph of *Penicillium*. You can see a photograph of a colony of this fungus on page 50. It produces penicillin and secretes it outside its cells.

Antibiotics

On pages 50–51 you learnt about the use of antibiotics to treat certain infections. Antibiotics are produced naturally by microorganisms. For example, the fungus *Penicillium* produces the antibiotic penicillin.

In optimum conditions *Penicillium* can double its mass every six hours. When the fungus grows in a tank of nutrient solution, the antibiotic is secreted into the solution. It is then a simple task to extract the antibiotic.

Microorganisms are usually grown in batches using huge industrial tanks called **fermenters**. The main difficulty is keeping the conditions inside the fermenter right for the microorganism. Fast-growing microorganisms take in a lot of oxygen and nutrients, and produce waste products and heat. The solution inside the fermenter must be carefully monitored and controlled.

Early stages of cheese production – churn milk.

Harvesting enzymes

Microbial enzymes are very important in food production. They are used to control the flavour, aroma, texture, and rate of production for many food products. **Rennin** is one example of an enzyme used in food production. It is a very important enzyme in cheesemaking.

Rennin is made in the stomachs of all young mammals. It causes the milk they feed on to lump together. This slows the food down as it passes through the gut, giving more time for enzymes to digest the food and for useful molecules to be absorbed.

Originally rennin for cheesemaking was taken from calf stomachs, but these are in short supply. Now most rennin comes from industrially grown fungi. Cheese made using rennin from fungi is labelled 'vegetarian cheese'.

Rennin causes the milk to form into solid lumps – the start of cheese.

Growing microorganisms for food

Microbial cells contain the same building blocks as cells from other organisms – carbohydrates, fats, and proteins. Bacterial proteins are similar to those in fish. Protein in yeast, a fungus, is similar to soya protein. Eating some microorganisms can provide a useful source of protein. Microbial cells are also low in fat and high in fibre from their cell walls. These nutrients add to their value as a food source. Microorganisms can be grown from simple starting nutrients, and they reproduce rapidly. Microbial biomass grown for food is called **single-cell protein (SCP)**.

Surprisingly, the technology for producing SCP was not developed by food companies. It was developed by oil companies. They wanted to use low value fuels as food for microorganisms, and make a more valuable product – SCP. The process works, but it has not been a commercial success. Oil prices have risen, and other protein sources, such as soya, have become cheaper.

Several types of SCP are used in animal feed, but only one type has been cleared for human consumption. Quorn is made from a fungus. The fungus isn't really a single cell. It grows as a cluster of interwoven fungal **hyphae**. When these are pressed together, the fungus looks like pastry. It is treated to match the taste and texture of meat and sold as mince and pieces for cooking. It is also made into convenience foods such as burgers and sausages, and used in ready-made meals.

Quorn provides a high-protein food grown by microbial cells.

Not all microorganisms can be eaten. Eating just a small amount of this fungus would be fatal.

Key words

fermenter
rennin
single-cell protein (SCP)
hyphae

Questions

1 Name three types of product which can be produced by the fermentation of microorganisms. Give one example of each.

2 Produce a flow chart to explain the main steps in the fermentation of microorganisms to produce antibiotics.

3 Why is it important to control the conditions inside a fermenter?

④ Why do microorganisms produce antibiotics naturally?

Pharmacogenetics – a look into the future

'In five to ten years' time a visit to your GP could routinely involve a genetic test before you are given a prescription,' says Dr Alice Rebeck, a scientist working in pharmacogenetics for GlaxoSmithKline (GSK). This is already happening for some medicines. Pharmacogenetics is the study of how people's responses to medicines vary due to differences in their genes. By understanding these differences, we will in the future be better able to predict how people will respond to a particular medicine.

Large amounts of data are collected from clinical trials. Alice reviews and analyses this data and talks to other research scientists and physicians about how they can put the data to good use. Their aim is to develop safer and more effective medicines to meet the needs of patients.

Why do people respond differently to medicines?

People with the same disease will not all respond to the same medicine in the same way. For example, imagine four patients suffering from an illness such as asthma. Each patient has similar symptoms. They are all prescribed the same medicine at the exact same dosage, but they respond differently:

- The first person recovers from the illness.
- The second does not notice any improvement in how they feel.
- The third person recovers, but develops an adverse reaction to the medicine – an itchy rash.
- The fourth person only feels better after another visit to their GP, who increases the dosage of the medicine.

It is possible that these different outcomes for each patient are the result of differences in their genes. On pages 142–3 you read about the relationship between genes and proteins. Genes are the coded instructions for the production of proteins in a cell. Each gene is the code for a different protein. If two people have different versions of a gene (called alleles) they may not make exactly the same protein. For example, if a gene codes for a specific enzyme, differences in this gene between people could result in differences in this enzyme.

- Some people may not produce it at all.
- Others may produce it, but in different amounts.
- Others may make a non-working form of the enzyme.

These differences in the structure or amount of a protein may change the way a particular medicine works in the body.

Clinical trials

Like all pharmaceutical companies, GSK conducts clinical trials to investigate how safe and effective new medicines are. (You can read more about clinical trials on page 53.) During the trial, blood samples are taken from each participant and different measurements are made to monitor their response to the new medicine. This sometimes includes extracting DNA from the blood samples to look at variation in the participants' genes. Scientists investigate whether these genetic variations are associated with different responses to the new medicine. This is **pharmacogenetics**.

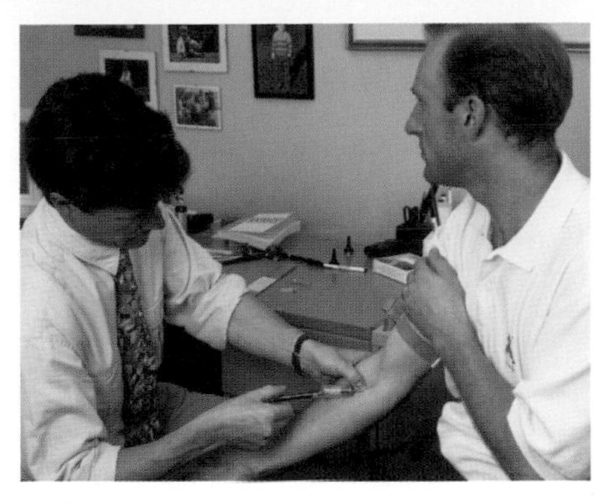

A clinical trial.

Alice describes a recent example. 'We tested a new medicine being developed to treat obesity in the US. At first there didn't seem to be any difference in weight loss between those taking the new medicine and the control group taking the placebo. However, some of the patients receiving the new medicine did show a significant weight loss. We looked at genes that are related to the way this new medicine was thought to work in the body. We were able to show that most of the patients who achieved weight loss had a version of a gene that was different from patients who did not lose weight. This data suggested that we could use pharmacogenetics to predict which patients would obtain the most benefit from taking this new medicine.'

GSK has had similar results with another new medicine under investigation for treating Alzheimer's disease. Alzheimer's is a form of dementia which can affect older people. People with the disease gradually become very forgetful, making everyday tasks very difficult. They may not recognize familiar faces, and have problems speaking, understanding, reading, or writing. A new medicine was tested on patients with Alzheimer's. Unfortunately these patients did not improve.

Could pharmacogenetics help? Scientists working on this new medicine knew that a gene called APOE is associated with Alzheimer's disease. People with different versions of the APOE gene have a different level of risk of developing Alzheimer's. Scientists looked at the APOE gene of the patients on the early clinical trials. Patients with particular versions of the APOE gene did improve when they were given the new medicine. Clinical trials are now continuing to confirm these findings.

'Pharmacogenetic studies like this will enable new medicines to be targeted at those people who are most likely to benefit from a medicine. I really enjoy my work and feel privileged to be part of a team that can make a difference to patients' lives,' says Alice.

Find out about:

- genetic modification of bacteria

3B Genetic modification

Bacterial cells make proteins

Bacterial cells are about one-tenth the length of typical animal cells. They have no nucleus, but they have one large circular chromosome in their cytoplasm. Most bacteria also have rings of DNA, which contain extra genes. These DNA rings are called **plasmids**.

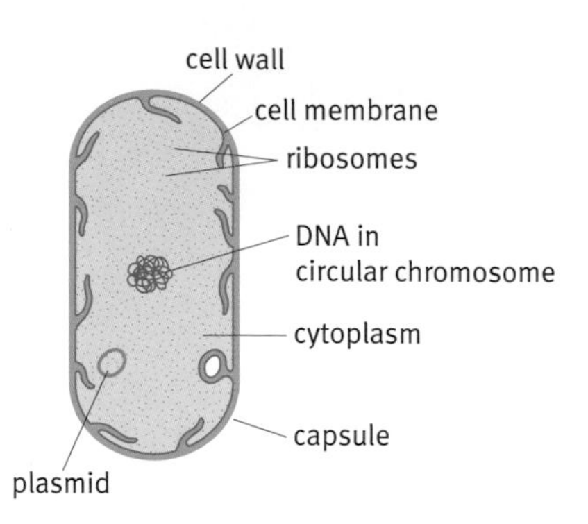

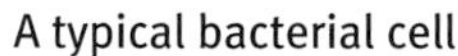

A typical bacterial cell

Bacteria produce a wide variety of proteins. Some of these do similar jobs to proteins in human cells, e.g. proteins in cell membranes. However, they are not identical to the equivalent human protein. Bacteria don't make many proteins which human cells produce. How can we make bacteria produce a protein which they don't normally make? Any cell can only make a protein if it has the gene which codes for it. Scientists can add a human gene to bacterial cells so that they make the human protein.

Genetic modification

Changing the genes of an organism is called genetic modification. Bacterial cells can be modified by adding genes from other microorganisms, plants, or animals.

Plants are sprayed with chemicals to reduce disease and pest damage. The chemicals are expensive to make and can cause pollution.

Many drugs used to treat diseases are proteins that can be made using these genetically modified (**GM**) bacteria. An example of such a drug is human insulin. Before genetic modification these proteins used to be extracted from animals. For example, pig insulin was given to people with diabetes. This worked, but there could be unwanted harmful effects. The insulin produced from GM bacteria does not cause these side effects.

Genetic modification of plants

GM varieties of crops could transform farming by cutting the use of chemical sprays.

Another use of genetic modification is to add new genes to plants. An estimated one-third of all crops are lost to pests, disease, and weeds. Plant diseases account for at least 10% of crops lost. Selective breeding has been used to produce crop plants which have more resistance to diseases. Genetic modification takes this one stage further. Plants can be given genes from other species, including genes from microorganisms. These genes often code for proteins that will protect the plant from common diseases.

Page 28 explains another type of genetic modification – gene therapy.

H Putting new genes into cells

Getting new genes into the cells is the most difficult step of genetic modification. A **vector** is needed to carry the gene into the cell. To modify bacteria scientists make use of bacterial plasmids. Plasmids are easier to manipulate than a bacterial cell's main chromosome. They are small and are designed to move in and out of cells.

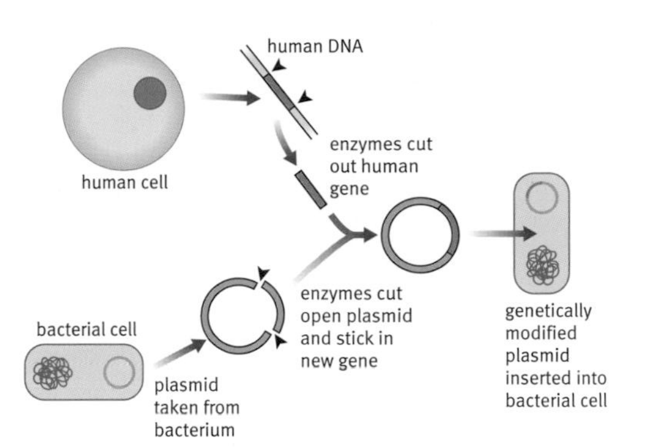

Plasmids are used as vectors in genetic modification of bacteria.

How can you tell which cells have been modified?

Not all the bacterial cells in a population will take up the modified plasmid. Scientists need a way of identifying which cells have been genetically modified. They do this by attaching a second gene to the plasmid. For example, this could be a gene from jellyfish which codes for a green fluorescent protein which is very easy to spot. Alternatively a gene for antibiotic resistance could be used. This gene can be used to separate out the GM bacteria from cells which have not taken up the plasmid:

- Make a modified plasmid that contains the human insulin gene and also a gene for antibiotic resistance.
- Add the modified plasmid to a bacteria population.
- Treat the population with the specific antibiotic.
- The bacteria which survive must contain the plasmid, so they will also make insulin.
- Grow these bacteria and harvest the insulin.

Large genes are too big to fit into a plasmid. Fortunately plasmids aren't the only things that can get through bacterial cells walls. Bacteria are infected by viruses called **bacteriophages**. Scientists use bacteriophages to carry larger genes into bacterial cells. This may sound odd, because viruses usually cause disease. However, the bacteriophages are disabled before they are used for genetic modification.

Vectors are also used to get new genes into plant cells. You can read more about this in the following Biology in action pages.

Key words

plasmids
GM
vector
bacteriophage

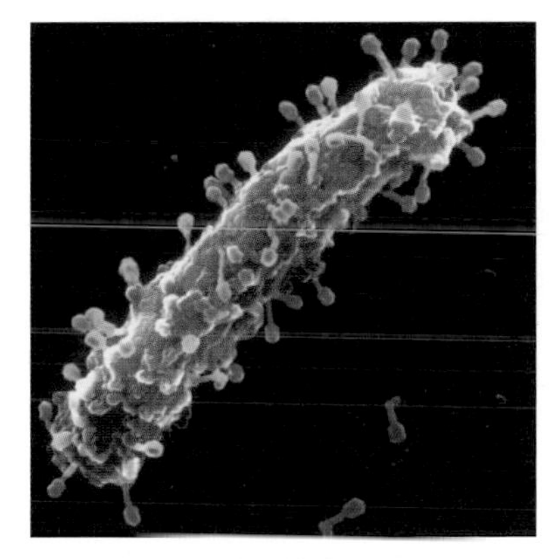

Bacteriophages attacking a bacterium (× 15 000)

Questions

1 Draw a labelled diagram of a bacterial cell.

2 Write a flow chart to explain the main stages in genetic modification of bacteria.

3 Give one example of genetic modification in

a bacteria **b** plants

④ Explain how vectors are used to transfer new genes into

a bacterial cells **b** plant cells

5 When the insulin gene is attached to a plasmid, a gene for antibiotic resistance may also be added. Explain why this second gene is needed.

Philippe Vain

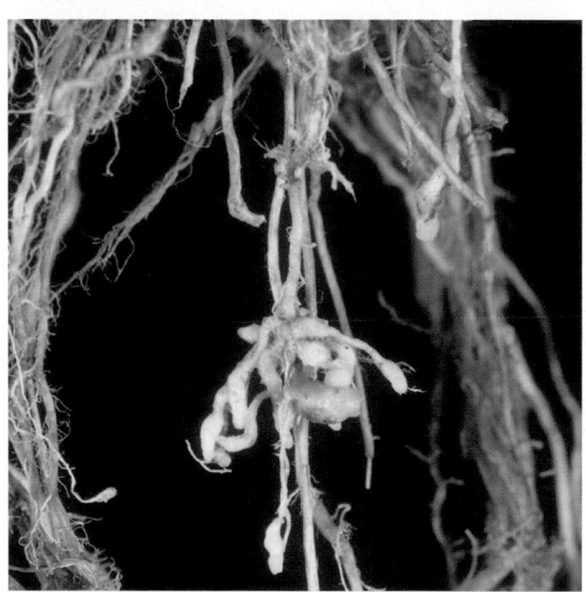
Philippe works on a species of root-knot nematode worm called *Meloidogyne* sp.

Designing for life

Philippe Vain is a plant biotechnologist at the John Innes Centre in Norwich. His team, along with researchers at the University of Leeds, has been designing genetically modified crops aimed at helping many of the world's poorest farmers. 'Our goal is to improve the pest resistance of key crops – rice, bananas, and potatoes – for developing countries,' says Philippe. 'In ten years' time there will still be more than half a billion people in the world without a reliable source of food. It is a much better strategy to give these people the means of food production instead of supplying food aid all the time.'

Nematode worms reduce crop yields

The target of his research is nematodes, microscopic worms that live in the soil. These worms attack the roots of crops, taking nutrients from the plant and laying their eggs inside the tissues. 'If you have a small infestation, you're going to get a reduced yield, a large infestation and you'll lose most of the crop.' For a poor farmer this can be a matter of life and death.

Farmers could kill the worms by spraying the crops using chemicals called nematicides, but these are expensive and highly toxic to humans and the environment. Instead, it was decided to develop a crop that was resistant to the pests.

Adding an extra gene

The plants already have genes for natural substances called cystatins. The cystatin genes are active in certain parts of the plant, for example, their seeds. Cystatins affect insect digestion, so insects cannot eat parts of the plant which contain them. Cystatins have no effect on humans. In fact we eat them all the time, in seeds from crops such as rice and maize.

When a particular gene is active in a cell we say that it is being expressed. This means that the protein which it codes for is being made in the cells. Philippe's team added another copy of the cystatin gene to the plants, which is expressed in the root cells.

Agrobacterium species cause cancer in plants. This plant is infected with *Agrobacterium tumifaciens*, which causes crown-gall disease.

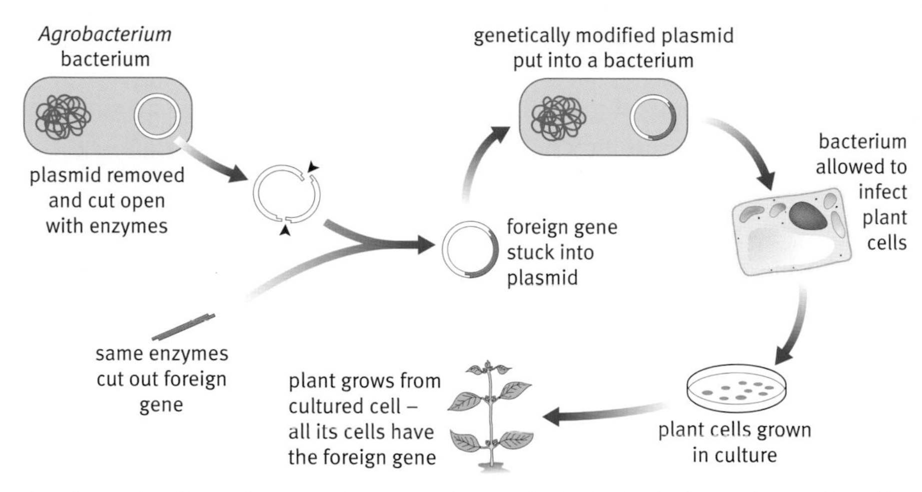

Agrobacterium is used to transfer a gene into a plant's DNA. The bacterium's plasmid acts as the vector in this example of genetic modification.

This would make the roots indigestible to the nematodes. The researchers used the bacteria *Agrobacterium* as a vector to carry the extra gene into the plant's gentic material.

The only difference between the final genetically modified plant and the original is an extra copy of the cystatin gene. Nevertheless, by law, any genetically modified plant has to go through extensive testing and safety trials before it can be released into the environment.

The resulting plants show a high level of resistance to the nematode and are ready to be offered to farmers as part of a government aid project. Phillipe Vain: 'You want to make a contribution. It's very rare to have a crop improvement strategy that really works, so it's very exciting to see the outcome.

'For us, the best result will be people trying the crop and it making a difference to their lives.'

Much of Philippe's work is on banana plants and rice.

Find out about:

- concerns about GM plants

3C How risky are GM crops?

In the US a wide variety of genetically modified (GM) crops are grown and sold. Some of the genes which have been added to crops are for disease resistance, but other characteristics have also been modified. For example, tomatoes have been modified to grow larger and with a sweeter flavour to meet customer demands.

Opponents of GM crops are concerned about the possible effects that any genetically modified organism (GMO) could have. As a result of such concerns, European countries have stricter regulations controlling the release of GMOs.

Anyone who wishes to introduce a GMO into the UK environment must apply to Defra (the Department for Environment, Food and Rural Affairs). Each application is looked at by the Advisory Committee on Releases to the Environment (ACRE), an independent scientific expert committee. ACRE advises Defra on every GMO application. Very few GM crops are currently grown in the UK.

Why are some people concerned about GMOs?

GMOs are living things. They reproduce between themselves, and interbreed with non-GM organisms. It is impossible to predict with absolute certainty how GMOs will interact with other species. The potential risks of introducing GMOs into the ecosystem need to be balanced against the potential benefits.

The table outlines some of the arguments about GMOs:

Arguments against the release of genetically modified organisms	Counter arguments
The added genes could make 'safe' plants produce toxins or allergens.	Food safety organizations can check for these.
Marker genes for antibiotic resistance could be taken up by disease organisms.	These antibiotics are not used in medicine, so it wouldn't matter.
Pesticides may 'leak' out of GM plant roots, and damage insects and microbes they were not designed to kill.	Insect-resistant plants reduce pesticide application so they benefit the environment.
GM crops may cause ecosystem changes that cannot be reversed.	Farmers may benefit from healthier crops and lower costs of production.
Multinationals will increase their domination of world markets.	Some GM technology has already been shared with developing nations.
Many consumers in EU countries refuse to buy GM products so farmers may lose markets.	Consumers in most countries are prepared to buy GM crops.
Poor farmers will not be able to afford the GM seeds.	Gene technology can develop more nutritious and higher-yielding plants that will benefit developing countries.

Questions

1 A disease-resistance gene is added to a crop plant. What advantages could this bring for the

- **a** plant
- **b** farmer
- **c** environment
- **d** consumer?

2 What possible disadvantage could there be for each of these groups?

3 Give one argument which you agree with either for or against GM crops which is

- **a** social
- **b** environmental
- **c** economic
- **d** ethical

3D Genetic testing

Find out about:

- how DNA technology is used for genetic testing

Genetic tests are used to identify faulty alleles in adults, fetuses, and embryos. On pages 20–25 you found out about different views people may hold on genetic testing.

The first genetic tests were developed in the 1980s, and are now available for several genetic disorders, such as cystic fibrosis, haemophilia, and Huntington's disorder. More recently tests have been developed to provide information about a person's risk of developing certain **multifactorial** diseases. A multifactorial disease is caused by a number of factors. Genes may be part of the cause, but they are not the only reason the disease develops. Environmental factors also play a major part in determining whether a person develops the disease.

The technology behind genetic testing

Genetic tests use artificially made pieces of DNA called **gene probes**. A gene probe is a short piece of single-stranded DNA which has complementary bases to the allele being tested for. So the probe will stick to this allele. If the probe sticks to a person's DNA, then they have the disease allele. Scientists use two different techniques to find out whether the probe has stuck to a DNA sample:

- **UV** – a fluorescent molecule is attached to the gene probe when it is made. These molecules glow under ultraviolet (UV) light.
- **Autoradiography** – the gene probe is made from radioactive DNA bases. These blacken X-ray film.

The technique of using gene probes to identify disease alleles is very similar to the process of **DNA fingerprinting**. The only difference is in the gene probes which are used. You can read more about DNA fingerprinting in the next Biology in action pages.

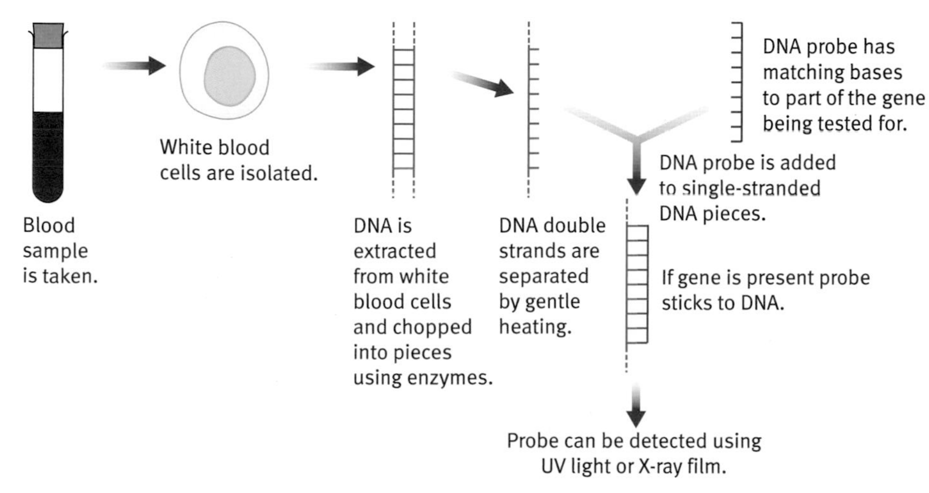

Using a gene probe for a genetic test.

Key words

multifactorial disease
gene probes
UV
autoradiography
genetic fingerprinting

Questions

1. Suggest why white blood cells are used as the source of a DNA sample.
2. Describe the structure of a gene probe.
3. Why are DNA probes useful for genetic testing?
4. How do scientists find out if a gene probe has stuck to a person's DNA?

DNA fingerprinting

How are genes copied?

Genes are copied using a technique called **polymerase chain reaction (PCR)**. PCR uses enzymes to replicate the gene's DNA. The process is similar to DNA replication in cells, when chromosomes are copied before cell division. It takes just a few minutes for PCR to copy a piece of DNA, and 20 cycles will turn one DNA molecule into a million.

PCR is very useful in forensic DNA work. Often the amount of DNA found at a crime scene is very small. Before the PCR technique was introduced, this DNA was often not enough to give a reliable DNA fingerprint. PCR is also used to copy a gene for genetic modification (page 214).

Discovering DNA fingerprinting

Sir Alec Jeffreys in the 1980s

'My life changed on Monday morning at 09.05 am 10 September 1984. In science it is unusual to have such a "Eureka moment". We were getting extraordinary variable patterns of DNA including from our technician and her mother and father. My first reaction to the results was 'this is too complicated'; and then the penny dropped and I realized we had genetic fingerprinting."

Sir Alec Jeffreys is describing the discovery that made him one of the most famous scientists in the world. He had detected regions along the DNA molecule known as **minisatellites**. These are sections of DNA that are not genes. A unit of about 30 base pairs is repeated over and over again tens or even thousands of times. The same minisatellite can be different lengths in different people. DNA fingerprinting looks at the lengths of many different minisatellites throughout a person's DNA.

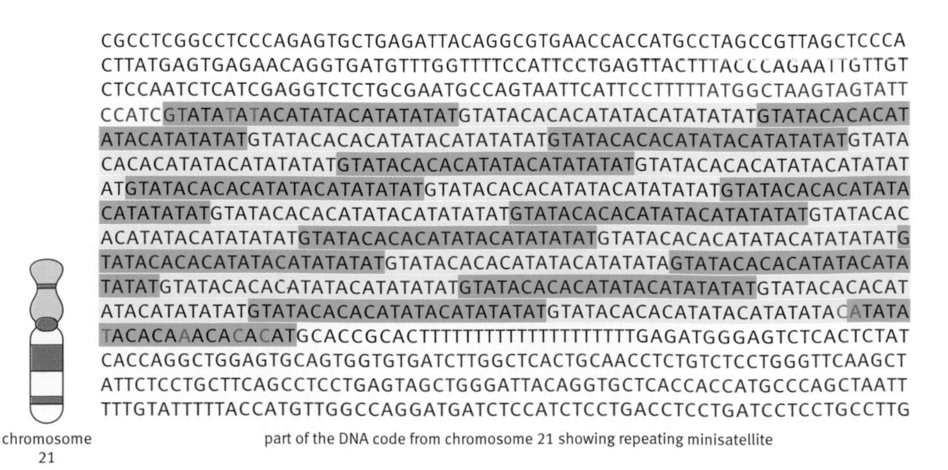

chromosome 21

part of the DNA code from chromosome 21 showing repeating minisatellite

The first human minisatellites were described in 1980. Four years later Sir Alec Jeffreys realized how they could be used in DNA fingerprinting.

How the technique works

Dr Tim Slingsby is a young scientist working with Sir Alec at Leicester University. He explains the basic techniques of DNA fingerprinting:

- *Extract DNA from the tissue sample:* break open the cells and purify the DNA.
- *Cut the DNA up into pieces:* **restriction enzymes** cut DNA at particular base sequences. One or two restriction enzymes that cut DNA at non-minisatellite sequences are added to the DNA sample. The pieces of cut-up DNA are different lengths in different people – because their minisatellites are different lengths.
- *Separate the fragments out:* gel electrophoresis separates the pieces of DNA out according to size. The cut-up DNA is put at one end of the gel and an electric current moves the pieces through the gel. Shorter pieces move faster than longer pieces, so they move further along the gel.
- *Make the pieces visible:* after gel electrophoresis the pieces of DNA have been separated, but we cannot see the different lengths. The DNA is transferred from the gel to a piece of membrane and a DNA probe is added that binds to the minisatellite sequences. The probe has a radioactive chemical attached to it. The membrane is placed next to an X-ray film. Where the probe has stuck to the DNA it causes the X-ray film to go black. So the final DNA fingerprint is a series of black lines on an X-ray film.

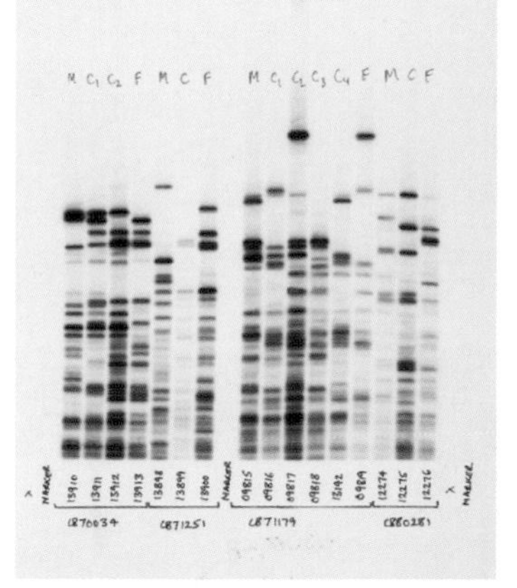

Each black line shows where pieces of DNA that contain minisatellite sequences have been separated out by gel electrophoresis.

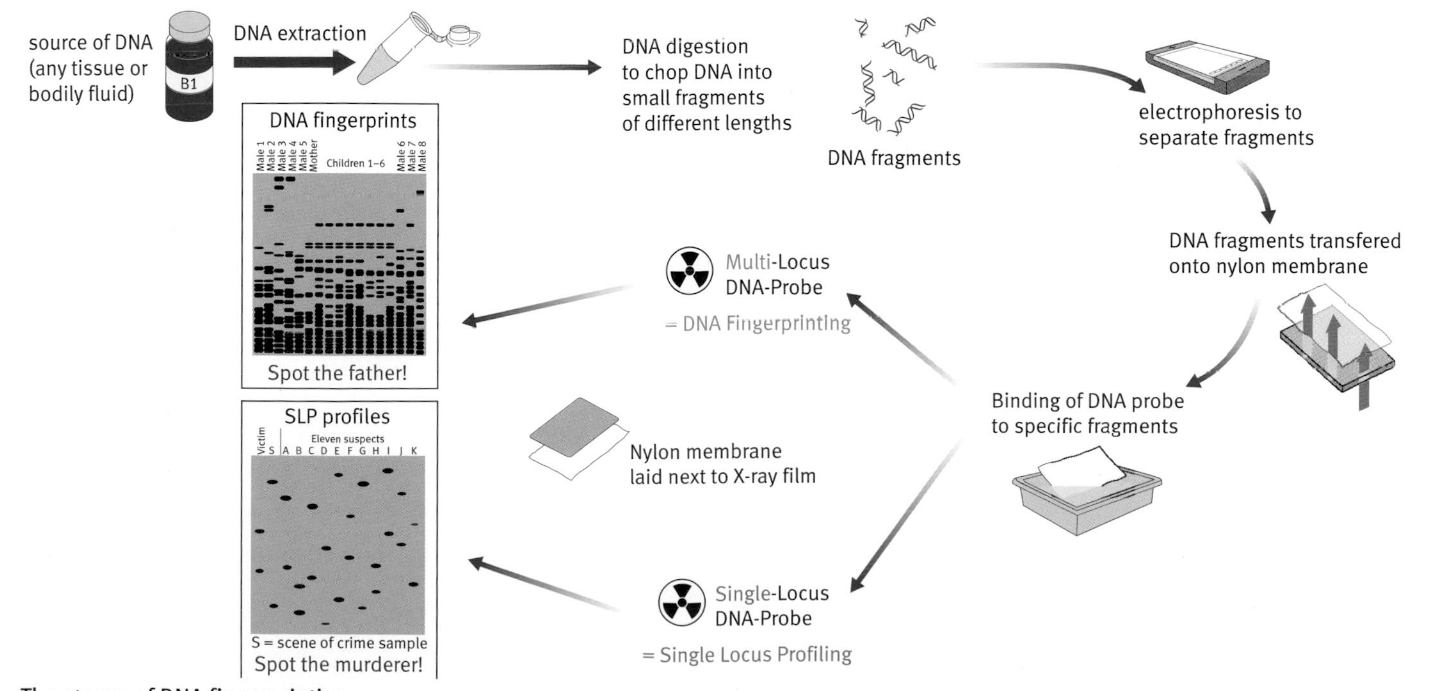

The stages of DNA fingerprinting

The impact of DNA fingerprinting

The first application of DNA fingerprinting was in an immigration dispute. In 1985 a young man was returning to the UK after a trip to Africa. He was refused re-entry to the UK on the grounds that he was not related to the people he claimed were his family. His solicitor had read about DNA fingerprinting. People from the same family share lots of their DNA, so there are similarities between their DNA fingerprint patterns. DNA fingerprinting proved that this man was a member of the family, and he was re-admitted to the UK. DNA fingerprinting is also used in paternity cases to show whether or not a man is the biological father of a child.

To begin with DNA fingerprinting was not used in criminal cases. A DNA fingerprint pattern was difficult to interpret, and needed too much high-quality DNA for the technique always to work. DNA from crime scenes is often only present in small amounts, e.g. a few drops of blood, or a single hair follicle. It may also be quite badly decomposed if the sample is old.

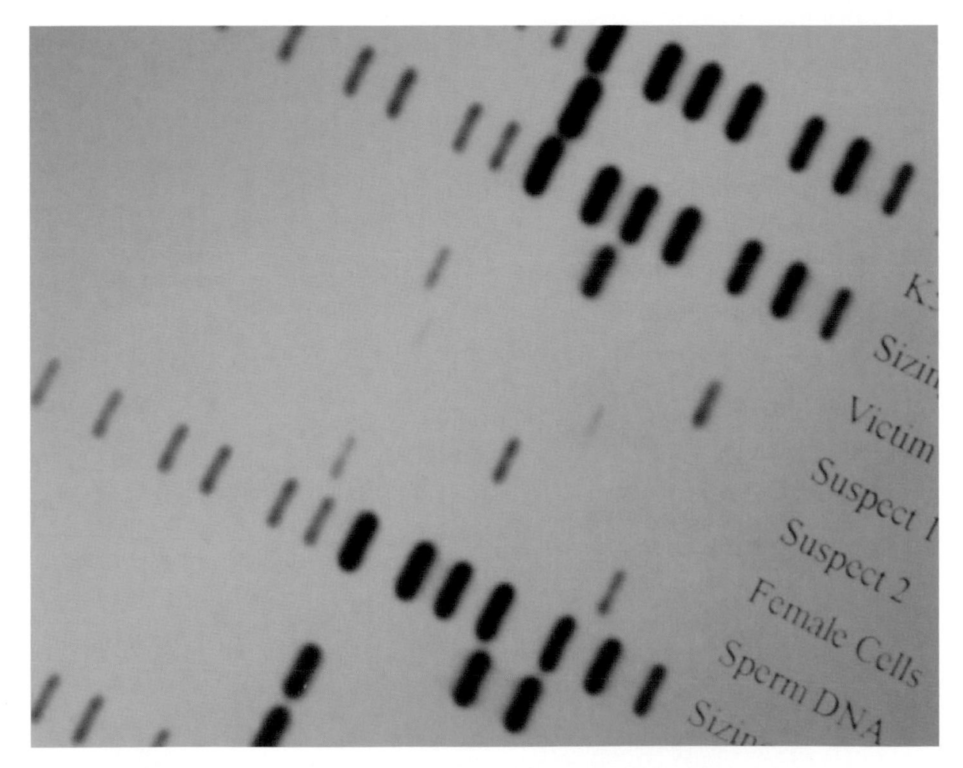

A DNA profile

So very quickly Sir Alec refined the technique and developed **DNA profiling**. This works in a very similar way to DNA fingerprinting, but only a few minisatellites are targeted. The pattern is much simpler, so it is easier to interpret. The technique can also be used with smaller samples of DNA. But Sir Alec points out that DNA profiling does not directly solve crimes. Just because a person's DNA is present at a crime scene does not necessarily mean they committed the crime. He says: 'It establishes whether sample X comes from person Y. It is then up to the court to interpret that in the context of other evidence in a criminal case.'

Perhaps surprisingly DNA profiling was first used to prove a person's innocence. In 1985 a man confessed to the brutal murder of a young girl. However, DNA profiling proved him innocent. DNA profiles were then taken of the local male population, and the man responsible for the crime was caught.

Looking to the future

Tim Slingsby is working on new applications of DNA technology: 'I am looking at DNA sequences in human sperm to investigate how variation is produced and maintained in humans from one generation to the next. I am particularly interested in **recombination**, a process that occurs during meiosis – cell division to produce sex cells. During recombination, pairs of chromosomes exchange sections of their DNA. This process leads to chromosomes in the sex cells that have a completely unique combination of alleles.'

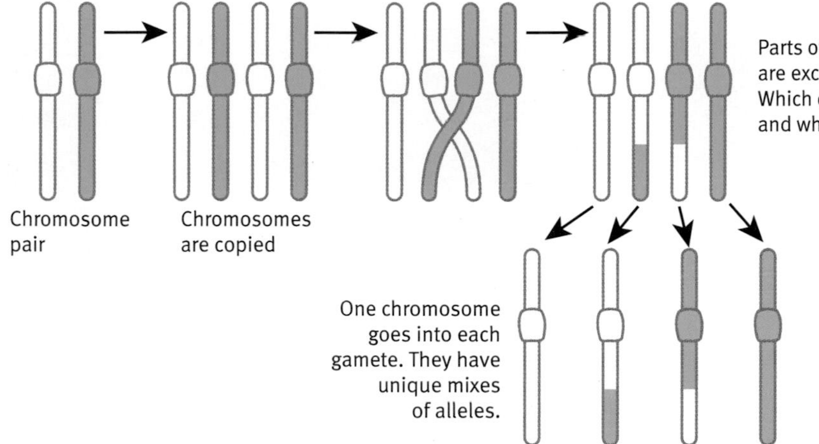

After recombination the chromosomes still have the same genes – but they may have different combinations of alleles.

'A better understanding of recombination will be valuable to scientists who are trying to identify genes that make people more likely to develop particular genetic diseases. It will also help the study of diseases that are caused when recombination goes wrong, for example, thalassaemia. Knowing more about recombination helps us to understand more about human evolution, and it could also be helpful for cancer research.'

So why did Tim choose to work at Leicester University? He explains: 'Alec is wonderfully enthusiastic about science and despite all of his success, he remains very approachable and down to earth. He is one of the very few scientists in his position who still works at the bench – and he churns out important results at a rate that is both inspiring and intimidating! He recognises the importance of public understanding of science, and often gives up his time to present his work outside of the science community. Alec's success shows that he is clearly more than an ordinary scientist, but one of the first things that I picked up from him is that every scientist needs one basic skill – curiosity.'

Topic 4 Blood and circulation

In your body, every cell matters. Blood delivers nutrients and oxygen to each group of cells and takes away their waste. Your heart pumps throughout your life to keep the blood moving.

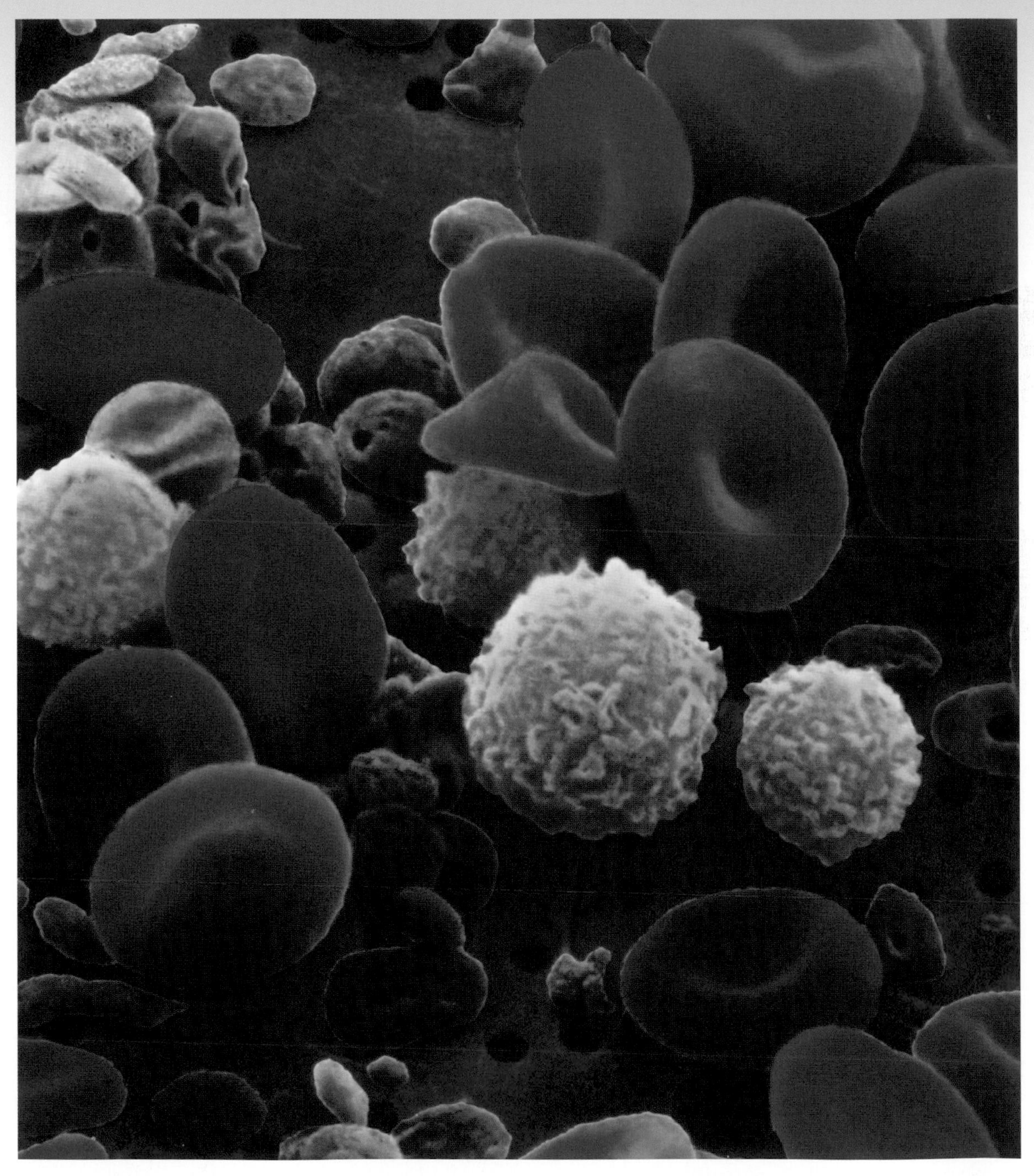

Blood cells (the colours in this photograph are computer-generated (false colour)). (× 2500)

Your heart

Your heart is a hollow sac of muscle. If you are at rest it beats about 60–80 times per minute, forcing blood through a network of blood vessels. Blood vessels carry blood to and from your body's cells. At the cells useful molecules pass from the blood into the cells, and waste products are taken away.

Blood

Blood tissue is made of several different types of cells floating in a clear liquid called **plasma**. Each type of cell has an important role in maintaining life. The plasma carries many dissolved chemicals around your body.

Blood groups

Blood donations save thousands of lives in the UK every day. Some people lose blood in accidents, and others need different parts of blood to treat particular illnesses. If you do have a blood transfusion, doctors and nurses will check carefully that you are given blood of a matching blood group. Your blood group is an inherited feature, determined by your genes.

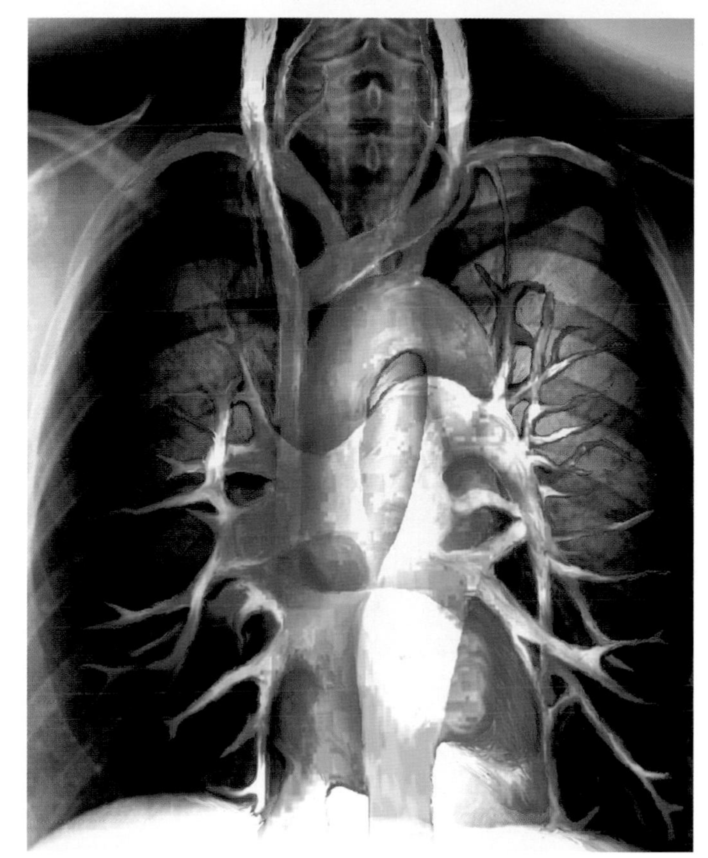

Artwork showing the heart and major blood vessels of the human chest.

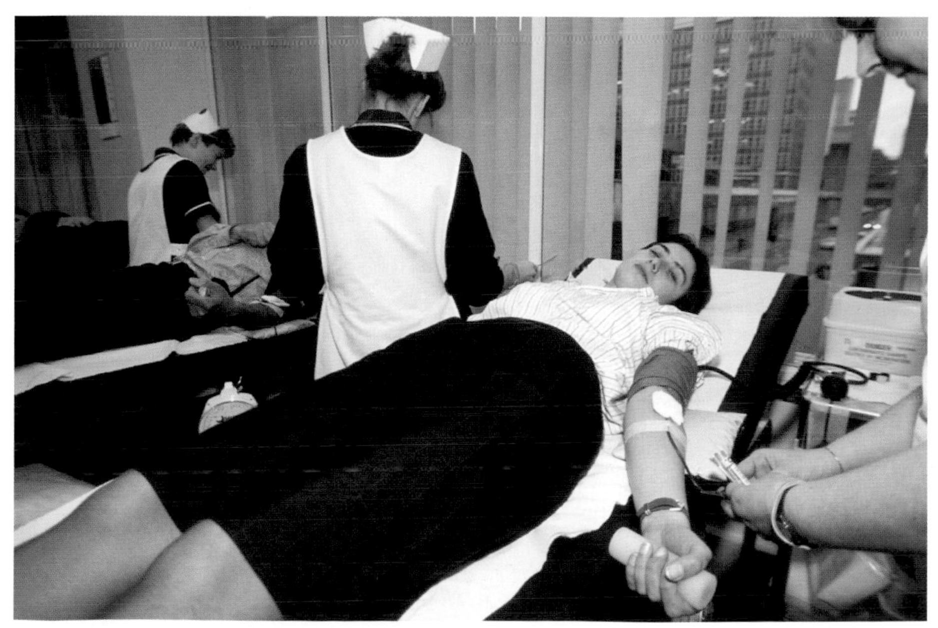

This woman is giving blood. Donated blood helps about one million people in the UK every year.

Key words

plasma

Find out about:

- the different parts of blood
- blood transfusions

4A Blood

Systems for moving molecules

All living things need a way of moving materials around their body. Small organisms, for example a single-celled *Amoeba*, can rely on simple diffusion.

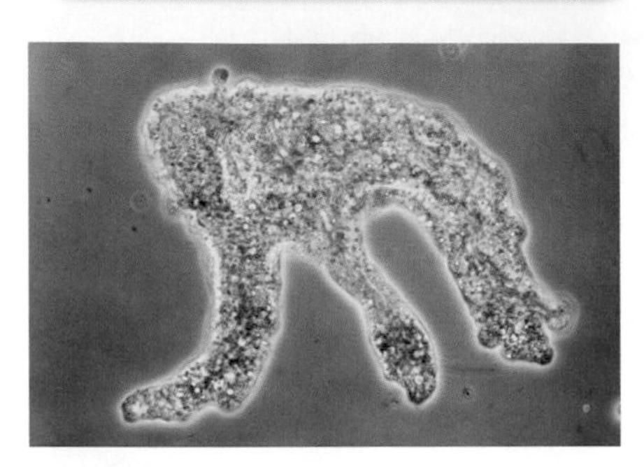

Amoeba

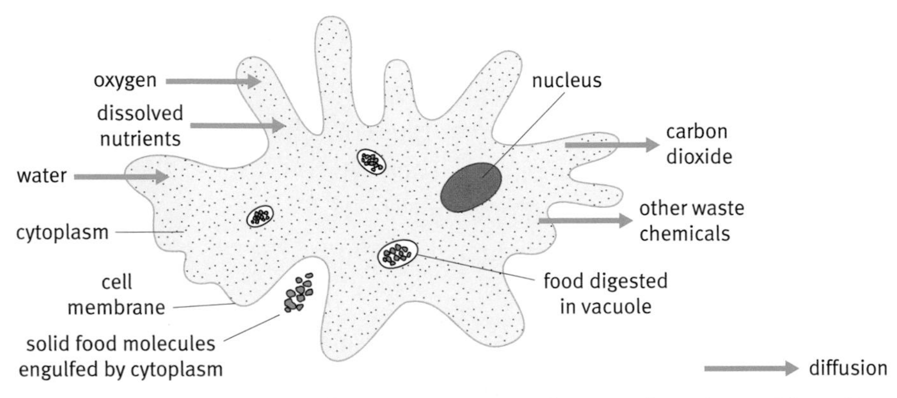

Amoeba is small enough for diffusion to transport substances in and out of the cell.

Diffusion is a slow process, so larger organisms need special transport systems. In most animals this is a **circulatory system**.

The human circulatory system

You have between 5 and 7 litres of blood circulating around your body. A sample of blood looks completely red. A closer look shows that it contains cells floating in a pale yellow fluid called plasma. Plasma is mainly water. It carries a wide range of dissolved materials including food molecules, hormones, and waste products from cells. Plasma also helps to distribute heat around the body.

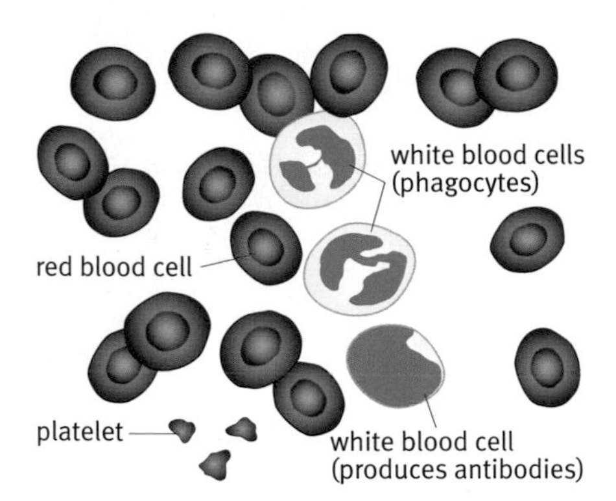

This diagram shows the cells in blood, but not in the correct proportions. Each 1 mm^3 of blood contains approximately 5 million red blood cells, 250 000 platelets, and 7000 white blood cells.

There are three types of cell floating in plasma:

- **red blood cells** – to transport oxygen
- **white blood cells** – to fight infection
- **platelets** – which play an important role in blood clotting at an injury site

Red blood cells

Red cells are the most obvious blood components because of their colour. They are packed with the protein **haemoglobin.** Haemoglobin binds oxygen as blood passes through the lungs. The oxygen is released from haemoglobin as blood circulates through the tissues of the body.

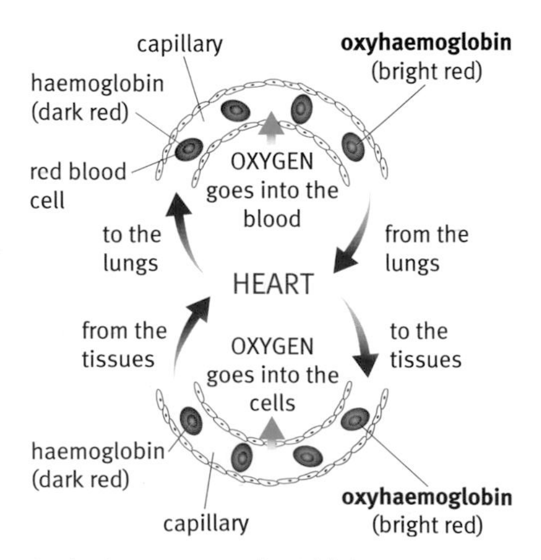

In the lungs oxygen is at high concentrations and binds to haemoglobin. At low oxygen concentrations in body tissues the oxygen is released. It diffuses into body cells which use the oxygen for respiration.

Red blood cells have no nucleus, which allows more space for haemoglobin. The biconcave shape gives the cells a large surface area. This means that oxygen can diffuse in and out of the cells more rapidly. This shape also gives cells flexibility to squeeze through capillaries, which are only one cell wide.

White blood cells

White blood cells protect the body from infection by disease-causing microorganisms. They produce antibodies, and engulf and digest microorganisms by phagocytosis. You can read more about how white blood cells work on page 42.

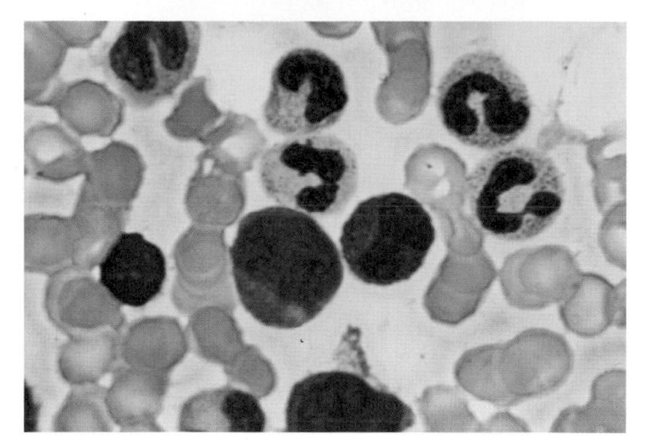

White blood cells

Platelets

Platelets are fragments of cells which are made from the cytoplasm of large cells. When a blood vessel is damaged, for example when you are cut, platelets stick to the cut edge. They send out chemicals which trigger a series of reactions that form a clot at the cut site. Clotting helps to stop too much blood being lost from the body.

If you do suffer a major blood loss you will need a **blood transfusion**. There is a huge demand for donated blood – 9 000 **donors** a day are needed to keep UK hospitals fully supplied. The blood is stored in bags mixed with an anticoagulant to stop it clotting. The large capital letter on each bag shows the blood type. Getting the wrong blood type in a transfusion can be very serious.

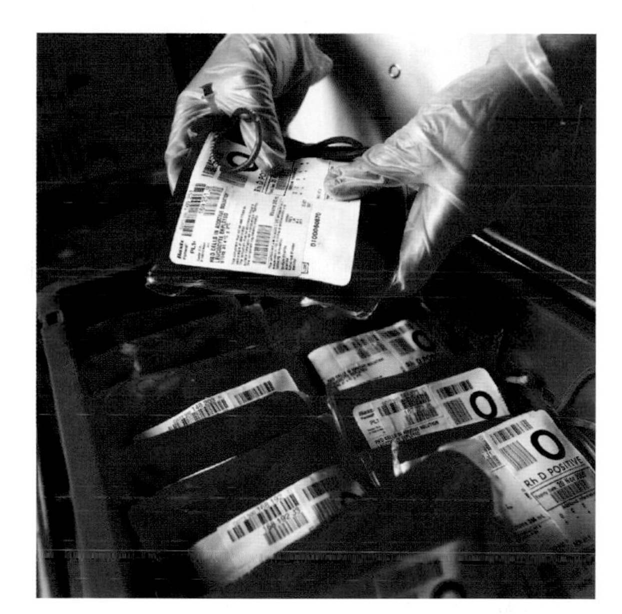

Donated blood

The ABO blood types

Every cell carries markers on the outside. These markers are called antigens. Antigens are what white blood cells use to detect foreign cells. If your white blood cells detect foreign antigens on a cell, they produce antibodies to destroy it. Red blood cells can carry two different antigens, A and B. This determines your **ABO blood type**:

- just A antigens – blood type A
- just B antigens – blood type B
- both A and B antigens – blood type AB
- neither type of antigen – blood type O

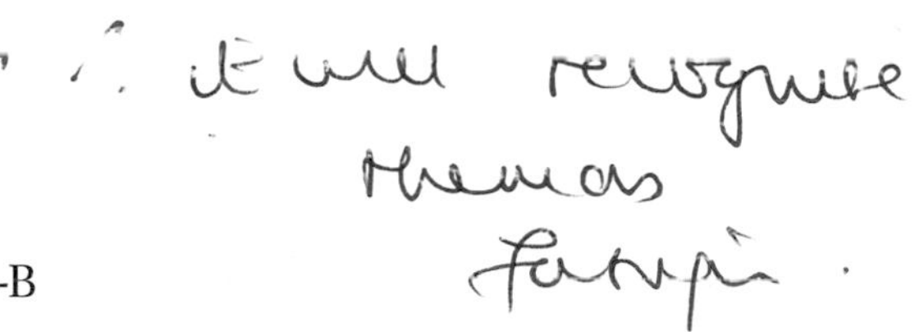

Your plasma contains antibodies against the red blood cell antigens you do not have. There are two types of antibody, anti-A and anti-B. So for example, someone with A antigens on their red blood cells will have anti-B antibodies in their blood plasma.

Blood type	Antigens	Antibodies
A	A	anti-B
B	B	anti-A
AB	A and B	none
O	neither	anti-B and anti-A

ABO blood types

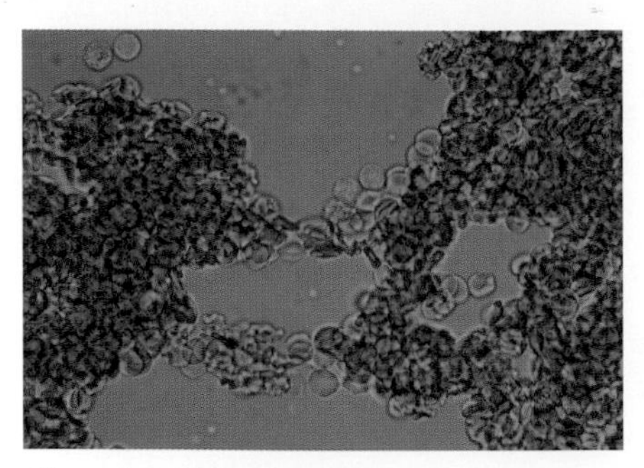

Clotted red blood cells

Blood transfusions

A person who gets a blood transfusion is called the **recipient**. It is essential to know the recipient's blood type. Anti-A antibodies have no effect on red blood cells carrying antigen B, but they will cause cells carrying antigen A to clot together in the bloodstream. These clots block blood vessels, stopping supplies of glucose and oxygen from reaching cells.

Before a blood transfusion the recipient's blood type is matched to suitable donated blood. Doctors check that the recipient's antibodies will not react with the donor's antigens. The antibodies in the donor's blood are present in much smaller amounts, and do not cause clotting in the recipient.

Recipient / Donor	O anti-A + anti-B	A anti-B	B anti-A	AB none
O anti-A + anti-B	–	–	–	–
A anti-B	clotting	–	clotting	–
B anti-A	clotting	clotting	–	–
AB none	clotting	clotting	clotting	–

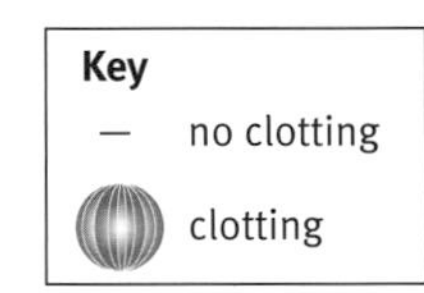

People with blood type AB don't have anti-A or anti-B antibodies. They can safely be given a transfusion of any blood type, so they are known as a **universal recipient**. Anyone with blood type O doesn't have A or B antigens. Their blood can be given to anyone – so they are known as a **universal donor**.

Questions

1 Write one sentence to describe the function of the four main components of blood.

2 List the different blood types in the ABO system.

3 For each type say which

a antigens are present on the red blood cells

b antibodies are present in the blood plasma

4 A patient has blood group O. Explain what would happen if they were given a transfusion of blood type A.

5 What blood type or types can a person with blood type O be given in a transfusion? Explain your answer.

6 Explain the meaning of the terms universal recipient and universal donor.

(7) When blood is donated it is checked for blood type. Find out what other checks are made.

Key words

circulatory system
red blood cells
white blood cells
platelets
haemoglobin
oxyhaemoglobin
blood transfusion
donor
ABO blood type
recipient
universal recipient
universal donor

4B How blood types are inherited

Find out about:

- how blood types are inherited

Your ABO blood type is an inherited feature. It is determined by a single gene, like other features such as dimples, earlobe shape, and hairy ring fingers. You studied the inheritance of simple features such as these on page 14. For example, the dimples gene has two versions, called alleles. The allele for dimples is dominant, and the allele for no dimples is recessive. A person who inherits one of each of these alleles from their parents will have dimples.

There is an important difference in the control of blood type. The gene for ABO blood type has three different alleles:

- allele I^A – red cells have A antigens
- allele I^B – red cells have B antigens
- allele I^o – red cells have neither antigen A nor antigen B

Any person will only inherit two ABO type alleles – one from each parent. The table shows the possible combinations and the blood group these produce.

Alleles	Blood type
I^AI^A	A
I^AI^o	A
I^BI^B	B
I^BI^o	B
I^AI^B	AB
I^oI^o	O

A person with the alleles I^A and I^B has both types of antigen on their red blood cells. These alleles are **codominant**. Allele I^o is recessive to both the I^A and I^B alleles.

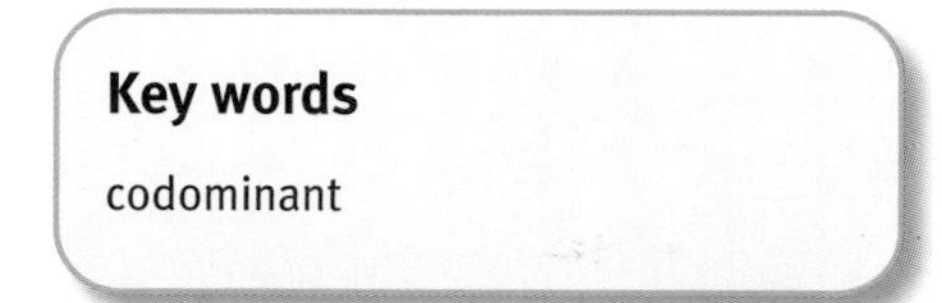

Key words

codominant

Predicting blood groups

You can use a genetic cross diagram to predict a child's possible ABO blood type. For example, a man with blood type AB and a woman with blood type O have a child. What blood type could the child inherit?

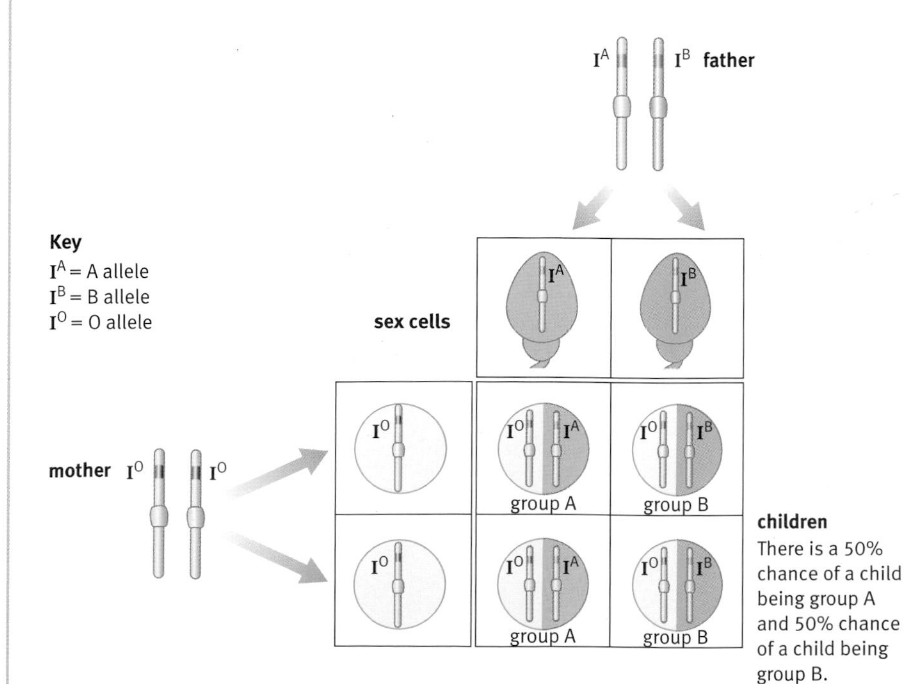

children
There is a 50% chance of a child being group A and 50% chance of a child being group B.

Questions

1 What blood type would a person with these alleles have:

a I^A and I^o **b** I^A and I^B
c I^B and I^B **d** I^B and I^o
e I^o and I^o?

2 The I^A and I^B alleles are codominant. Explain what this means.

3 Use a genetic cross diagram to show the blood types that a child could have if their parents were

a type AB and type A (with alleles I^A and I^o)

b type O and type B (with alleles I^B and I^o)

Karen Sugden

Genes and environment

Almost all of our characteristics are affected by both our genes and our environment. Many scientists, such as Karen Sugden, are investigating the links between these two factors.

Does depression have a genetic cause?

Karen works in the new hi-tech laboratories of the Social, Genetic and Developmental Psychiatry Centre at King's College, London. She is part of a team looking at the genetic basis for depression.

'Depression is one of the most important disorders in terms of worldwide health,' says Karen. 'It's vital to get a better understanding of the disease.' About two out of every three people in the UK will suffer from depression at some point in their lives. It may be mild, or last just a few weeks. For some people the condition is very serious. They need medication and other treatment to help them recover.'

What is depression?

People often say that they feel 'depressed' when they're fed up because of something that has happened to them. For example, they may have done badly in an exam. These ups and downs are quite normal, and people usually get over them fairly quickly.

When a person has depression they feel a low mood which may last for several weeks. Everyday tasks, such as going to the shops or tidying up, may feel too much to manage. A person with depression may also suffer from physical symptoms, such as tiredness and headaches.

'I know now that depression affects different people in different ways. To start with I felt very down, but there were other things like I'd lose my temper easily, sometimes for nothing really. I couldn't focus on anything, nothing seemed worth doing, and I couldn't imagine being happy. I've had different treatments, some medication from my doctor, but also he's helped me to look after myself better generally and that really does help. I make sure that I eat properly, and do get out of the house. It's important for me to be active, because that helps me sleep. My doctor has really supported me, and helped me to get better.'

What causes depression?

Karen's experiments involve looking at a gene called 5-HTT, which helps to regulate a chemical in the brain known as serotonin. Serotonin acts as a chemical messenger between nerve cells.

The 5-HTT gene comes in two different alleles – a long version and a short version. We all inherit two copies of the gene, one from our mother and the other from our father.

'We knew from other studies that the 5-HTT gene was involved in depression and we also knew that stress causes depression,' explains Karen. 'What we didn't understand was why if bad things happen in people's lives, some people get depressed and some people don't.'

The scientists studied a group of 1000 people living in New Zealand. 'We got cheek swabs from the individuals,' says Karen. 'From these cheek cells we then extracted the DNA.' The DNA samples contained each person's entire genome – all their genes.

Using the latest DNA analysis equipment, the researchers were able to take a sample from each person's DNA and examine their 5-HTT genes. Karen's team compared this information against whether they had suffered stressful life events or depression.

They discovered that those people who had the short allele of the 5-HTT gene were more likely to get depression if they suffered stressful events in their lives.

This suggests that depression is a disease that is influenced by both genetic and environmental factors. Just because someone has the short 5-HTT allele does not mean that they will get depression. But if they suffer an extremely stressful event, it makes depression more likely.

Karen says that making the discovery made the hard work involved in the research worthwhile. 'To see the results come out, particularly when you have a theory and that theory is proved right, is very rewarding.' But when it comes to understanding depression, this work is only the start: 'There's still a huge way to go, lots and lots of things to find out yet!'

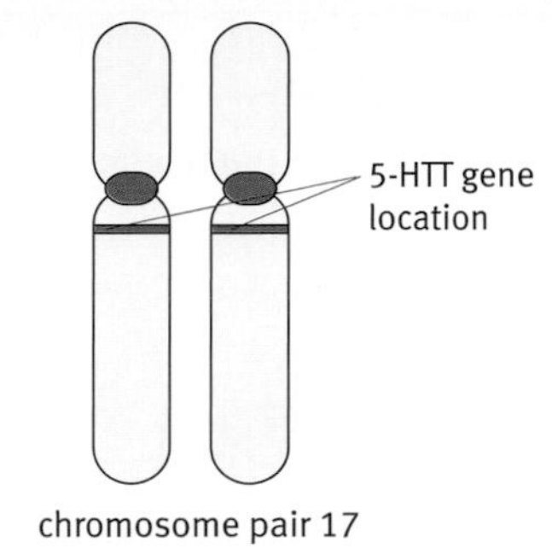

The 5-HTT gene is found on human chromosome 17. Each person has two copies of the gene, one from each of their parents.

Find out about:

- the structure of the human heart

4C The heart

Your heart is a hollow muscle. It is a divided into two separate halves by a wall down the middle. Each half is about the size of a clenched fist. The heart pumps blood around the body, even when you are resting. On average your heart beats 100 000 times a day, which makes 35 million beats a year. To get an idea of the force each heartbeat produces, squeeze a tennis ball as hard as possible.

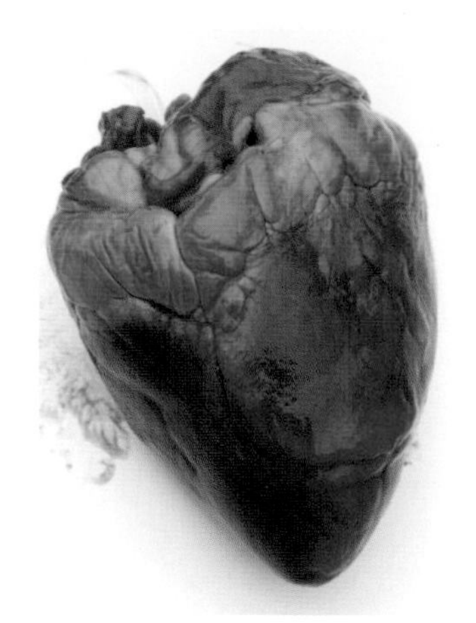

A sheep's heart is similar in structure to the human heart.

Inside the heart

Each side of the heart has two chambers – an **atrium** and a **ventricle**. Blood enters the atria which act as holding areas. The thin walls of the atria allow them to swell as blood arrives from the veins. The atria muscle then contracts, and blood is pushed into the ventricles. The ventricle walls have much thicker muscle. They contract with enough force to push blood out and away from the heart.

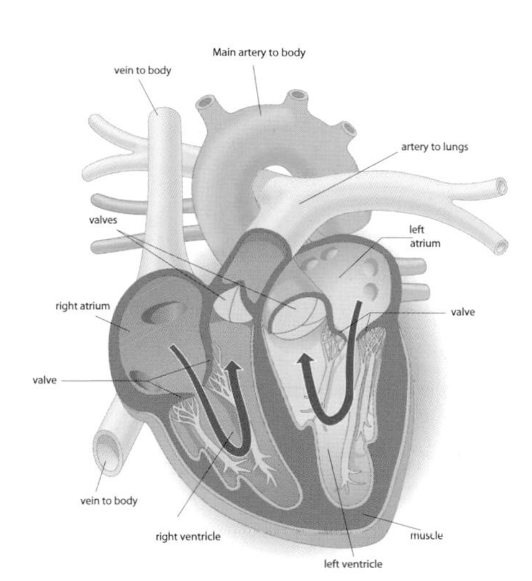

Diagram of a human heart. The artery leaving the right-hand side of the heart has been coloured blue to show that it is carrying blood which is short of oxygen.

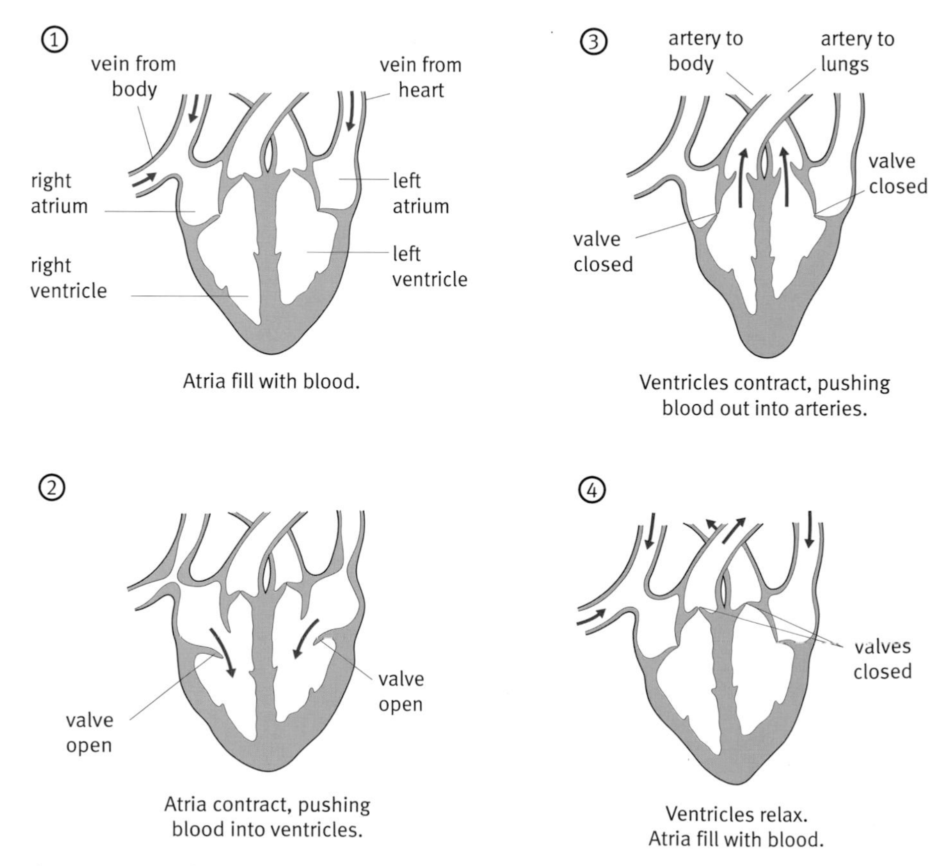

The sequence of events in a heartbeat. Both sides of the heart contract at the same time.

A double circulation

Blood from the body enters the right atrium of the heart. It is pumped out of the right ventricle towards the lungs to pick up oxygen. The blood is now **oxygenated**. It returns to the left atrium and passes into the left ventricle. Here it gets another, harder pump which carries it around the rest of the body. The left ventricle has a thicker wall of muscle than the right, because it has to pump blood to the whole of the body. The right ventricle only pumps blood as far as the lungs. As the blood passes around the body it gradually gives up its oxygen to the cells. It becomes **deoxygenated**. The blood then returns to the right atrium again. So blood passes through the heart twice on every circuit of the body. This is called a **double circulation**.

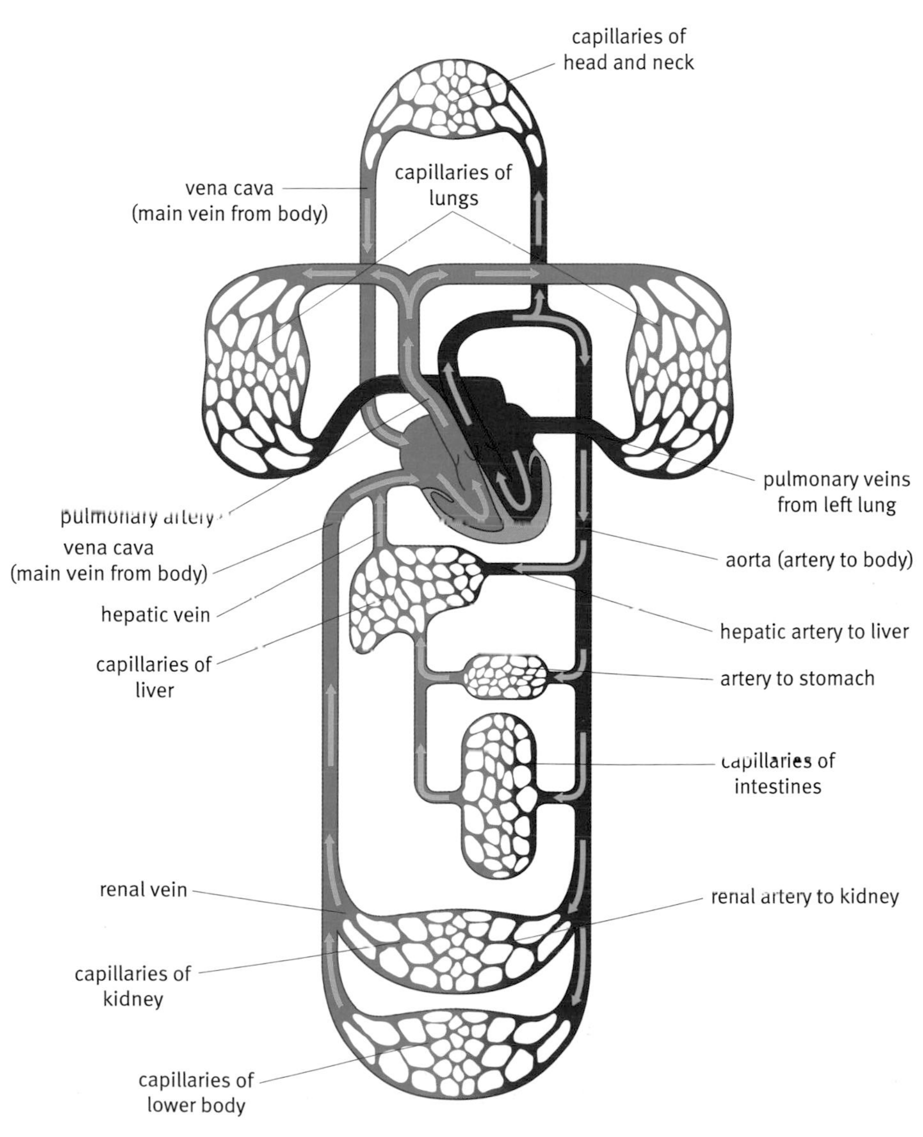

Blood is carried away from the heart by arteries, and towards the heart in veins. You can read more about the structure and function of blood vessels in Module B2 *Keeping healthy*.

Key words

atrium
ventricle
oxygenated
deoxygenated
double circulation

Questions

1 Explain what is meant by a double circulatory system.

2 Explain the difference in wall thickness between
  a atria and ventricles
  b the right and left ventricles

Find out about:

- the function of valves in the heart and veins
- why tissue fluid is important

4D Valves and tissue fluid

Valves – a one-way system

When the ventricles contract they push blood out of the heart. But what stops blood going backwards? This is the job of the heart **valves**. They act like one-way doors to keep the blood flowing in one direction. There are two sets of valves in the heart:

- Between each atrium and ventricle. These valves stop blood flowing backwards from the ventricles into the atria.
- Between the ventricle and the arteries leaving the heart. These valves stop blood flowing backwards from the arteries into the ventricles.

Valves are also found in veins. Blood pressure is lower in veins than in arteries. Valves stop blood from flowing backwards in the veins in between each pump from the heart.

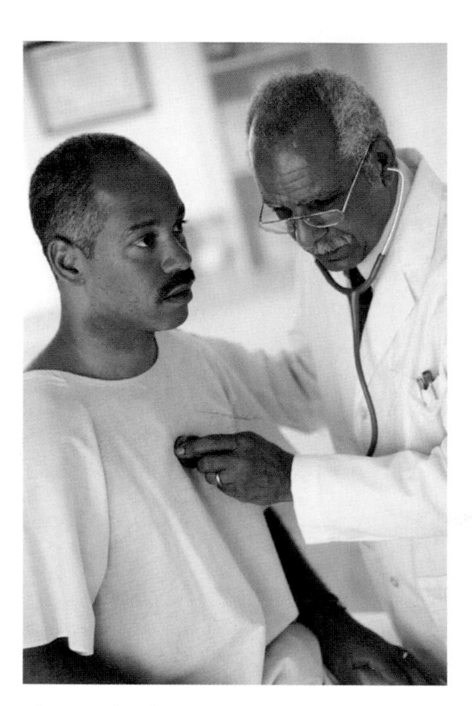

The comforting 'lub-dub' sound of a heartbeat is the sound of heart valves snapping shut. The 'lub' sound is caused by the valves between the atria and ventricles shutting. The 'dub' sound is made as the valves between the ventricles and the arteries close.

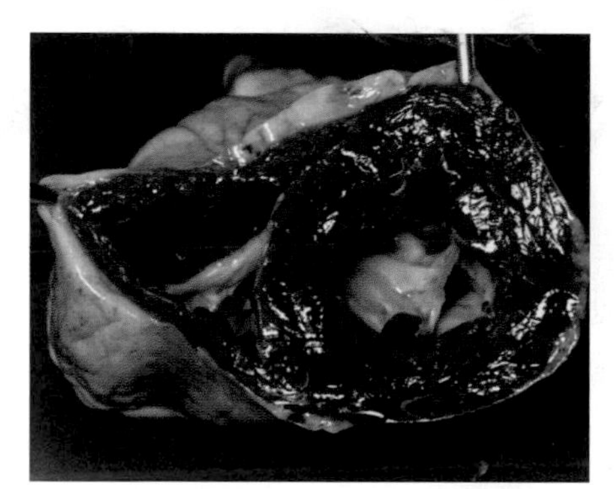

A valve between a ventricle and an artery. Strong tendons hold the valve flaps in place, preventing blood from flowing backwards.

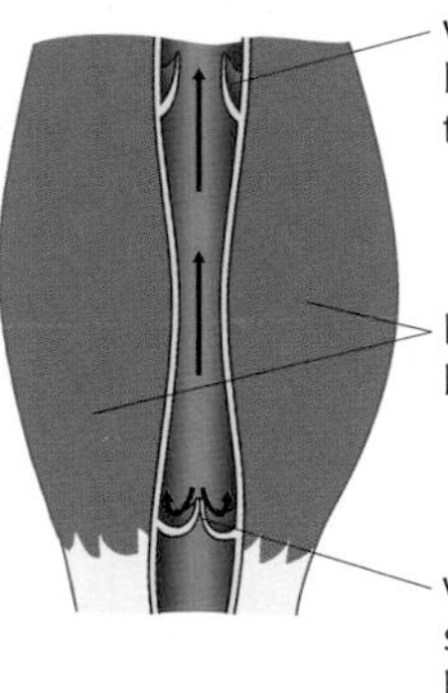

Valves stop the blood flowing backwards in the veins.

Why is a double circulation important?

The lungs and other body organs contain a **capillary network**. As blood squeezes through these capillaries its pressure drops. It has less force moving it along. If blood went straight from the lungs to the rest of the body, it would move too slowly to provide enough oxygen for the body's cells. The double circulation gives the blood a pump to get through each network of capillaries – the right ventricle pumps it through the lungs, and the left ventricle pumps it through the rest of the body.

Key words

valves
capillary network

Capillary networks

On average you will have 6 litres of blood in your body which is all pumped through the heart three times each minute. But the blood spends most of that time in capillary networks. Capillaries are where chemicals in the body's cells and in the blood are exchanged. The structure of capillaries makes them ideally suited for this function.

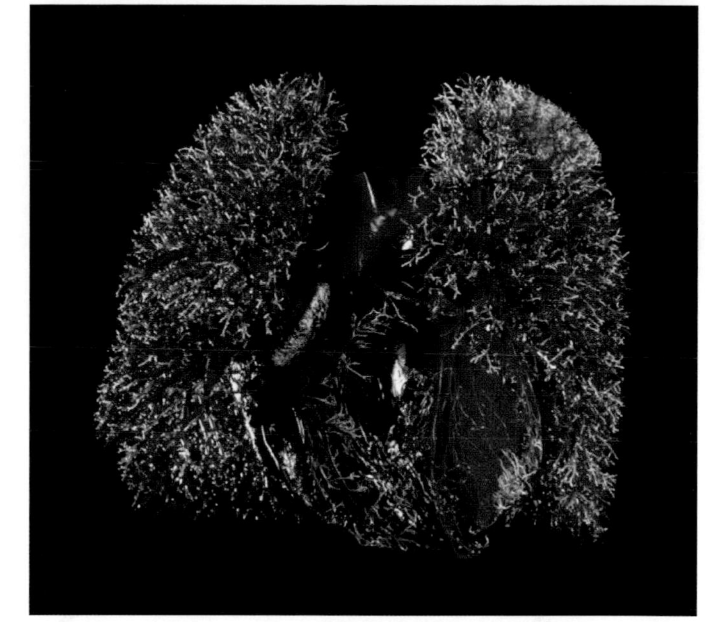

A resin cast of the capillary network of an adult human's lungs.

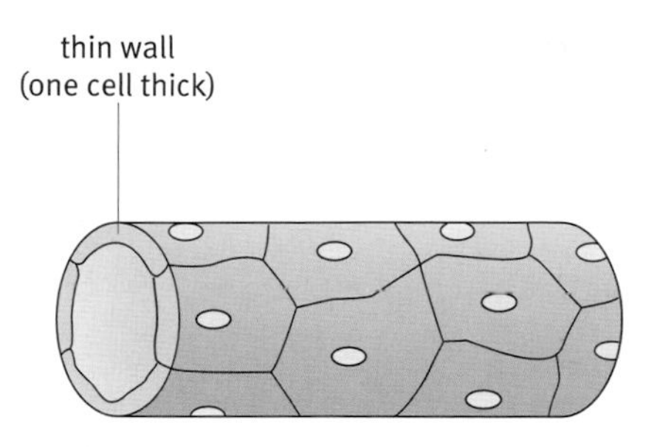

Capillary walls are very thin and porous.

When blood enters a capillary network from an artery it is at high pressure. Blood plasma is squeezed out of the capillary. It forms a liquid called **tissue fluid**, which bathes all of your cells.

Tissue fluid contains all the dissolved raw materials being carried by blood plasma. These chemicals diffuse from the tissue fluid into cells. Waste products from cells diffuse out into the tissue fluid.

As blood passes through the capillary network its pressure drops. Plasma stops being squeezed out, and tissue fluid with waste products from cells moves back into the capillaries.

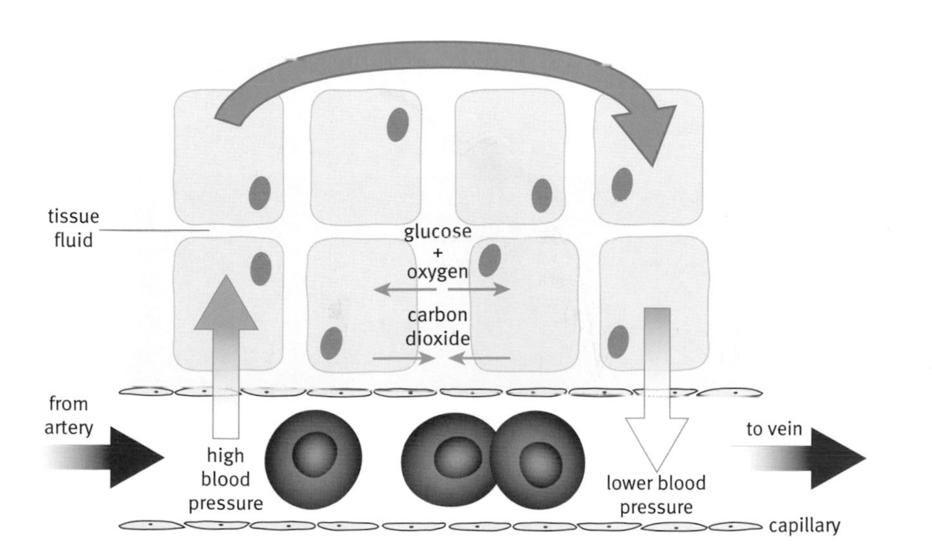

Capillary actions

Questions

1 Draw a flow chart to show the route blood takes on its journey around the body. Include all the key words on pages 230–234 in your description. Start with:

Blood leaves the left ventricle ➔

2 Describe the job of valves in the heart and veins.

3 What is tissue fluid made from?

4 Explain why tissue fluid leaks out at the start of capillaries, and moves back in towards the end of the capillary network.

5 Explain how tissue fluid helps the exchange of chemicals between the capillaries and body tissues.

6 Name four chemicals which are exchanged between body cells and tissue fluid.

Topic 5 Energy for life

Animals take in energy in the food they eat. This energy has to be transferred into a useful form to power movement, growth, or to keep the body warm. Energy is released from food by a process called **respiration**.

Muscle cells are using large amounts of energy to produce this movement.

Respiration

Respiration is a carefully coordinated series of chemical reactions. It happens in all plant and animal cells. Most living things need oxygen for respiration. When you exercise, respiration happens faster to release more energy. You must get more oxygen into your body, so you breathe faster.

Respiration without oxygen

Many organisms can release energy from food without oxygen. This is called **anaerobic** respiration. Most organisms can only survive in this way for a short time. A few organisms can survive either with or without oxygen, for example, yeast cells. There are a small number of bacteria species which cannot survive in oxygen, for example, *Clostridium tetani*, which causes the disease tetanus.

Key words

respiration
anaerobic

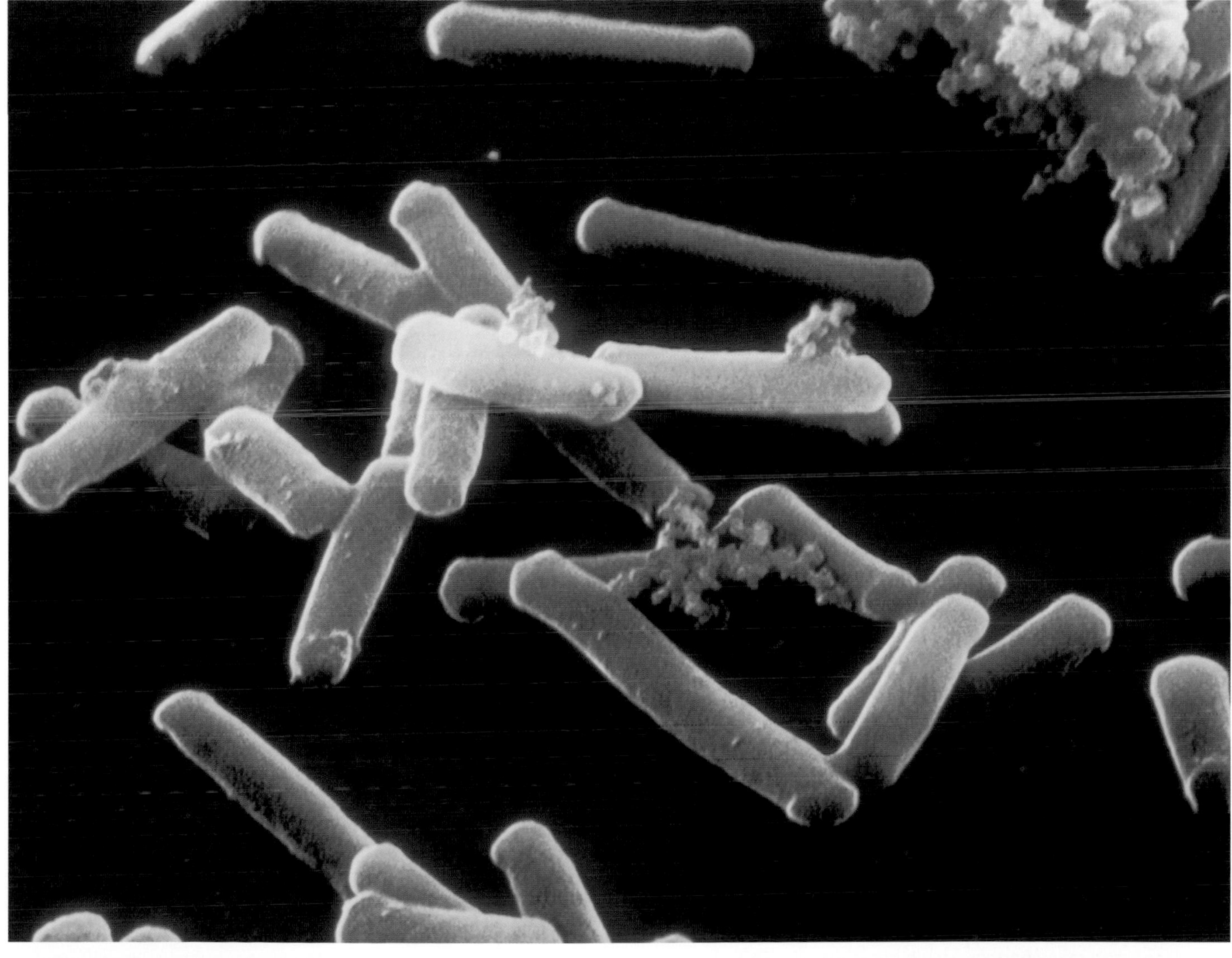

Clostridium tetani bacteria live in soil. They produce spores which protect them from contact with oxygen in air. If the spores enter a human body in a deep cut, the bacteria can grow. (× 13 000)

Find out about:

- respiration

5A Energy for life

Your body is a demanding animal. Billions of cells each carry out thousands of chemical reactions every second to keep you alive. These cells need a constant supply of energy to drive these reactions. The food you eat provides you with molecules to make new cells. But it is also a store of chemical energy. This is converted by respiration into energy your cells can use.

Most of your energy comes from **aerobic** respiration. During aerobic respiration glucose from your food reacts with oxygen. The reactions release energy from the glucose. Respiration can be summarized by this equation:

This girl is using energy for many different actions – such as processing information in her brain, moving, maintaining a constant internal environment, and growing and repairing her tissues.

$$C_6H_{12}O_6 + 6O_2 \longrightarrow 6CO_2 + 6H_2O \quad \text{(+ energy released)}$$

glucose + oxygen → carbon dioxide + water

Respiration is a long series of reactions. It is summarized by this equation.

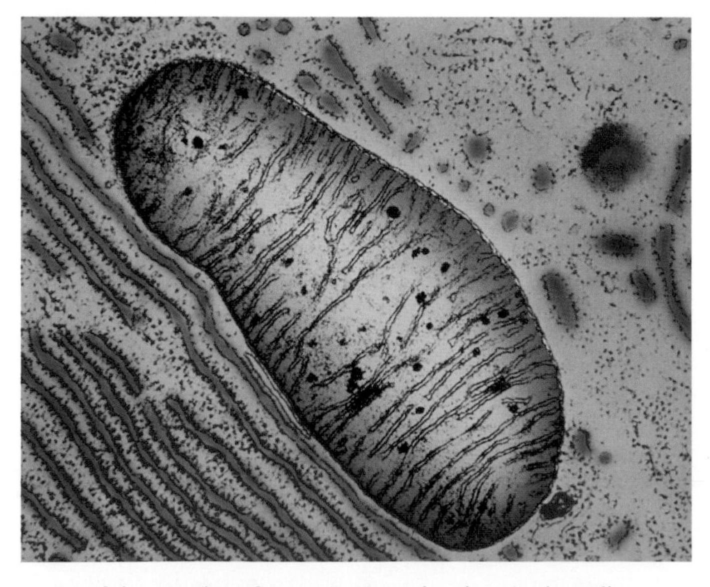

Some of the reactions for respiration take place in the cell cytoplasm, and many happen inside **mitochondria**. This electron micrograph shows a single mitochondrion (magnification × 64 000).

This fuel is combining with oxygen in a violent chemical reaction. Fuel in the body also reacts with oxygen, but in a very different reaction – not at all like combustion.

Maximizing diffusion

Oxygen moves from the air in your lungs into your blood by diffusion. Carbon dioxide diffuses the opposite way. This is called **gas exchange**. Gas exchange happens at the alveoli in the lungs.

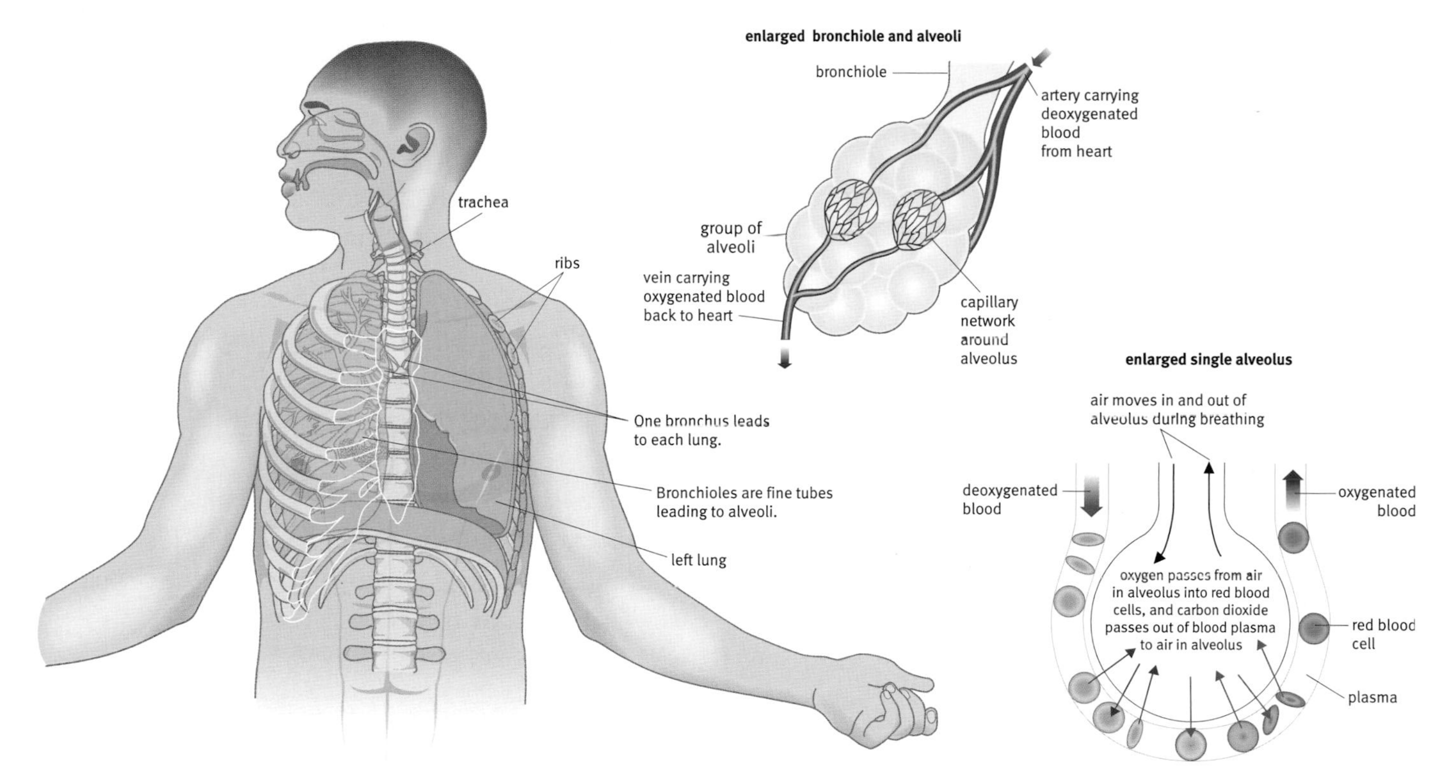

Gas exchange at the alveoli

Diffusion is the movement of molecules from an area of their high concentration to an area of their lower concentration. The greater the difference in concentration, the faster diffusion happens. (You can remind yourself about diffusion on page 108.)

Faster and deeper breathing refreshes the air in your lungs more often. A faster blood flow also carries oxygen away from the lungs as quickly as possible. Both of these features keep the biggest possible difference between the oxygen concentration in the alveoli and that in your blood. So gas exchange in the lungs happens at its fastest rate.

Key words

aerobic
mitochondria
gas exchange

Questions

1 Write down three processes your body needs energy for.

2 a Write down the word equation for aerobic respiration.
  b Annotate the equation to show where the reactants come from, and what happens to the products.

3 Describe what happens during gas exchange in the lungs.

(4) Explain how a large concentration gradient for oxygen is maintained between the lungs and the blood.

Find out about:

- your body's response to exercise

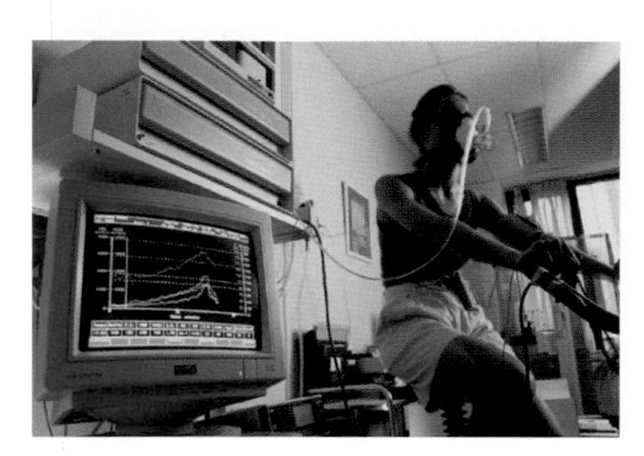

This athlete is having a fitness test. The equipment measures the maximum volume of oxygen she consumes in one minute as she cycles harder and harder. High volumes of oxygen consumption show high levels of endurance.

5B The effect of exercise

Responding to exercise

Working muscles need more energy than relaxed muscles. When you exercise, aerobic respiration in muscle cells must happen faster to provide this extra energy. To do this the muscle cells need more oxygen and glucose. The cells also need to get rid of more waste carbon dioxide made during respiration.

Like other chemicals, oxygen, glucose, and carbon dioxide are transported around the body in your blood. So two things happen when you exercise:

- *Your breathing rate increases* – more oxygen is brought into your body, and more carbon dioxide can be got rid of. Just sitting around, you breathe about 12 times a minute. Each breath takes in about 0.5 litres of air. During exercise you can breathe three times faster, and four times deeper.
- *Your heart rate increases* – oxygen and glucose are transported to your muscle cells faster, and carbon dioxide is removed faster. A resting heart pumps about 5 litres of blood per minute, but this volume can easily be tripled during exercise. Some blood is also redirected to the muscles, from tissues that can reduce their activity, e.g. from the digestive system.

Anaerobic respiration

When you exercise hard your muscle cells need large amounts of energy. The oxygen demand for aerobic respiration may be greater than the body can provide. If this happens the muscle cells can use anaerobic respiration. This releases energy from glucose without oxygen. This can be an advantage if your muscles need extra energy very quickly.

glucose ⟶ lactic acid (+ energy released)

Anaerobic respiration in animals is summarized by this equation.

Anaerobic respiration can only be used by muscles for a short period of time. It releases much less energy from each gram of glucose than aerobic respiration does. Also, the waste product **lactic acid** is toxic if produced in large amounts. It builds up in muscles, making them feel sore and tired.

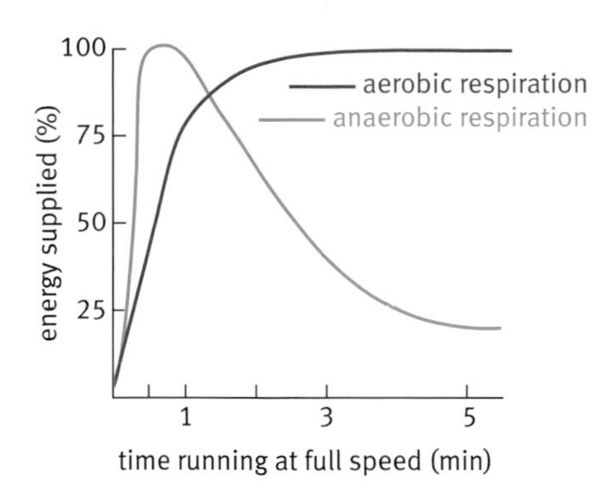

The 100 m sprint takes about 10–11 seconds. The heart and lungs cannot increase the oxygen supply to the muscles fast enough, so most of the energy required during the short race comes from anaerobic respiration.

H Getting rid of lactic acid

Toxic lactic acid cannot be left in muscle cells. After anaerobic respiration it is broken down into carbon dioxide and water. This reaction needs oxygen. So in human beings anaerobic respiration only provides energy without oxygen for a short period of time. The oxygen must be 'paid back' eventually. The amount of oxygen needed to get rid of the lactic acid is called the **oxygen debt**.

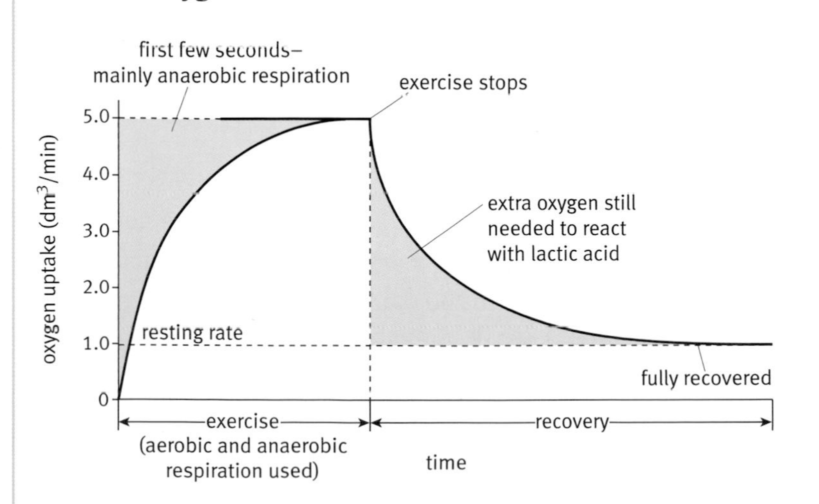

During exercise the heart rate increases, and so does the amount of blood pumped per beat. Both these factors combine to pump more blood per minute. Once exercise has finished you carry on breathing deeply. Your breathing rate returns to normal when all the lactic acid has been broken down.

Key words

lactic acid
oxygen debt

Questions

1. Write down the word equation for anaerobic respiration in human cells.
2. Explain why human cells cannot use anaerobic respiration for more than a short period of time.
3. Write a flow chart to explain why your heart and breathing rates increase when you exercise.
4. Explain what is meant by the term 'oxygen debt'.

Find out about:

- anaerobic respiration in plants and microorganisms
- ATP as an energy store

5C Anaerobic respiration and ATP

Anaerobic respiration in other organisms

Many animals use anaerobic respiration for short bursts of energy. For example, both a predator running after prey, and the prey running for its life, will use anaerobic respiration. Other organisms can also use anaerobic respiration, for example:

- parts of plants, e.g. roots in waterlogged soils, germinating seeds
- some microorganisms, e.g. yeast, lactobacilli (used in cheese and yogurt production)

In these organisms the product of anaerobic respiration is **ethanol** instead of lactic acid:

glucose ⟶ ethanol + carbon dioxide (+ energy released)

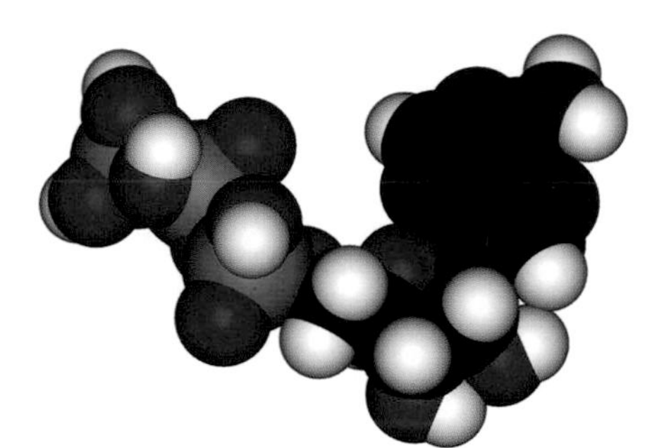

Model of ATP

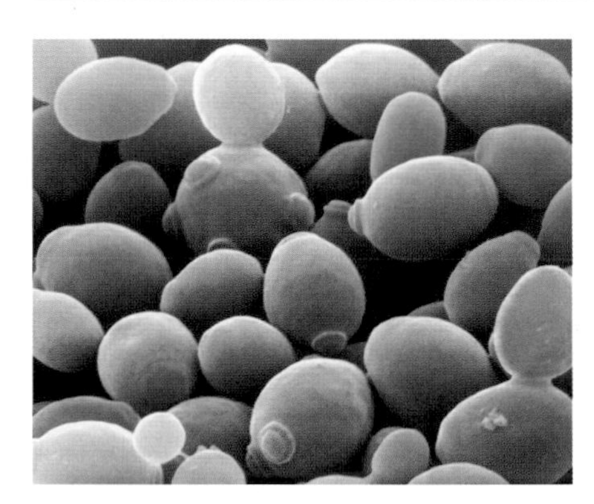

Yeast cells can respire anaerobically until the ethanol builds up and becomes too toxic.

Germinating seeds respire anaerobically.

What happens to the energy from respiration?

All respiration releases energy from glucose. This energy will be needed by many processes in the cell. There must be a way of storing the energy temporarily, so that it can be moved to the part of the cell that needs it. This is achieved by a very important chemical – **ATP**.

ATP is made by the cell using energy released during respiration. ATP molecules move freely around the cell. When energy is needed for reactions in the cell, some ATP is broken down, releasing energy. So ATP can be described as the 'energy currency' of the cell.

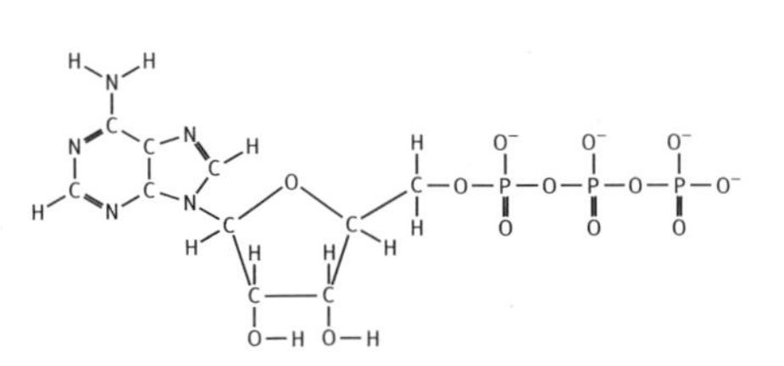

Diagram of an ATP molecule. ATP stands for adenosine triphosphate.

H ATP in muscle cells

Skeletal muscles may make up to 40% of a person's body weight. They are used for holding the body upright, breathing, and moving. A whole muscle is made up of hundreds of muscle fibres. Each muscle fibre is full of fine protein filaments. These can slide past each other, shortening the whole muscle. This happens when a muscle contracts to move part of the body.

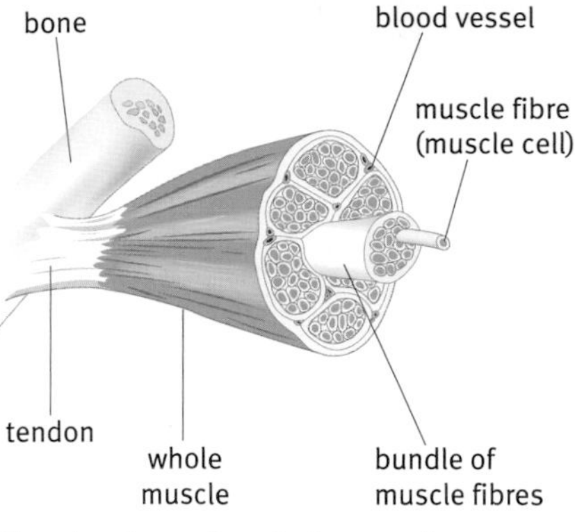

The structure of a whole muscle.

Muscle tissue is packed with thin protein filaments. ATP allows the filaments to slide past each other, contracting the muscle.

Muscle contraction needs a large supply of ATP. At any moment each of your muscle cells has about a million molecules of ATP. This is enough to keep your skeletal muscles going for just 10 seconds. So when ATP is broken down to release energy, it must be remade immediately. Even if you did no exercise, you'd need to break down and rebuild each ATP molecule about 800 times a day.

Questions

1 Describe conditions where anaerobic respiration is an advantage to:

  a human beings

  b another organism

2 Give an example of anaerobic respiration which is used by human beings to make a useful product.

③ Describe how the chemical ATP is produced.

④ Why is ATP sometimes called the 'energy currency' of the cell?

⑤ Describe how the structure of muscles enables them to contract.

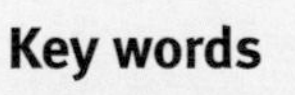

Key words

ethanol
ATP

Topic 6 Get moving

You have an internal skeleton to support your body. This makes you a **vertebrate.** Around your skeleton is a complex system of different tissues which hold your bones together and move them. Altogether they form your **skeletal-muscular system.**

The skeletal-muscular system in action.

Muscles

Muscles pull on your bones to move your skeleton. Different tissues join muscles to bones, and strap bones together. Each of these tissues is designed for the particular job it has to do.

Joints

Sports can be tough on your joints. The ends of the bones are under great force when you run or land from a jump. The tissues that hold bones together can be twisted by a sudden change of direction, or loss of balance. A really bad twist can even pull a muscle away from its bone. Physiotherapists can speed recovery from injury but it is better to be aware of the risks and take steps to avoid them.

Getting fit

Your muscles and joints need to be worked to keep them healthy. This applies to your skeletal-muscular system and your heart muscle. People are different and one way of keeping fit does not suit everyone. Some people prefer to play team sports, others go for a brisk walk every day, and some people like to visit a gym. If you join a gym your fitness is assessed before you begin training. It is useful if the gym keeps a record of your data so that anyone on the team can monitor your progress, and change your training plan if this is needed.

Warming up before a game can help avoid injury.

Key words

vertebrate
skeletal-muscular system

Find out about:

- the skeletal system

6A The skeleton

Your **skeleton** provides a tough, flexible framework for the rest of your body. It supports the soft tissues of the body. Without your skeleton you would be a jellyfish-like mass. Even when you are standing still, muscles pull on your bones to maintain your posture.

As well as supporting your body your skeleton:

- stores minerals such as calcium and phosphorus
- makes red blood cells, platelets, and some white blood cells in bone marrow
- forms a system of levers with muscles attached, which allows the body to move
- protects internal organs

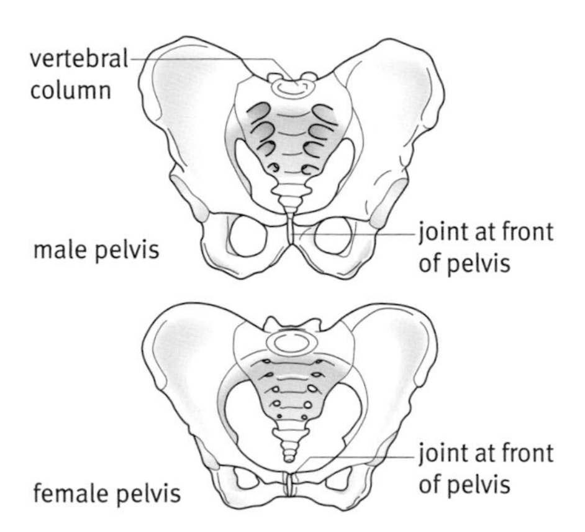

One function of the skeleton is to provide protection for internal organs. The pelvis protects the reproductive organs. The female pelvis is shallower and wider for childbearing.

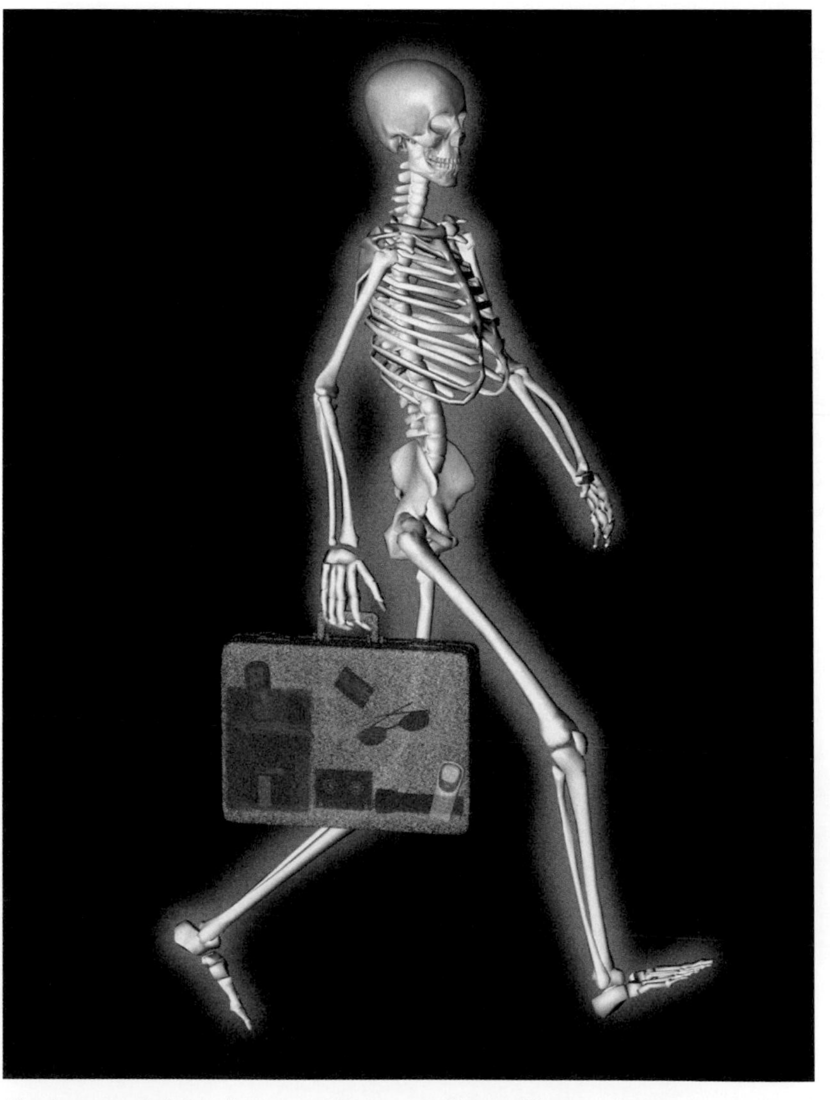

The human skeleton has over 200 bones. Most will move, but some are fixed in position, e.g. those in the skull. Skull bones are flexible during early development, but fuse together soon after birth.

Living bone

The skeleton is not just dry bone. Its tissues, such as bone and cartilage, are made of living cells. Blood brings nutrients and oxygen to the cells.

Bone is continually broke down and rebuilt. Even an adult's skeleton is continually changing. Weight-bearing exercises such as jogging stimulate bone growth, increasing its density. Inactivity makes bone less dense and weaker.

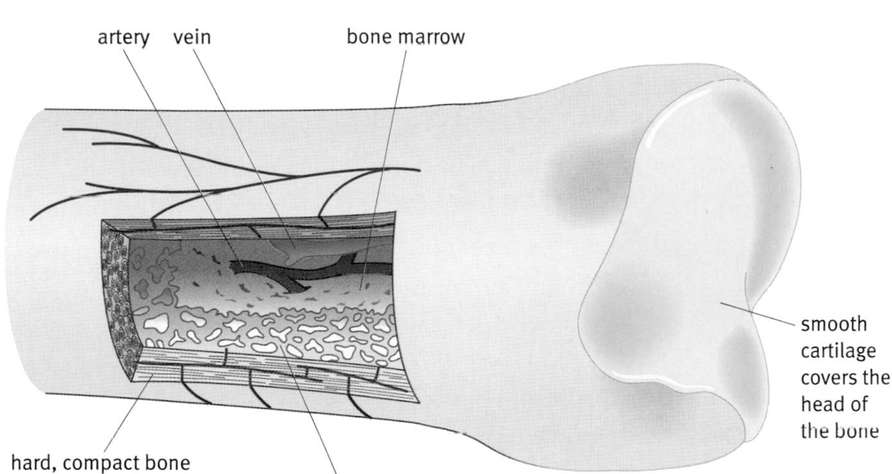

The structure of bone

Key words

skeleton

Questions

1 List four functions of the skeleton.
2 Give features from the photographs on this page which show that bone is a living tissue.
3 Describe how exercise changes bones.

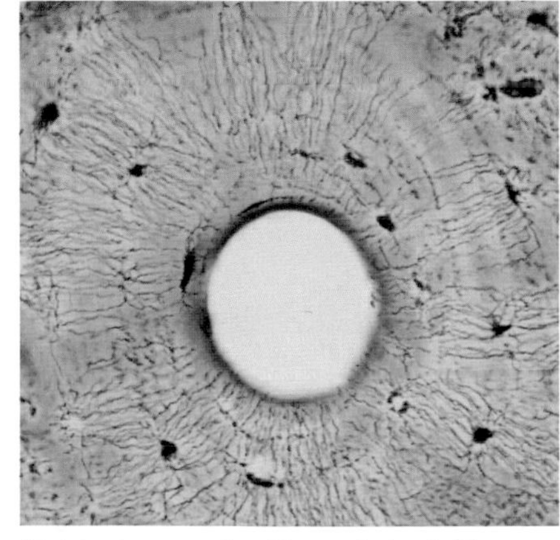

Light micrograph of bone (×600). You can see the central canal surrounded by circular bands. The dark dots are cavities containing bone cells.

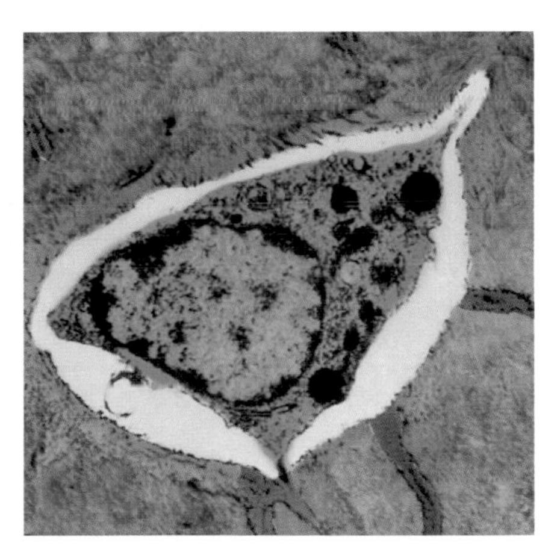

An electron micrograph showing a section through one cavity and its bone cell (×4000)

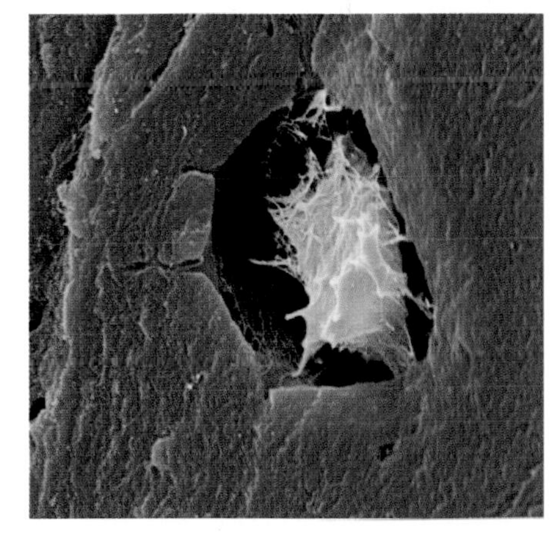

A scanning electron micrograph showing the surface of a single bone cell in its cavity (×4000)

How films can help fractures

You might be surprised to find an engineer working with an anatomy book by his side. But that is what Alessio Murgia does in the Department of Cybernetics at Reading University. Alessio has been using the technology that makes virtual characters in films to help people with mobility problems. This is called kinematics.

Creating virtual characters

To make characters like Gollum in *Lord of the Rings* an actor is covered in special reflective markers. Large reflectors are put onto the body, and smaller ones are used on the face. The light reflected from these markers is picked up by cameras and fed into a computer. The actor is filmed and a computer programme plots the positions of the markers for each frame. This produces a video which tracks the actor's every move and expression. The film is used as a base for the animators to make realistic virtual characters.

Studying how people move

Alessio explains the process in more detail. 'In the lab we used twelve cameras because of the complexity of human movement. Each reflective marker must be picked up by at least two cameras to pinpoint its position.

'This technology can be used to study gait – that's the way people walk – and upper limb movements, for example, how people who have fractured their wrists change the way they use their arms to cope with everyday tasks.' Alessio's work focuses on the human arm, but other scientists have also studied gait to help young people with restricted mobility. These photographs show a patient with celebral palsy filmed using this technology.

Kinematic technology can be used to help people with cerebral palsy. The environment in which these experiments are done is important. Here the lab is painted to look like an underwater world, so the wires and backpack seem like part of the adventure for the child.

If the patient keeps walking in this way he might develop bone deformities as his bones are still growing. On his back, the patient has a box that collects the electrical signals from the muscles – this shows when each muscle contracts during the gait pattern as the subject walks. People involved in this kind of study can compare this with a control pattern from someone with normal gait.'

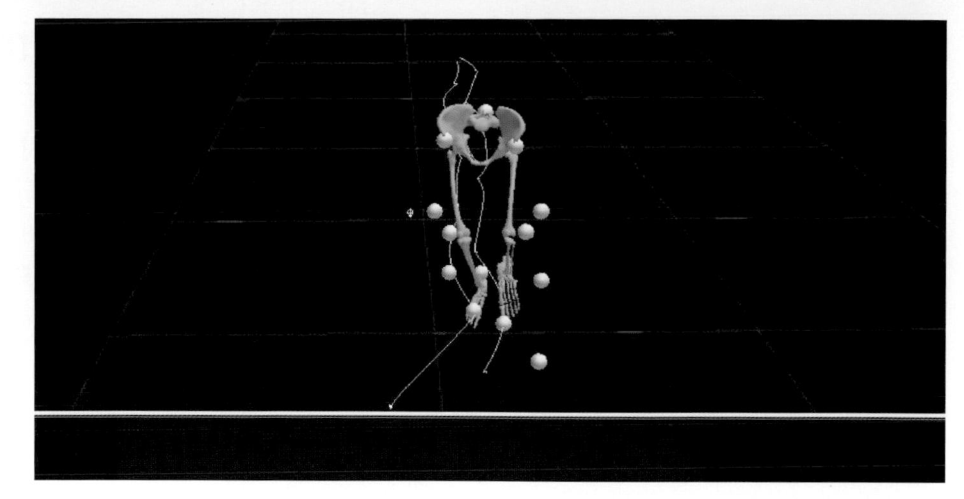

The computer simulates the child's gait pattern.

Alessio does not treat patients himself. His work is used by a physiotherapist or surgeon to make decisions about the best treatment for a patient. 'In cases like this it might be possible for the surgeon to lengthen or shorten tendons to correct the gait.'

Fractured wrists

Alessio's own research involves the study of shoulder, elbow, and hand kinematics to help people that are recovering from wrist fractures. 'The way we move our limbs is actually very complex – take the arm for example, it's more than just triceps and biceps contracting. There are muscles attached by tendons to sites on the shoulders. If someone has fractured their wrist they will use these muscles to turn their hands instead of the muscles round the wrist. There are two important wrist movements: *pronation* when you rotate your palm downwards, and *supination* when the palm is rotated upwards. These movements are very important in daily living, for example, every time you turn a key, put a cup to your lips, pick something up, or stroke a cat. If you have fractured your wrist you compensate by using other muscles in the elbow and shoulder.

'Because of the many degrees of freedom in the human body there are many ways of achieving a particular movement. In the past physiotherapists have relied on test movements in clinics to monitor how fracture patients are coping with moving, but that is not the same as real tasks at home. Using the movement information extracted from this film, the physiotherapist knows where to intervene to restore the right combination of muscle movement.'

Find out about:

- how movement is produced at your joints

6B Joints and movement

Holding the bones together

Two or more bones meet at a **joint**. Different types of joint allow different sorts of movement. Ball-and-socket joints, at your hip and shoulder, are the most versatile. These joints move in every direction, like a computer joystick. Hinge joints, such as the knee and elbow, move in just two directions – back and forwards.

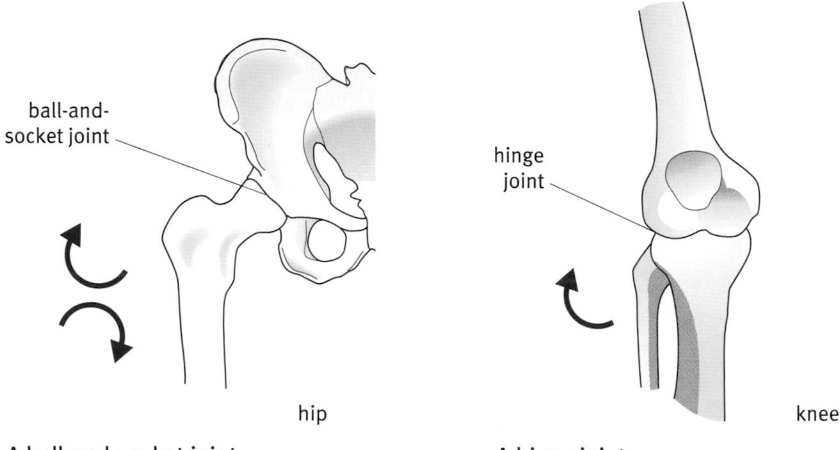

A ball-and-socket joint

A hinge joint

Tough, fibrous bands called **ligaments** hold the bones in place, and limit how far the bones can move. **Cartilage** stops bones from knocking against each other as they move. It forms a rubbery shock-absorbing coat over the end of each bone. This stops the bones from damaging each other. Cartilage is smooth, but friction could still wear it down. To reduce this, as far as possible, the joint is lubricated with oily **synovial fluid**.

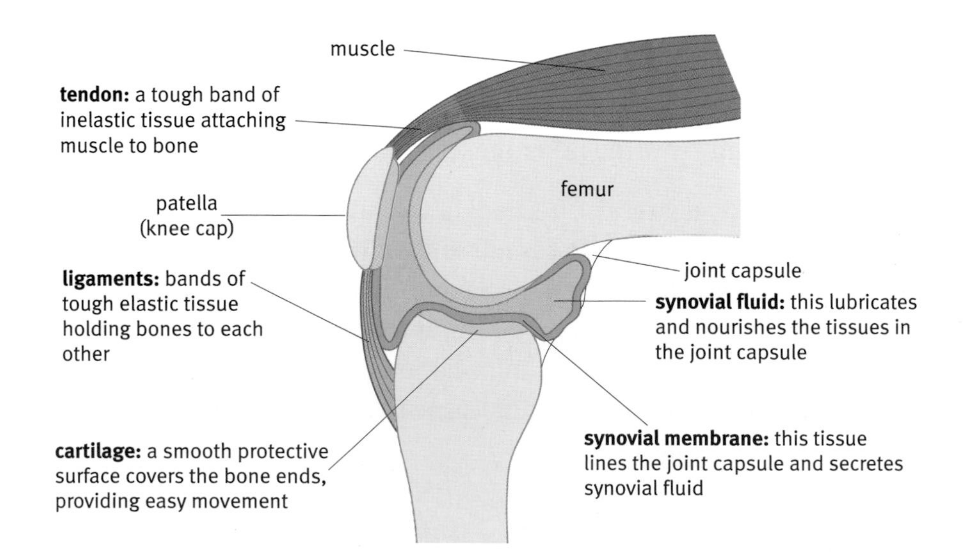

The knee joint. Like most joints in the body, this is a synovial joint.

How muscles move bones

Muscles can only pull a bone for movement. They cannot push it. A muscle contracts to pull on a bone and move it at a joint. After contracting the muscle is only stretched again when the bone is pulled back by another muscle. So at least two muscles must act at every joint:

- One contracts to bend the joint.
- The other contracts to straighten it.

Muscles which work opposite each other are called an **antagonistic pair**.

There are over 600 muscles attached to the human skeleton. They make up almost half the total body weight.

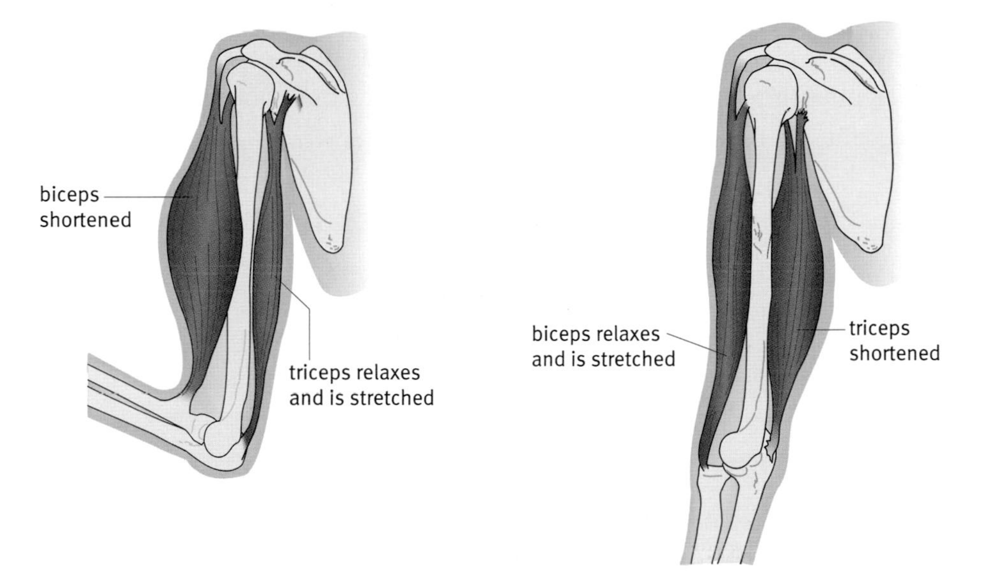

The biceps and triceps muscles contract to move the elbow joint.

Questions

1 Describe the difference between a tendon and a ligament.

2 Explain why this difference is important.

3 Name the parts of a synovial joint and explain their function.

4 Name the muscle which:

  a bends the arm

  b straightens the arm

5 Explain what is meant by an antagonistic pair of muscles.

(6) Professional dancers and gymnasts often develop osteoarthritis in their knee joints in later life. The surface of their cartilage becomes rough and wears away. Suggest what symptoms this will cause.

Key words

joint
ligament
cartilage
tendon
synovial fluid
antagonistic pair

Find out about:

- injuries caused by excessive exercise

6C Sports injuries

Joint injuries

If you follow a sports team you will know how often players may get injured. It is an occupational hazard. Joints are tough and well designed, but there is a limit to the force they can withstand. Common injuries include **sprains**, **dislocations**, torn ligaments, and torn tendons.

Football is particularly hazardous. There are lots of stops, starts, and changes of direction, and perhaps some bad tackles. It is not just professional footballers who suffer – 40% of knee injuries happen to under-15 footballers.

Joint injuries are common in football players.

Sprains

The most common sporting injury is a sprain. This usually happens when you overstretch a ligament by twisting your ankle or knee. Often people will say that they have 'torn a muscle', when they have actually sprained a ligament. There are several symptoms:

- redness and swelling
- surface bruising
- difficulty walking
- dull, throbbing ache or sharp, cramping pain

The usual treatment for sprains is **RICE** – rest, ice, compress, elevate.

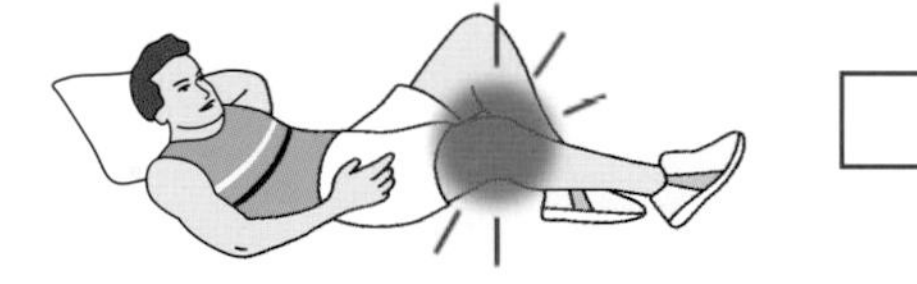

REST means immobilizing the injured part (e.g. keeping the weight off a torn muscle).

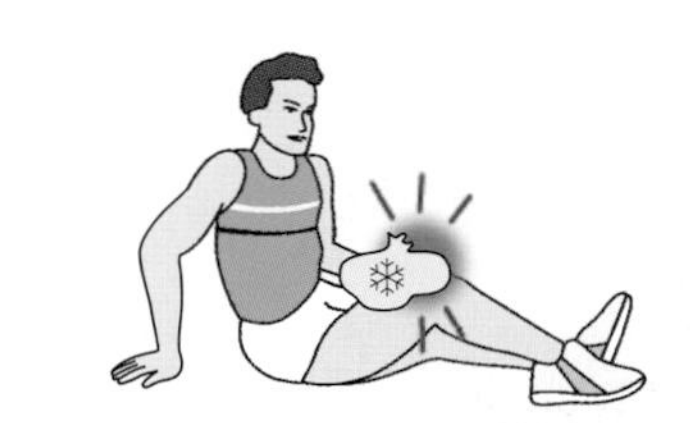

ICE acts as an anaesthetic, reduces swelling, and slows the flow of blood to the injured area. To avoid damaging the tissue, the ice is applied indirectly (e.g. in a tea towel or plastic bag) for up to 20 minutes at a time with 30 minutes between applications.

COMPRESSION usually involves wrapping a bandage round the injured part to reduce swelling. The bandage should be snug but not too tight.

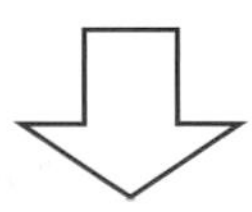

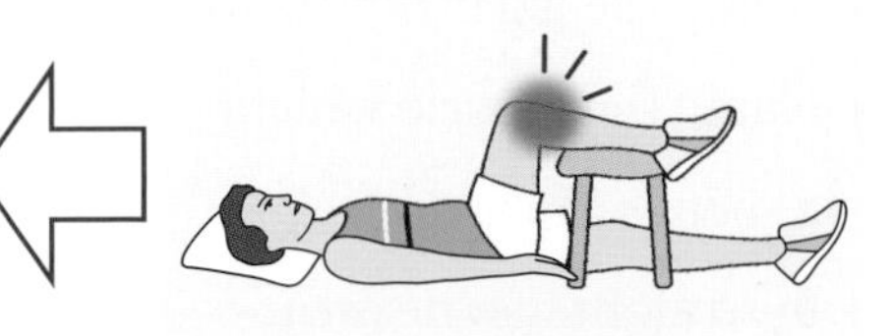

ELEVATION means raising the injured limb. This reduces swelling by helping to keep excess fluid away from the damaged area.

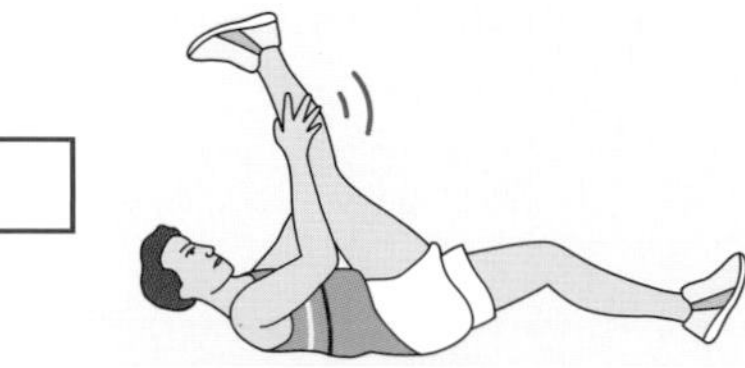

SIMPLE STRETCHING ROUTINES help to regain mobility, but only when swelling stops.

AEROBIC EXERCISING of the injured part is not restarted until it has regained at least 75% of the previous level of strength, and then only moderately. This exercise helps build muscle and return the athlete to peak fitness.

Recovery from a sports injury often involves RICE followed by stretching and strengthening exercises. RICE stands for rest, ice, compression, and elevation.

After 72 hours of RICE treatment, heat and gentle massage can be used to loosen the surrounding muscles. If the injury keeps occurring, physiotherapy can be used to strengthen the surrounding muscles. The diagram shows a set of exercises which a physiotherapist might suggest for this type of injury.

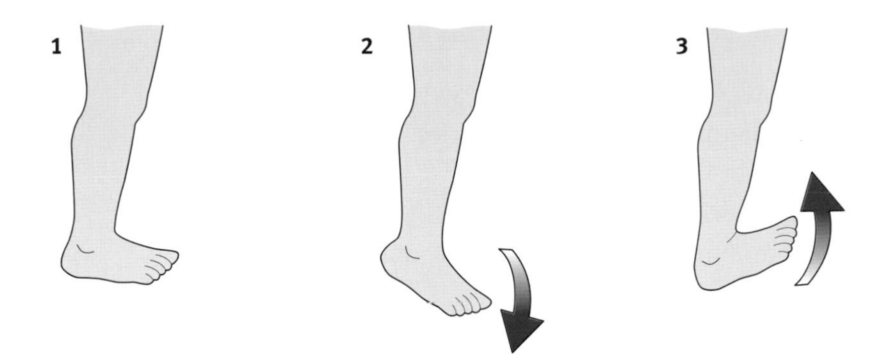

Physiotherapists prescribe different exercises for different injuries. If you suffer a joint injury, your GP may refer you to a physiotherapist for treatment.

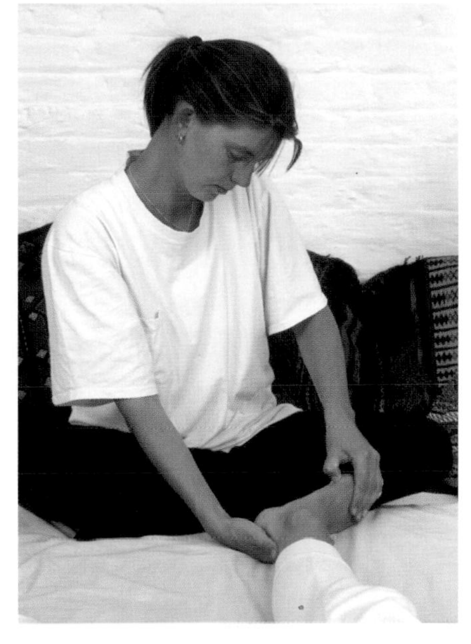

Physiotherapist treating a patient's ankle joint.

Dislocations

Gymnasts also suffer from joint injuries, often at the knee joint. Cartilage in the knee is an excellent shock absorber, but floor routines put a lot of force on the joints. If a gymnast lands off balance, their kneecap can become dislocated. This happens when the bone slips out of the joint. In contact sports such as rugby, dislocations of the shoulder are extremely common. Dislocations are very painful.

Gymnasts land with great force. It can be equivalent to carrying ten times their body weight.

Questions

1 Describe

a the symptoms and **b** RICE treatment for a sprain.

2 Describe two other types of common joint injury.

3 Describe a set of exercises which a physiotherapist might give someone to treat a particular joint injury.

Key words

sprain
dislocation
RICE

Find out about:

- fitness training

6D Following a training programme

Health and fitness

Regular exercise is very important for a healthy lifestyle. Simple brisk walking is a very effective way to exercise. Before someone begins a fitness training programme, it is important to have a health check. This can be with a GP or fitness trainer. They check factors such as:

- general health checks, e.g. blood pressure
- whether they take any medicines
- whether they smoke
- if they drink alcohol, how much?
- the amount of exercise they normally do
- family medical history
- previous treatments for injury or illness

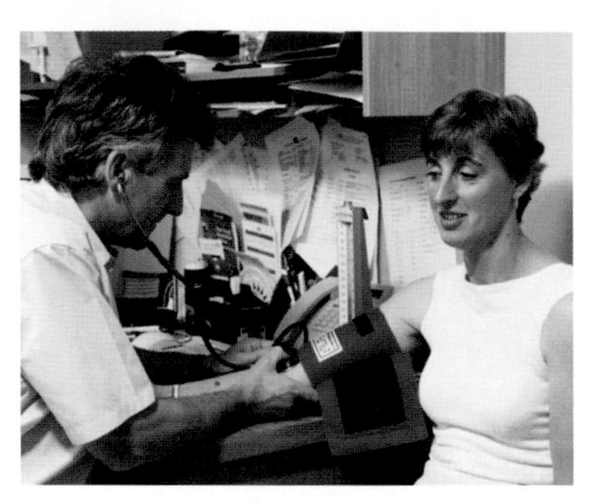

Blood pressure is the force of blood as it flows through the blood vessels.

What is 'normal' blood pressure?

Factors such as your blood pressure and heart rate vary during the day. They are lowest when you are at rest, and higher when you are moving quickly. Two people both sitting at rest will have different heart rates and blood pressure. So there is not a single 'normal' value for heart rate or blood pressure. Doctors use a range to describe the value for blood pressure which most people are within.

How is this information used?

The background information on a person's general health is used to decide on the best fitness programme. There are different types and levels of exercise. Information about a person's current health and fitness levels helps a trainer decide on the best programme for them.

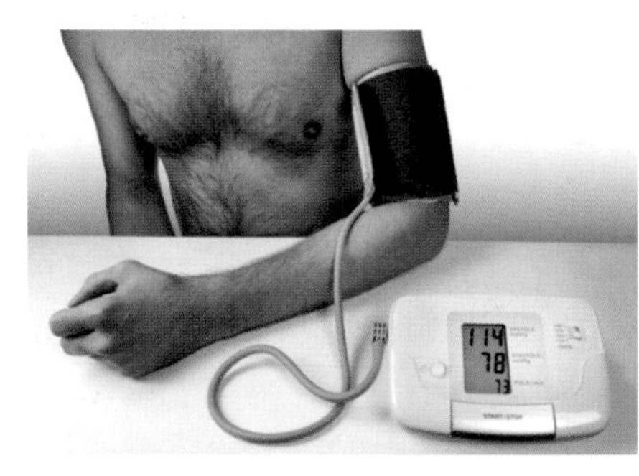

The higher number measures the **systolic** pressure. This is the pressure when the left ventricle contracts to push blood into the arteries. The lower number is the **diastolic** pressure. This is the pressure in the arteries between pumps, when the heart is relaxed and refilling with blood.

These same checks will also be made by a doctor before they prescribe a patient any medical treatment. All medical treatments can cause unwanted harmful effects. Before prescribing any treatment a doctor weighs up the potential side effects against the benefits they expect the treatment will bring. The patient's current condition is important information when making this decision.

Making progress

If you are following a fitness programme your trainer will check your progress regularly. If your fitness improves faster than they had expected, they may increase the intensity of your exercise programme. If you get an injury, they will suggest that you cut out exercises which use the injured joint or muscle until you have recovered.

It is also very important to monitor patients on a course of medical treatment. If they do not respond to a particular treatment, doctors may wish to adjust the dosage of a medicine, or try a different treatment.

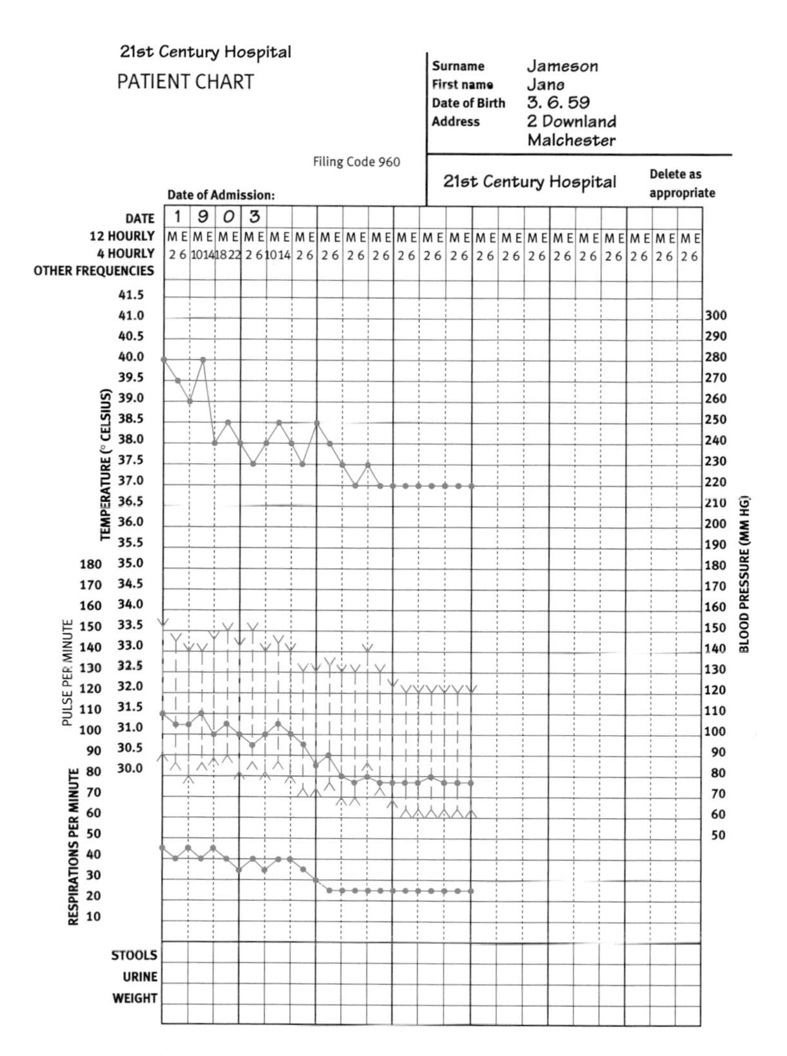

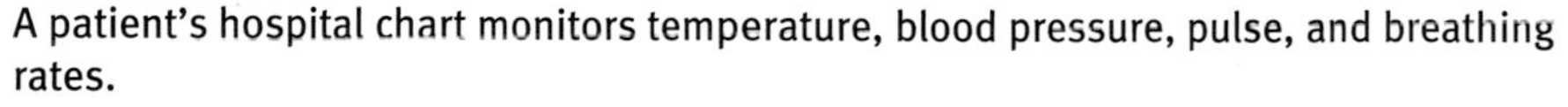
A patient's hospital chart monitors temperature, blood pressure, pulse, and breathing rates.

Keeping records

Information is stored in a person's medical or fitness records. These records must always be available. If the person does not make as much progress as expected, the records will help decisions about changing the treatment or training programme to be made quickly. The information must be stored carefully so that anyone on the medical or fitness team can access them easily. For example, if a patient in hospital needs treatment during the night, staff on duty must have access to records made during the previous day.

You can read more about an athlete's training programme on the next two pages.

Key words

systolic
diastolic

Questions

1. List four pieces of medical or lifestyle history that should be checked before a person begins medical treatment or a fitness programme.
2. Explain why this information must be checked.
3. What are the two readings given for blood pressure? Explain the difference between them.
4. Explain why record-keeping is important during a fitness or treatment programme.
5. When a GP prescribes a course of treatment for a patient, they often book regular follow-up appointments. Explain why these appointments are important.
6. Suggest two reasons why a course of medical treatment or a fitness training programme might be changed before it has been completed.
7. Write a short paragraph to describe the stages a physiotherapist would work through to treat someone with a sporting injury.

Anna Bevan winning the 800 metres

Sports training

A new athlete hits the headlines

In 2003 Anna Bevan was an average runner. In 2005, she stormed to victory in the 800 metres. Her coach, Dan Forde, describes how Anna's training programme was developed.

'I coach the athletics team at the university. Anna was studying sports science. During a physiology practical her maximal oxygen consumption (VO_2 max) was measured. She gave excellent results, so her lecturer asked me to do a bleep test and step test on Anna.

The tests confirmed that Anna was aerobically very fit. She had been an all-round sportsperson at school, playing tennis, netball, hockey, and rugby. However, her first love was athletics. She had tried sprinting but wasn't very successful. I offered to train Anna and she accepted.'

Before training begins

'It's important for a coach and athlete to relate well and trust each other. Before training began we spent some time talking and got to know each other. Also, I checked Anna's health, because health and fitness are not the same thing. Although Anna performed well in fitness tests, she might have had a medical condition that made it dangerous for her to train seriously.'

Health check

'Anna completed a physical activity readiness questionnaire. I also asked our athletics team doctor to carry out a more detailed health screening.

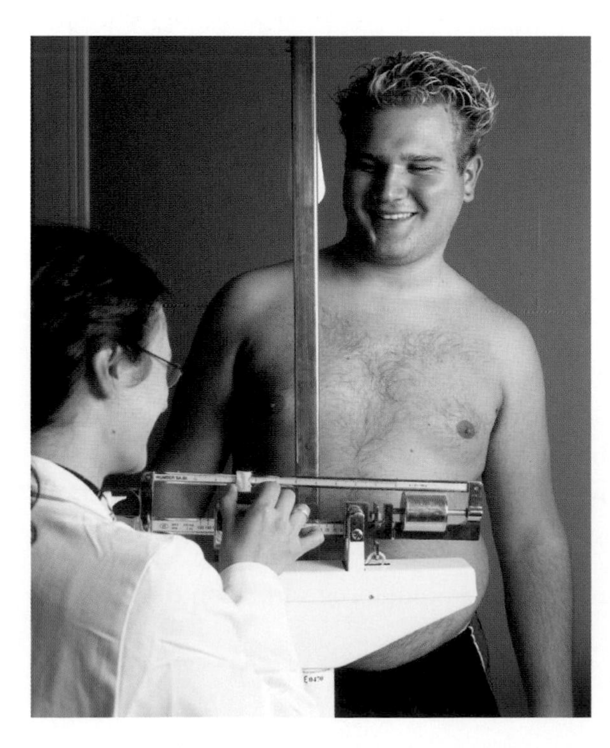
This man is being weighed and measured so that his body mass index can be calculated.

If Anna was aiming to be an elite athlete, she would have to put her body under extreme physical stress. The doctor asked questions about Anna's lifestyle and medical history. Then he took blood and urine samples, and measured her body temperature, resting heart rate, and blood pressure. Anna had an ECG to make sure that her heart was healthy.

'The doctor also measured Anna's weight and height to make a rough calculation of her body mass index. He took skinfold measurements to estimate her body fat content.'

Choosing the right event

'Before Anna started training I wanted to find out why she had not done well in school athletics competitions. I suspected that she might be trying the wrong distance, so I asked for a muscle biopsy. The results showed that Anna's proportion of fast-twitch to slow-twitch fibres suited her better to middle-distance events rather than sprinting. Anna agreed to give this a try, and we designed a training programme for her.'

Starting the programme

'Anna did several sprints with rest, walking, or light jogging in between. She also did one long, slow run each week. Before Anna trained each day her resting heart rate and body temperature were measured. She also weighed herself to make sure she was well hydrated. Water is lost during intensive exercise, and must be replaced before another intensive session.

'Half-way through the first season Anna strained her hamstring muscle. The start of a race was delayed too long and she got cold. When she sprinted out of the blocks part of her muscle snapped, like an overstretched elastic band. A physiotherapist was on hand to perform the **RICE** routine almost immediately. He also gave Anna a rehabilitation programme. We monitored her progress over a number of weeks to ensure she did not push herself too hard too soon. Fortunately she could run at full fitness again before the end of the season.'

B7 Biology across the ecosystem

Summary

Interdependence

- Life depends on solar energy absorbed during photosynthesis and stored in chemicals like starch.
- Autotrophs make food from minerals, carbon dioxide, and water but heterotrophs need ready-made food.
- Chlorophyll absorbs light for photosynthesis. The equation is:

  carbon dioxide + water → glucose + oxygen
- Some glucose is used for respiration and to make cellulose, protein, starch, and chlorophyll.
- Nitrates are absorbed and added to glucose to make proteins.
- They are absorbed by active transport in the roots. H
- Starch has little effect on the osmotic balance of the cell, so it is used for energy storage. H
- Photosynthesis may be limited by low temperatures and low carbon dioxide or light levels.
- Human activity is increasing atmospheric carbon dioxide levels.
- Compensation points occur when respiration makes as much carbon dioxide as photosynthesis uses.
- Energy is transferred when living things are eaten or broken down by decomposers.
- Pyramids of biomass show the food available to the herbivores and carnivores in an ecosystem.
- Only about 10% of the energy in biomass is passed on at each stage in a food chain. The rest is lost to decomposers or as waste heat.
- Soil contains living organisms, decaying material, sand, clay, air, water, and dissolved minerals.

Further up the food chain

- Mutualistic relationships benefit partner species; commensalism benefits one without harming the other; and parasitism harms the host.
- Parasites cause human disease and reduce farmers' yields.
- Parasites evolved with their hosts, so they can evade the host's immune system. For example, malaria parasites hide in red blood cells. They produce large numbers of offspring to ensure transfer to new hosts.
- Sickle-cell anaemia is caused by a faulty recessive allele. It makes red cells jam in small capillaries, causing pain, organ damage, and early death. H
- Natural selection has increased the allele's frequency where malaria exists because having one copy of the allele protects you. H

New technologies

- Bacteria have a cell wall, cell membrane, circular DNA chromosome, and plasmids.
- Bacteria and fungi are grown in fermenters to produce antibiotics, single-cell protein, and enzymes like rennin.
- Genetic modification involves isolating and copying genes and putting them in new cells.
- This is done by adding a virus or plasmid as vector. H
- GM bacteria make drugs, and hormones such as insulin. GM plants have added characteristics like disease resistance which lead to higher yields.
- Economic, social, and ethical implications have to be considered before genetically modified organisms are released.
- Genetic tests are carried out by isolating DNA, adding gene probes, and using UV or autoradiography to locate them.

Circulation

- Blood contains red cells to carry oxygen, white cells to fight infection, and platelets to seal cuts.
- A gene with three alleles, decides your blood type.
- Of the three alleles, I^A, I^B, I^O, two – I^A and I^B – are codominant.
- The chance of inheriting each blood group can be predicted using genetic diagrams. H
- The recipient of a blood transfusion must not have antibodies to the donor's red cells.
- The heart has an atrium on each side, to collect blood from veins, and a ventricle on each side to pump it into the arteries.
- We have a double circulatory system. The right ventricle sends blood to the lungs for gas exchange, and the left ventricle sends it around the body.
- Valves in the heart and veins keep blood flowing in the right direction.
- Tissue fluid bathes cells and aids diffusion of oxygen, carbon dioxide, glucose, and urea between capillaries and tissues.

Respiration

- Aerobic respiration releases large amounts of energy.
- This energy is used to synthesize ATP, the cell's energy currency. H
- The equation for aerobic respiration is:

  glucose + oxygen → carbon dioxide + water
- During exercise, respiration increases to provide more energy (ATP) for muscle contraction.
- Heart and breathing rates increase to supply oxygen and glucose faster, and remove carbon dioxide faster, during exercise.
- 'Normal' measurements for factors such as heart rate and blood pressure vary.
- Anaerobic respiration releases less energy. The equation is:

  glucose → lactic acid
- Anaerobic respiration is used to provide short bursts of energy when aerobic respiration cannot meet the demand. H
- The 'oxygen debt' is the amount of oxygen used to break down lactic acid after anaerobic exercise. H

Skeletal system

- Vertebrate skeletons support the body and allow movement.
- Bones are held together by slightly elastic ligaments.
- Tough inelastic tendons attach muscles to bone.
- Layers of cartilage and synovial fluid reduce friction at joints.
- Muscles operate in antagonistic pairs because they can only contract.
- Health information is needed before medical treatment or an exercise regime starts.
- Regular contact with health or fitness practitioners and accurate records are essential.
- These records may be needed by other practitioners. H
- The benefits of treatments need to be weighed against their side effects and there can be more than one way to achieve targets.
- Training can be monitored by measuring changes in the heart rate's response to exercise.
- Assessments of progress need to take the accuracy of the monitoring technique, and the reliability of the data, into account. H
- A fitness programme can be modified in response to injury or to promote further improvement.
- Excessive exercise can dislocate a bone from its socket, tear a ligament or tendon, or sprain a ligament.
- Sprains are treated by rest, ice, compression, and elevation (RICE). When the swelling has gone down, stretching routines can be used to regain mobility.
- Physiotherapists prescribe exercises to build muscles and counteract joint problems.

Glossary

23 pairs Human body cells have 23 pairs of chromosomes in the nucleus.

ABO blood type All people can be divided into four groups depending on the antigens which are carried on their red blood cells. These may be type A, type B, both A and B, or neither. This way of classifying people's blood type is the ABO system.

absorb (digestion) Absorption, during digestion, happens when small molecules pass from the small intestine into the blood. They can then be carried to all your body cells.

active 'working memory' One explanation for how the human memory works.

active site The part of an enzyme where the reacting molecules fit into.

active transport Molecules are moved in or out of a cell using energy. This process is used when transport needs to be faster than diffusion, and when molecules are being moved from a region where they are at low concentration to where they are at high concentration.

ADH A hormone making kidney tubules more permeable to water, causing greater re-absorption of water.

adrenaline A hormone which has many affects on the body, for example, increasing heart rate, increasing breathing rate.

aerobic respiration Respiration which uses oxygen.

AIDS Acquired Immune Deficiency Syndrome, a disease caused by the HIV virus. The body's immune system is attacked by the virus and gradually becomes weakened.

allele Different versions of the same gene.

amino acids The small molecules which are joined in long chains to make proteins. All the proteins in living things are made from 20 different amino acids joined in different orders.

anaerobic respiration Respiration which does not use oxygen.

antagonistic effectors Antagonistic effectors have opposite effects.

antagonistic pair Two muscles which work to move the same bone in opposite directions, e.g. the biceps and triceps muscles.

antibiotic Drugs that kill or stop the growth of bacteria and fungi.

antibiotic resistant Microorganisms that are not killed by antibiotics.

antibodies A group of proteins made by white blood cells to fight dangerous microorganisms. A different antibody is needed to fight each different type of microorganism. Antibodies bind to the surface of the microorganism, which triggers other white blood cells to digest them.

antigen Proteins on the surface of a cell. A cell's antigens are unique markers.

artery Blood vessels which carry blood away from the heart.

asexual reproduction When a new individual is produced from just one parent.

ATP ATP (adenosine triphosphate) is a chemical used by living things to store and transfer energy during chemical reactions.

atrium (plural atria) One of the upper chambers in the heart. The two atria pump blood to the ventricles.

autoradiography Gene probes are often made using radioactive DNA bases. The radioactivity blackens X-ray film, which shows whether the gene probe has bound to the DNA sample.

autotroph An organism which produces its own organic compounds.

autotrophic bacteria Bacteria which produce their own organic compounds. Some use light energy for photosynthesis, others use simple inorganic molecules as a source of energy.

auxins A plant hormone that affects plant growth and development. For example, auxin stimulates growth of roots in cuttings.

axon A long, thin extension of the cytoplasm of a neuron. The axon carries electrical impulses very quickly.

bacteriophage A type of virus which infects bacteria.

bacterium One type of single-celled microorganism. They do not have a nucleus. Some bacteria may cause disease.

base pairing The bases in a DNA molecule (A, C, G, T) always bond in the same way. A and T always bond together. C and G always bond together.

behaviour Everything an organism does; its response to all the stimuli around it.

biodiversity The great variety of living things, both within a species and between different species.

blind trial A clinical trial in which the patient does not know whether they are taking the new drug, but their doctor does.

blood pressure The pressure exerted by blood pushing on the walls of a blood vessel.

blood transfusion Transfer of blood from one person to another.

capillary network Large numbers of narrow blood vessels which pass through each organ in the body. Capillaries receive blood from arteries and return it to veins. Capillary walls are only one cell thick.

carbohydrate A natural chemical made of carbon, hydrogen, and oxygen. The hydrogen and oxygen are present in the proportions as water. An example is glucose $C_6H_{12}O_6$. Carbohydrates includes sugars, starch, and cellulose.

carbon cycle The cycling of the element carbon in the environment between the atmosphere, biosphere, hydrosphere, and lithosphere. The element exists in different compounds in these spheres. In the atmosphere it is mainly present as carbon dioxide.

carrier Someone who has the recessive allele for a characteristic or disease but who does not have the characteristic or disease itself.

cartilage Tough, flexible tissue found at the end of bones and in joints. It protects the end of bones from rubbing together and becoming damaged.

cause When there is evidence that changes in a factor produce a particular outcome, then the factor is said to cause the outcome, for example, increases in the pollen count cause increases in the incidence of hay fever.

cellulose The chemical which makes up most of the fibre in food. The human body cannot digest cellulose.

central nervous system In mammals the brain and spinal cord.

cerebral cortex The highly folded outer region of the brain, concerned with conscious behaviour.

chlorophyll A green pigment found in chloroplasts. Chlorophyll absorbs energy from sunlight for photosynthesis.

chloroplast An organelle found in some plant cells where photosynthesis takes place.

chromosomes Long, thin, thread-like structures in the nucleus of a cell made from a molecule of DNA. Chromosomes carry the genes.

circulatory system The heart and blood vessels. The circulatory system transports useful chemicals and waste products around the body.

clinical trial When a new drug is tested on humans to find out whether it is safe and whether it works.

clone A new cell or individual made by asexual reproduction. A clone has the same genes as its parent.

codominant Some genes have two alleles which are neither dominant or recessive. If a person has a copy of both these alleles, they will both be expressed and show up in that person. These alleles are co-dominant.

commensalism A relationship between two organisms of different species where one organism gains, and the other neither gains or loses from the relationship.

common ancestor A species which two or more other species both evolved from.

compensation point Respiration uses glucose in a plant, photosynthesis produces glucose in a plant. When respiration and photosynthesis are taking place in a plant at the same rate, there is no net gain or loss of glucose. This is the compensation point.

consciousness The part of the human brain concerned with thought and decision making.

consumers Organisms which eat others in a food chain. This is all the organisms in a food chain except the producer(s).

control In a clinical trial, the control group is people taking the currently used drug. The effects of the new drug can then be compared against this group.

core (of the body) Central parts of the body where the body temperature is kept constant.

correlation A link between two things. For example, if an outcome happens when a factor is present, but not when it is absent. Or if an outcome increases or decreases when a factor increases. For example, when pollen count increases hayfever cases also increase.

cuttings A shoot or leaf taken from a plant, to be grown into a new plant.

cystic fibrosis An inherited disorder. The disorder is caused by recessive alleles.

decomposers Organisms which feed on dead organisms. They break down the complex organic chemicals in their bodies, releasing nutrients back into the ecosystem to be used by other living organisms.

deoxygenated Blood in which the haemoglobin is not bound to oxygen molecules.

development How an organism changes as it grows and matures. As a zygote develops, it forms more and more cells. These are organised into different tissues and organs.

diastolic The blood pressure when all parts of the heart muscle are relaxed and the heart is filling with blood.

digestion Breaks down large food molecules into smaller ones. This is needed so that they can pass into your blood.

dislocation An injury where a bone is forced out of its joint.

DNA The chemical that makes up chromosomes – deoxyribonucleic acid. DNA carries the genetic code, which controls how an organism develops.

DNA fingerprinting A DNA fingerprint uses gene probes to identify particular sequences of DNA bases in a person's genetic make-up. The pattern produced in a DNA fingerprint can be used to identify family relationships.

DNA profiling A DNA profile is produced in the same way as a DNA fingerprint, but fewer gene probes are used. DNA profiling is used in forensic science to test samples of DNA left at crime scenes.

DNA technology Any process which uses our knowledge of DNA to solve a problem or make a new product.

dominant Describes an allele that will show up in an organism even if a different allele of the gene is present. You only need to have one copy of a dominant allele to have the feature it produces.

donor A person who gives blood to another person.

double circulation A circulatory system where the blood passes through the heart twice for every complete circulation of the body.

double helix The shape of the DNA molecule, with two strands twisted together in a spiral.

double-blind A clinical trial in which neither the doctor nor the patient knows whether the patient is taking the new drug.

Ecstasy A recreational drug that increases the concentration of serotonin at the synapses in the brain, giving pleasurable feelings. Long-term effects may include destruction of the synapses.

effector The part of a control system that brings about a change to the system.

embryo The earliest stage of development for an animal or plant. In humans the embryo stage lasts until age two months.

embryo selection A process where an embryo's genes are checked before the embryo is put into the mother's womb. Only healthy embryos are chosen.

endangered Species which are at risk of becoming extinct.

environment Everything that surrounds you. This is factors like the air, the Earth, and water, as well as other living things.

environmental Things in your environment that affect the way you develop.

enzyme A protein that catalyses (speeds up) chemical reactions in living things.

ethanol Waste product from anaerobic respiration in plants and yeast.

ethics A set of principles which may show how to behave in a situation.

excretion The removal of waste products of chemical reactions from cells.

extinct A species is extinct when all the members of the species have died out.

extremities Parts of the body away from the core, for example fingers.

false negative A wrong test result. The test result says that a person does not have a medical condition but this is incorrect.

false positive A wrong test result. The test result says that a person has a medical condition but this is incorrect.

fatty sheath Fat wrapped around the outside of an axon to insulate neurons from each other.

feral A child who has not learned to speak because they have not been exposed to language early enough in life.

fermenter A large vessel in which microorganisms are grown to make a useful product.

food chain In the food industry this covers all the stages from where food grows, through harvesting, processing, preservation and cooking to being eaten.

food web A series of linked food chains showing the feeding relationships in a habitat – 'what eats what'.

fossil The stony remains of an animal or plant that lived millions of years ago, or an imprint of its mark (for example, a footprint) in a surface.

fungus A group of living things, including some microorganisms, that cannot make their own food.

gametes The sex cells that fuse to form a zygote. In humans, the male gamete is the sperm and the female gamete is the egg.

gas exchange The exchange of oxygen and carbon dioxide that takes place in the lungs.

gene probe A short piece of single-stranded DNA used in a genetic test. The gene probe has complementary bases to the allele which is being tested for.

gene switching Genes in the nucleus of a cell switch off and are inactive when a cell becomes specialized. Only genes which the cell needs to carry out its particular job stay active.

gene therapy Replacing faulty alleles with normal alleles. The aim is to cure genetic disorders.

genes A section of DNA giving the instructions for a cell about how to make one kind of protein.

genetic Factors that are affected by an organism's genes.

genetic modification (GM) Altering the characteristics of an organism by introducing the genes of another organism into its DNA.

genetic screening Testing a population for a particular allele.

genetic variation Differences between individuals caused by differences in their genes. Gametes show genetic variation – they all have different genes.

glands Parts of the body that make enzymes, hormones, and other secretions in the body, for example sweat glands.

habitat The place where an organism lives.

haemoglobin The protein molecule in red blood cells. Haemoglobin binds to oxygen and carries it around the body. It also gives blood its red colour.

heart attack The coronary arteries become blocked and the supply of blood to the heart muscle is interrupted, damaging the heart muscle.

heat stroke A life-threatening rise in body temperature where the body temperature control system fails.

heterotroph An organism which must eat other organisms for its source of organic compounds.

HIV Human Immunodeficiency Virus, the virus that causes AIDS.

homeostasis Keeping a steady state inside your body.

hominid Animals more like humans than apes that lived in Africa millions of years ago.

hormone A chemical messenger secreted by specialised cells in animals and plants. Hormones bring about changes in cells or tissues in different parts of the animal or plant.

host An organism who's body is infected with a parasite.

Huntington's disorder An inherited disease of the nervous system. The symptoms do not show up until middle age.

hyphae A network of fine threads which form the body of a fungus.

hypothalamus The part of the brain that controls many different functions, for example body temperature.

hypothermia A fall in body temperature to below 35°C.

immune Able to react to an infection quickly, stopping the microorganisms before they can make you ill, usually because you've been exposed to them before.

immune system A group of organs and tissues in the body that fight infections.

indirectly When something humans do affects another species, but this wasn't the reason for the action. For example, a species habitat is destroyed when land is cleared for farming.

infectious A disease which can be caught. The microorganism which causes it is passed from one person to another through the air, through water, or by touch.

infertile An organism that cannot produce offspring.

influenza A disease caused by a particular virus. Symptoms include a very high temperature, sweating, aching muscles. In some cases 'flu' is fatal.

inherited A feature that is passed from parents to offspring by their genes.

insulin A hormone produced by the pancreas. It is a chemical which helps to control the level of sugar (glucose) in the blood.

interdependence The relationships between different living things which they rely on to survive.

involuntary An automatic response made by the body without conscious thought.

joint A point where two or more bones meet.

kidneys Organs in the body which remove waste urea from the blood, and balance water and salt levels.

lactic acid Waste product from anaerobic respiration in animals.

life cycle The stages an organism goes through as it matures, develops, and reproduces.

lifestyle diseases Diseases which are not caused by microorganisms. They are triggered by other factors, for example, smoking diet, lack of exercise.

ligament Tissue which joins two or more bones together.

limiting factor The factor which prevents the rate of photosynthesis from increasing at a particular time. This may be light intensity, temperature, carbon dioxide concentration, or water availability.

long-term memory The part of the memory that stores information for a long period, or permanently.

malaria A disease caused by a protozoan called *Plasmodium*.

match Some studies into diseases compare two groups of people. People in each group are chosen to be as similar as possible (matched) so that the results can be fairly compared.

meiosis Cell division that halves the number of chromosomes to produce gametes. The four new cells are genetically different from each other and from the parent cell.

memory The storage and retrieval of information by the brain.

meristem cells Unspecialized cells in plants that can develop into any kind of specialized cell.

microorganism Living organisms that can only be seen by looking at them through a microscope. They include bacteria, viruses, and fungi.

mitochondria An organelle in plant cells where respiration takes place.

mitosis Cell division that makes two new cells identical to each other and to the parent cell.

models of memory Explanations for how memory is structured in the brain.

motor neuron A neuron that carries nerve impulses from the brain or spinal cord to an effector.

mRNA Messenger RNA, a chemical involved in making proteins in cells. The mRNA molecule is similar to DNA but single stranded. It carries the genetic code from the DNA molecule out of the nucleus into the cytoplasm.

multicellular An organism made up of many cells.

multifactorial disease A disease caused by several different factors, including genetic and environmental factors.

multi-store model One explanation for how the human memory works.

muscles Muscles move parts of the skeleton for movement. There is also muscle tissue in other parts of the body, for example in the walls of arteries.

mutation A change in the DNA of an organism. It alters a gene and may change the organism's characteristics.

natural selection When certain individuals are better suited to their environment they are more likely to survive and breed, passing on their features to the next generation.

mutualism An association where both organisms seem to benefit.

nerve cell A cell in the nervous system that transmits electrical signals to allow communication within the body.

nerve impulses Electrical signals carried by neurons (nerve cells).

nervous system Tissues and organs which control the body's responses to stimuli. In a mammal it is made up of the central nervous system and peripheral nervous system.

neuron Nerve cell.

neuroscientist A scientist who studies how the brain and nerves function.

newborn reflexes Reflexes to particular stimuli that usually occur only for a short time in newborn babies.

nitrate ions An ion is an electrically charged atom or group of atoms. The nitrate ion has a negative charge, NO_3^-.

one-gene–one-protein theory The idea that each gene on a chromosome controls the production of one protein in the cell.

optimum temperature The temperature at which enzymes work fastest.

organelles The specialized parts of a cell, such as the nucleus and mitochondria. Chloroplasts are organelles that occur only in plant cells.

organic compound A chemical which contains carbon in its molecules.

organs Parts of a plant or animal made up of different tissues.

osmosis The diffusion of water across a partially permeable membrane.

osmotic balance If a cell contains too high a level of dissolved chemicals it will gain too much water by osmosis. If a cell's level of dissolved chemicals is too low, it will lose too much water by osmosis. When the cell has the correct level of dissolved chemicals, it is osmotically balanced.

oxygen debt After a period of anaerobic respiration the body uses oxygen to break down the lactic acid. The amount of oxygen which is needed to do this is the oxygen debt.

oxygenated Blood in which the haemoglobin is bound to oxygen molecules (oxyhaemoglobin).

oxyhaemoglobin A haemoglobin molecule which has bound to oxygen molecules. Oxyhaemoglobin gives up its oxygen to body tissues as red blood cells are carried around the body.

pancreas An organ in the body which produces some hormones and digestive enzymes. The hormone insulin is made here.

parasite An organism that lives in or on another organism.

partially permeable membrane A membrane that acts as a barrier to some molecules but allows others to diffuse through freely.

peer review The process whereby scientists who are experts in their field critically evaluate a scientific paper or idea before and after publication.

penicillin An antibiotic made by one type of fungus.

peripheral nervous system The network of nerves connecting the central nervous system to the rest of the body.

pharmacogenetics Investigating whether genetic variations between people cause different responses to a medicine.

phloem A plant tissue which transports sugar throughout a plant.

photosynthesis The process in green plants which uses energy from sunlight to convert carbon dioxide and water into the sugar glucose.

phototropism The bending of growing plant shoots towards the light.

physiotherapy Treatment of injuries of the skeletal-muscular system.

phytoplankton Single-celled photosynthetic organisms found in an ocean ecosystem.

pituitary gland Part of the human brain which coordinates many different functions, for example release of ADH.

placebo Occasionally used in clinical trials, this looks like the drug being tested but contains no actual drug.

plasma The clear straw-coloured fluid part of blood.

plasmids Small circle of DNA found in bacteria. Plasmids are not part of a bacterium's main chromosome.

platelets Cell fragments found in blood. Platelets play a role in the clotting process.

polymer A material made up of very long molecules. The molecules are long chains of smaller molecules.

polymerase chain reaction (PCR) A technique used to make many copies of a sample of DNA.

population A group of animals or plants of the same species living in the same area.

predator An animal that kills other animals (its prey) for food.

pre-implantation genetic diagnosis (PGD) This is the technical term for embryo selection. Embryos fertilised outside the body are tested for genetic disorders. Only healthy embryos are put into the mother's uterus.

processing centre The part of a control system that receives and processes information from the receptor, and triggers action by the effectors.

producers Organisms found at the start of a food chain. Producers are autotrophs, able to make their own food.

protein Chemicals in living things that are polymers made by joining together amino acids.

protozoan A type of single-celled organism.

Prozac A brand name for an antidepressant drug. It increases the concentration of serotonin at the synapses in the brain.

pyramid of biomass A chart which shows the relative amount of living mass (biomass) at different levels in a food chain.

pyramid of numbers A chart which shows the relative number of organisms at different levels in a food chain.

receptor The part of a control system that detects changes in the system and passes this information to the processing centre.

recessive An allele that will only show up in an organism when a dominant allele of the gene is not present. You must have two copies of a recessive allele to have the feature it produces.

recipient A person who receives blood from another person.

recombination Recombination happens during meiosis. Pairs of chromosomes exchange sections of DNA. This produces chromosomes in sex cells that have a unique mixture of alleles.

red blood cells Blood cells containing haemoglobin, which binds to oxygen so that it can be carried around the body by the bloodstream.

reflex arc A neuron pathway that brings about a reflex response. A reflex arc involves a sensory neuron, connecting neurons in the brain or spinal cord, and a motor neuron.

rennin An enzyme which acts on a protein in milk, causing it to form solid clumps. Rennin is used in cheese-making. Traditionally rennin is obtained from the stomachs of young mammals, but now over half the rennin used in cheese-making is produced by a genetically engineered yeast.

reproductive cloning Making new individuals that are genetically identical to the parent.

respiration A series of chemical reactions in cells which release energy for the cell to use.

response Action or behaviour that is caused by a stimulus.

restriction enzymes A group of natural enzymes which cut DNA at particular places. Each restriction enzyme cuts DNA at a specific base sequence.

ribosomes Organelles in cells. Amino acids are joined together to form proteins in the ribosomes.

RICE RICE stands for Rest, Ice, Compression, Elevation. This is the treatment for a sprain.

risk factor A variable linked to an increased risk of disease. Risk factors are linked to disease but may not be the cause of the disease.

rooting powder A product used in gardening containing plant hormones. Rooting powder encourages a cutting to form roots.

selective breeding Choosing parent organisms with certain characteristics and mating them to try to produce offspring that have these characteristics.

sensory neuron A neuron that carries nerve impulses from a receptor to the brain or spinal cord.

serotonin A chemical released at one type of synapse in the brain, resulting in feelings of pleasure.

sex cell Cells produced by males and females for reproduction - egg cells and sperm cells. Sex cells carry a copy of the parent's genetic information. They join together at fertilisation.

shivering Very quick muscle contractions. Releases more energy from muscle cells to raise body temperature.

short-term memory The part of the memory that stores information for a short time.

sickle-cell anaemia A disease in which a large number of red blood cells are sickle shaped and cannot carry oxygen properly.

simple reflex An automatic response made by an animal to a stimulus.

single-cell protein (SCP) A microorganism grown as a source of food protein. Most single-cell protein is used in animal feed, but one type is used in food for humans.

skeletal-muscular system All the bones and muscles which work together to move the body.

skeleton The bones that form a framework for the body. The skeleton supports and protects the internal organs, and provides a system of levers that allow the body to move. Some bones also make red blood cells.

specialized A specialized cell is adapted for a particular job.

species A group of organisms that can breed to produce fertile offspring.

sprain An injury where ligaments are located.

starch A type of carbohydrate found in bread, potatoes, and rice. Plants produce starch to store the energy food they make by photosynthesis. Starch molecules are a long chain of glucose molecules.

stem cells Unspecialized animal cells that can divide and develop into specialized cells.

stimulus A change in the environment that causes a response.

structural proteins Proteins which are used to build cells.

symbiosis Any association between two organisms.

symptom What a person has when they have a particular illness, for example, a rash, high temperature, or sore throat.

synapses A tiny gap between neurons that transmits nerve impulses from one neuron to another by means of chemicals diffusing across the gap.

synovial fluid Fluid found in the cavity of a joint. The fluid lubricates and nourishes the joint, and prevents two bones from rubbing against each other.

systolic The blood pressure when the blood is pumped from the left ventricle to the rest of the body.

tapeworm A parasitic worm which lives primarily in the gut of other living organisms, such as humans.

tendon Tissue that joins muscle to a bone.

therapeutic cloning Growing new tissues and organs from cloning embryonic stem cells. The new tissues and organs are used to treat people who are ill or injured.

tissue Group of specialized cells of the same type working together to do a particular job.

tissue fluid Plasma that is forced out of the blood as it passes through a capillary network. Tissue fluid carries dissolved chemicals from the blood to cells.

triplet code A sequence of three bases coding for a particular amino acid in the genetic code.

trophic levels The different steps in a food chain.

universal donor People with a blood type that can be given to any other person, whatever their blood type.

universal recipient People who can receive blood of any type in a transfusion.

unspecialized Cells which have not yet developed into one particular type of cell.

urea A waste product made by the liver from the breakdown of amino acids the body does not use.

urine Waste excreted from the body in the kidneys and stored in the bladder.

UV Ultraviolet (UV) radiation is a type of radiation that we cannot see. It is beyond the violet end of the visible spectrum.

vaccination Introducing to the body a chemical (a vaccine) used to make a person immune to a disease. A vaccine contains weakened or dead microorganisms, or parts of the microorganism, so that the body makes antibodies to the disease without being ill.

valves Flaps of tissue which act like one-way gates, only letting blood flow in one direction around the body. Valves are found in the heart and in veins.

variation Differences between living organisms. This could be differences between species. There are also differences between members of a population from the same species.

vasoconstriction Narrowing of blood vessels.

vasodilation Widening of blood vessels.

vector A method of transfer. Vectors are used to transfer genes from one organism to another.

vein Blood vessels which carry blood towards the heart.

ventricle One of the lower chambers of the heart. The right ventricle pumps blood to the lungs. The left ventricle pumps blood to the rest of the body.

vertebrate An animal with a spinal column (backbone).

virus Microorganisms that can only live and reproduce inside living cells.

white blood cells Cells in the blood that fight microorganisms. Some white blood cells digest invading microorganisms. Others produce antibodies.

XX The pair of sex chromosomes found in a human woman's body cells.

XY The pair of sex chromosomes found in a human man's body cells.

xylem A plant tissue which transports water through a plant.

yield The crop yield is the amount of crop that can be grown per area of land.

zooplankton Single-celled animals found in an ocean ecosystem. Zooplankton feed on phytoplankton.

zygote The cell made when a sperm cell fertilizes an egg cell in sexual reproduction.

Index

Acknowledgements

Publisher's Acknowledgments

Oxford University Press wishes to thank the following for their kind permission to reproduce copyright material;

p10l Liam Bailey/Photofusion Picture Library/Alamy; **p10r** Alamy; **p11** Dr Paul Andrews/University of Dundee/Science Photo Library; **p13** Mehau Kulyk/Science Photo Library; **p14l** David Crausby/Alamy; **p14r** Bsip Astier/Science Photo Library; **p15t** Dan Sinclair/Zooid Pictures; **p15b** Zooid Pictures; **p17** Mauro Fermariello/Science Photo Library; **p18l** Dopmine/Science Photo Library; **p18r** CNRI/Science Photo Library; **p19** Oxford University Press; **p20** Ian Miles-Flashpoint Pictures/Alamy; **p21** BSIP/Laurent/Science Photo Library; **p22** Ariel Skelley/Corbis UK Ltd.; **p26t** Bsip, Laurent/H.Americain/Science Photo Library; **p26b** Pascal Goetgheluck/Science Photo Library; **p27** Andrew Parsons/PA/Empics; **p28l** plainpicture/Alamy; **p28r** Mark Clarke/Science Photo Library; **p29t** Michael Stephens/PA/Empics; **p29b** AP Photo; **p30bl** Holt Studios International; **p30br** Claude Nuridsany & Marie Perennou/Science Photo Library; **p30t** David Scharf/Science Photo Library; **p31** Ph. Plailly/Eurelios/Science Photo Library; **p32t** Leo Mason/Corbis UK Ltd.; **p32b** Dr Yorgos Nikas/Science Photo Library; **p33** Yoav Levy/Phototake Inc/Alamy; **p38t** Science Photo Library; **p38b** Guzelian Photographers; **p39** Sipa Press/Rex Features; **p40** Guzelian Photographers; **p41** David Scharf/Science Photo Library; **p42** Guzelian Photographers; **p45** Guzelian Photographers; **p46l** Melanie Friend/Photofusion Picture Library/Alamy; **p46r** Melanie Friend/Photofusion Picture Library/Alamy; **p46b** John Dee/Rex Features; **p48** Detail Parenting/Alamy; **p49t** Getty Images; **p49b** Philip Wolmuth/Alamy; **p50t** Robert Pickett/Corbis UK Ltd.; **p50b** W. Eugene Smith/Time Life Pictures/Getty Images; **p51b** Paul A. Souders/Corbis UK Ltd.; **p52t** Erich Schrempp/Science Photo Library; **p52** Pete Saloutos/Corbis UK Ltd.; **p53l** Simon Fraser/MRC Unit, Newcastle General Hospital/Science Photo Library; **p53r** Ed Kashi/Corbis UK Ltd.; **p53b** Humphrey Evans/Cordaiy Photo Library Ltd./Corbis UK Ltd.; **p54** Dr P. Marazzi/Science Photo Library; **p56bl** Science Photo Library; **p56br** Biophoto Associates/Science Photo Library; **p56t** Guzelian Photographers; **p58l** Matt Meadows, Peter Arnold Inc./Science Photo Library; **p58c** Bettmann/Corbis UK Ltd.; **p58r** Sipa Press/Rex Features; **p60** Janine Wiedel Photolibrary/Alamy; **p66l** Michael Prince/Corbis UK Ltd.; **p66r** Wayne Bennett/Corbis UK Ltd.; **p67l** Kit Houghton/Corbis UK Ltd.; **p67c** /Corbis UK Ltd.; **p67r** Will & Deni McIntyre/Corbis UK Ltd.; **p67b** Ray Tang/Rex Features; **p69t** Oxford University Press; **p69c** Tom Brakefield/Corbis UK Ltd.; **p69b** Jeff Lepore/Science Photo Library; **p70t** Holt Studios International; **p70b** VVG/Science Photo Library; **p74** Mary Evans Picture Library; **p76** British Association for the Advancement of Science; **p77** Eddie Adams/Corbis UK Ltd.; **p79t** A. Barrington Brown/Science Photo Library; **p79b** Renee Lynn/Corbis UK Ltd.; **p80** Graham Neden/Ecoscene/Corbis UK Ltd.; **p81t** Geoscience Features Picture Library; **p81c** Robert Lee/Science Photo Library; **p81b** Jeffrey L. Rotman/Corbis UK Ltd.; **p82l** Alex Rakosy/Custom Medical Stock Photo/Science Photo Library; **p82r** ArendSmith/Robert Harding Picture Library; **p83l** Bsip, Chassenet/Science Photo Library; **p83r** Bsip, Chassenet/Science Photo Library; **p84** Daniel Cox /Photolibrary Group; **p85l** Christian Jegou/Publiphoto Diffusion/Science Photo Library; **p85r** John Reader/Science Photo Library; **p86bl** Niall Benvie/Corbis UK Ltd.; **p86br** Alex Bartel/Science Photo Library; **p86t** W. Perry Conway/Corbis UK Ltd.; **p88** Tim Davis/Science Photo Library; **p89** Julia Hancock/Science Photo Library; **p94** Dimitri Iundt/Corbis UK Ltd.; **p95t** Publiphoto Diffusion/Science Photo Library; **p95b** Martyn F. Chillmaid; **p97** Jim Varney/Science Photo Library; **p98** J.C. Revy/Science Photo Library; **p99** Eric and David Hosking/Corbis UK Ltd.; **p100t** Rick Price/Corbis UK Ltd.; **p100b** Simon Fraser/Science Photo Library; **p102t** James Fraser/Rex Features; **p102b** Alfred Pasieka/Science Photo Library; **p104** Martin Dohrn/Science Photo Library; **p105t** John Cleare Mountain Camera; **p105b** Dave Bartruff/Corbis UK Ltd.; **p106tl** Focus Group, LLC/Alamy; **p106bl** Stock Connection Inc./Alamy; **p106tc** Corbis UK Ltd.; **p106bc** David Stoecklein/Corbis UK Ltd.; **p106tr** Wartenberg/Picture Press/Corbis UK Ltd.; **p106br** Owen Franken/Corbis UK Ltd.; **p107** Mark Clarke/Science Photo Library; **p108l** Zooid Pictures; **p108c** Zooid Pictures; **p108r** Zooid Pictures; **p108b** Corbis UK Ltd.; **p111** Professors P.M. Motta & F.M. Magliocca/Science Photo Library; **p112** Martyn F. Chillmaid; **p114t** Blair Seitz/Science Photo Library; **p114b** Getty Images; **p116** Marco Cristofori/Corbis UK Ltd.; **p117** Don Mason/Corbis UK Ltd.; **p122tl** Oxford Scientific Films/Photolibrary Group; **p122bl** Michael & Patricia Fogden/Corbis UK Ltd.; **p122tc** Oxford Scientific Films/Photolibrary Group; **p122cc** Corel/Oxford University Press; **p122tr** Corel/Oxford University Press; **p122br** Corel/Oxford University Press; **p124t** Edelmann/Science Photo Library; **p124c** Alexander Tsiaras/Science Photo Library; **p124b** Science Photo Library; **p125t** Bob Rowan/Progressive Image/Corbis UK Ltd.; **p125c** M.I. Walker/Science Photo Library; **p125b** Leo Batten/Frank Lane Picture Agency/Corbis UK Ltd.; **p126t** Joe McDonald/Corbis UK Ltd.; **p126b** Astrid & Hanns-Frieder Michler/Science Photo Library; **p127t** M J Higginson/Science Photo Library; **p127b** Martyn F. Chillmaid/Science Photo Library; **p129** Rob Lewine/Corbis UK Ltd.; **p131l** Francis Leroy/Biocosmos/Science Photo Library; **p131r** Corel/Oxford University Press; **p132tl** Carl & Ann Purcell/Corbis UK Ltd.; **p132cl** Holt Studios International; **p132bl** CNRI/Science Photo Library; **p132tr** Martin Harvey/Corbis UK Ltd.; **p132cr** Holt Studios International; **p132br** Dr Jeremy Burgess/Science Photo Library; **p134t** Science Photo Library/Science Photo Library; **p134b** A. Barrington Brown/Science Photo Library; **p136l** Bob Gibbons/Holt Studios International; **p136r** Helen Mcardle/Science Photo Library; **p137tl** Anthony Bannister/Gallo Images/Corbis UK Ltd.; **p137bl** Russ Munn/Agstock/Science Photo Library; **p137tr** Marko Modic/Corbis UK Ltd.; **p137br** foodfolio/Alamy; **p138t** Dr. Y. Nikas/Phototake Inc/Alamy; **p138c** Dr Yorgos Nikas/Science Photo Library; **p138b** Biophoto Associates/Science Photo Library; **p139** Edelmann/Science Photo Library; **p140** Mauro Fermariello/Science Photo Library; **p142** Div. Of Computer Research & Technology/National Institute Of Health/Science Photo Library; **p144t** Oxford Scientific Films/Photolibrary Group; **p144b** John Kaprielian/Science Photo Library; **p150bl** Astrid & Hanns-Frieder Michler/Science Photo Library; **p150br** Eye Of Science/Science Photo Library; **p150t** Dennis Kunke/Phototake Inc./Alamy; **p151tl** Jeff Rotman/Nature Picture Library; **p151bl** Manfred Danegger/NHPA; **p151tr** Tobias Bernhard/Oxford Scientific Films/Photolibrary Group; **p151br** Clive Druett/Papilio/Corbis UK Ltd.; **p152l** Sheila Terry/Science Photo Library; **p152c** Owen Franken/Corbis UK Ltd.; **p152r** Laura Dwight/Corbis UK Ltd.; **p153l** BSIP Astier/Science Photo Library; **p153c** Jennie Woodcock/Reflections Photolibrary/Corbis UK Ltd.; **p153r** Corbis UK Ltd.; **p154t** Adam Hart-Davis/Science Photo Library; **p154c** Adam Hart-Davis/Science Photo Library; **p156l** Jeffrey L. Rotman/Corbis UK Ltd.; **p156c** Eric and David Hosking/Corbis UK Ltd.; **p156r** Photolibrary Group; **p159l** Lawrence Manning/Corbis UK Ltd.; **p159c** Corbis UK Ltd.; **p159r** Sally and Richard Greenhill/Alamy; **p159** Open University; **p160t** Corbis UK Ltd.; **p160b** Sipa Press/Rex Features; **p163t** S.Kramer/Custom Medical Stock Photo/Science Photo Library; **p163b** Mark Lythgoe and Steve Smith/Wellcome Trust; **p164** Steve Bloom Images/Alamy; **p165** Anthony Bannister/Gallo Images/Corbis UK Ltd.; **p167** Jerry Wachter/Science Photo Library; **p168** Karen Kasmauski/Corbis UK Ltd.; **p169** Roger Ressmeyer/Corbis UK Ltd.; **p172** Sipa Press/Rex Features; **p173** Clarissa Leahy/Photofusion Picture Library

Illustrations by IFA Design, Plymouth, UK and Clive Goodyer.